Contributions to Economics

More information about this series at
http://www.springer.com/series/1262

Peter Galbács

The Theory of New Classical Macroeconomics

A Positive Critique

Springer

Peter Galbács
Budapest Business School
Miskolc
Hungary

Completely revised edition of the original Hungarian edition with the title "Aktív szabályozás vagy gazdaságpolitikai nihilizmus?" published by Akadémiai Kiadó Zrt., 2012

ISSN 1431-1933 ISSN 2197-7178 (electronic)
Contributions to Economics
ISBN 978-3-319-17577-5 ISBN 978-3-319-17578-2 (eBook)
DOI 10.1007/978-3-319-17578-2

Library of Congress Control Number: 2015938182

Springer Cham Heidelberg New York Dordrecht London

Printed on acid-free paper

Springer International Publishing AG Switzerland is part of Springer Science+Business Media (www.springer.com)

To the memory of András Vigvári.

Foreword

Peter Galbács, a former student of mine, offered me the privilege to write a brief introduction to his work on The Theory of New Classical Macroeconomics.

Peter Galbács' book is an exciting and challenging work that induces even those readers who are well educated in—and sometimes indoctrinated by—mainstream macroeconomics to rethink and revise some of the fundamental ideas of our profession. I would not pretend that I agreed with all parts of his analysis and conclusions, but this is exactly why Galbács' book is so useful and interesting. It would have been tiresome and boring to read a text which triggers only "I agree" types of reactions. Galbács' work is like a huge canvas: it offers an elegant, colorful, and complex picture of the different schools of macroeconomic theory. His main conclusion that the neoclassical theory of economics is still alive and it provides a good framework of analysis and economic policy is well supported by the author's reasoning in the text. I would like to encourage the author to pursue this line of research in the future. And last but not least, I am really proud of the fact that I could be one of his professors.

Budapest, Hungary

Ivan Major

Preface

Finding the black cat is not rewarded.
(Norman Davies: Europe–A History. Foreword, Polish Edition)

My career as a researcher started under the spell of Keynes. I took a sincere interest in the ultimate questions of economic theory relatively early, surprising even myself. Attending lectures on sociology and social philosophy at the faculty of humanities and social sciences, I almost ran away from the philosophy lectures. The lack of firm answers and the different systems of thoughts contradicting and overwriting each other were too disturbing and what is more even frustrating for me. So, my interest turned to economic sociology first and then to economics itself, hoping to find a disciplined science built on mathematics. I did believe that schools of economic thoughts could harmonize in the most fundamental questions and have only minor differences in their opinions. I could hardly realize that economic theories suffer the same confusion. Moreover, having been armed with lectures given by Endre Nagy, Miklós Maróth, and György Miklós, all these impressions made it obvious and nearly inevitable that my interest as an economist would turn to economic theory and to the philosophy of economics. Today, I know that economic theory for me, due to these early experiences, has become a thesaurus of philosophical texts describing the economic aspects of processes going on in societies. Of course, it is true as well that I realized only as an economist that the questions raised by a branch of science itself cannot be answered on the grounds of its methodology and theory. If we want to consider the objects and methods of economics, the theory of marginal utility does not help us too much.

I made myself familiar with Keynes' theory as an undergraduate student. Fortunately, my first attempts were helped and controlled by professors who required me to study the most important texts in the history of economic thoughts very young. The impact that Keynes exerted on me was huge, since, on one hand, I could become not only acquainted with the principles that had governed operative economic policy for decades but also got an insight into the most significant and most extensive theoretical debate of economics of the twentieth century. My early interest was mainly focused on the points on which Keynes was opposed to the program of neoclassical economics that had been generally accepted up to that time. Back then, of course, I regarded all the scientific debates as black and white and, with all my strength, I tried to declare Keynes the winner, wherever and whenever it was possible. I often entered into indirect criticism (while doing so, its master, Prof. László Vígh, also helped and encouraged me), so the conflict I was inquisitive about was soon relocated between Keynes and Milton Friedman. The orthodox monetarism led by Milton Friedman was the first after-Keynes great theoretical system with which I got acquainted thoroughly. I think, I made quite a good job, since a serious antipathy to Friedman's theory took shape in my mind and it still exists even today. However, there is no doubt that I indulged in studying monetarism thanks to Friedman.

Previously, I tended to evaluate new classical macroeconomics similarly, in which I was encouraged by both the oversimplified sights formed on the new classical group and the vulgarization of the new classical theory itself. I do not know exactly what that turning point was after which I showed greater concerns to Lucas and his group than to my former master, Keynes. It was roughly at the end of my university years, so, as a fellow, I devoted my dissertation to new classical macroeconomics. I remember well, my initial purpose was to write a monographic comparison of Keynes and monetarism, but it was expected to be a monumental mission, so a Ph.D. dissertation could not have been able to embrace all the results I wanted to include. Finally, the dissertation was written on the monetary and fiscal policy recommendations of new classical macroeconomics—and it soon has become the basis of this book as well.

The genre of this book is *comparative critical history of economic thoughts*. It must be noted that I was aware of neither the ultimate aim nor the path leading to it, so I could not foresee where I would get to in the end. Curiosity was my only guiding principle, and the book as a whole should be considered an explanatory note to my readings. My interest was not governed by a single theorem to be proved or criticized. I just wanted to know where the economists labeled as new classicals came from and what they actually taught. Therefore, I based my investigation on the original texts, so the first level of comparison was made between these texts and their commentaries (i.e., reviews on economic theories). Of course, the comparison emerged in the context of the history of economic thoughts as well, since I had to realize at an early stage that, for example, either the Phillips curve elaborated by the new classicals or their comprehensive suggestions of economic policy can be judged only by comparing them recurrently with the theory of Friedman or Keynes. Moreover, I did not want to write a reading log. I wanted to avoid a simple reformulation of the theory of new classical macroeconomics, so I tried to carry out a *critical* analysis. During my work I was mainly supported by the methodological principles set out in Chap. 1. My determination to avoid writing a mere summary explains why I discussed topics (albeit, sketchily) such as automatic fiscal stabilizers or the conceptual differences between fiscal consolidation and structural reforms, to which new classicals had not paid any attention no matter how close these issues were to the problems being investigated.

This volume consists of seven chapters succeeding in birth order. Chapter 1 is the only exception as it was written only after the logical structure of the book had been ready. I had to face the recurrent critical remark made in debates that reproached me with the lack of a well-articulated methodological introduction. Although my declared purpose from the very beginning was to exert a constructive critique against new classical macroeconomics, the clarified methodological principles on which this book had been built were missing; I could give account of these fundamentals only in the course of private conversations at best, being satisfied with the belief that my interlocutors were familiar with them. The methodological and philosophical system detailed in Chap. 1 that later became the most important and interesting part of the book for me took form during these conversations. The ultimate goal of this chapter was to identify the characteristic aspect of mainstream economics and to clarify its methodological grounds, setting the discussion and the evolution of the methodology of (mainstream) economics in the context of general history and history of science. To my great surprise, the fundamental differences between orthodox monetarism and new classical macroeconomics emerge even at the level of the methodological principles.

Chapter 2 turns directly to the new classical doctrines. Although mostly implicit doctrines were examined in Chap. 3 (for instance, when discussing the conception of new classical labor market), rational expectations hypothesis being the most often cited and the most influential new classical theorem was studied here. Underlying assumptions and theoretical consequences were also reviewed here. These investigations were made especially gripping by the direct intention of new classicals to give a realist description of processes and mechanisms observed in reality. This circumstance induced and justified me to compare new classical macroeconomics with reality. It is argued below that estimations of market participants, since agents do not have access to information the theory requires and do not possess the necessary methodological skills either, can meet the requirements that were specified by Muth (1961) only if agents are supported by an institution in a central position. It could be, as a typical example, an inflation targeting central bank—so, after all, rational expectations hypothesis can effectively underpin this monetary policy regime. This chapter is very precious for me, since it is based on my first important publication.

Chapter 3 continues reviewing the fundamentals. Here, we can get an inside view of the underlying logic that determines the new classical way of thinking. We are going to be forced again and again to realize that the existence of (general) equilibrium is based on a set of fragile assumptions among which also the equilibrium itself stands as the most objectionable. The most of Chap. 3 was uncertain in the beginning. The text is based on notes that were made in the course of some fundamental reviews of John Weeks' "A Critique of Neoclassical Macroeconomics." Since Weeks laid his emphasis on other aspects and details than I had to concentrate, these notes advanced to be a chapter on their own as time was passing by. Weeks mainly focused on the analysis of aggregate reproduction while I was rather interested in the theory of macroeconomic cycles and countercyclical economic policies. I supposed (and suppose even today) that the most fundamental discrepancy separating rivaling macro-theories is determined by how these theories explained and answered the problem of business cycles. My interest in the theory of business cycles may be explained by the fact that the theories scrutinized below could and can start governing operative economic policies.

I could hardly deny the fundamental effects Weeks had on the tone and the methodology of this work. I think, Weeks' *Critique* can be referred to as the most essential and significant critical treatise on mainstream economics. Its impacts and importance are hard to estimate—moreover, in stylistic terms, it is an outstanding manifestation of the collection of the briefest and most disciplined works in economics. But, at the same time, it is a difficult and ponderous piece of reading. It is only for the third or fourth time when the inter- and intra-textual relations and the limitations of the reasoning can be revealed. The robust text of Weeks makes us feel as if everything that is possible has been told about neoclassical-mainstream economics. Actually it is almost completely true since it is very difficult to find those points where the issues discussed could be complemented. Weeks' work is indispensable: I believe that it is almost impossible to write about new classical macroeconomics without having Weeks' monograph among the references: breaking away from his works could have only impaired my efforts and I would have seemed to reinvent the wheel. I did my best to avoid simply echoing Weeks' discussion and I think I managed to do so. It would have been difficult indeed to refute his thoughts, so I had to agree with him and, wherever it was necessary for the sake of my reasoning, I strived for abundant completions.

Chapters 4 and 5 make up the central part of the book. After the necessary theoretical preliminaries had been set out, new classical thoughts on monetary and fiscal policy could be analyzed and evaluated here for the first time. I was deeply biased by the conventional view on monetarism, according to which fiscal policy is only of secondary importance for mainstream economics, while monetary policy is assumed to have potential only in setting the price level. I think, if this notion was relevant at all, it was true only of Friedman's system at best. For him, macroeconomic systems settle in stable (long-lasting) equilibrium state, as his accounts on the natural rate of unemployment highlight, from which they can be dislocated through monetary policy interventions only temporarily and only with considerable detriments and avoidable-unnecessary costs. I had to fumble in the dark for a while when scrutinizing the countercyclical potential of monetary policy on these grounds. In the beginning, I showed some inclination to think that Friedman had stressed the countercyclical impotency of monetary policy (and, as a consequence, of fiscal policy) through these doctrines of his—but, later, it became evident that business cycles originate in economic policy, as argued implicitly by him. Actually, Friedman talked about the possibilities of monetary policy in growth enhancement—however, it is completely and fundamentally impotent in these terms so the only consequences of its efforts are the business cycles it generates. So, the best option of monetary policy is to specify its goal of increasing the quantity of money at an appropriate rate—and, moreover, fiscal policy should abstain from all of these. The new classical system is much more elaborated. It is a significant step forward that the emergence of business cycles is allowed now, so the countercyclical potential of economic policy can be a sensible question. While new classical macroeconomics could not abandon the traditional and common interpretations when evaluating monetary policy, giving room for monetary policy only in price level setting at a maximum, new classical doctrines regarding countercyclical fiscal policy (its scope and potential) seemed to be mainly unclarified to me. On the grounds of the permanent income theory specified under rational expectations hypothesis or the Barro–Ricardo equivalence, it is really possible to declare that fiscal policy is predestined to be impotent in new classical macroeconomics as well. However, it has to be realized how thorough the theory is when considering the assumptions and conditions necessary for this ineffectiveness. The ineffectiveness of fiscal policy is dependent upon tight conditions in the new classical theory, and, in this context, underlying this conditionality proved to be the greatest merit of the school.

Chapter 6 summarizes some conclusions that were highlighted by new classical macroeconomics but were outside the core of the theory. Such a topic is the problem of economic policy credibility to which we get starting from the equivalence theorem. Credibility is discussed with Reaganomics as a background, the interpretation of which can be found in the literature though, but it was worth stressing the relations to the equivalence theorem as well. It is also Chap. 6 where a topic analyzed previously recurs: it is inflation targeting, scrutinized in Chap. 2 before, with regard to the strong definition of rational expectations hypothesis. Here, some technical issues emerge such as strict inflation and output-gap targeting or flexible inflation targeting. I made efforts to stress the advantages of economic policy offered by flexible inflation targeting and to lay those features of the system in the limelight that can be evaluated and understood more easily with the support of new classical macroeconomics or rational expectations hypothesis.

Chapter 7 gives a short summary and raises some further questions to which I could not even try to find ultimate and satisfying answers, not only for the reason that the chosen topic of this work or certain size constraints would not have made it possible at all. These are problems the solutions to which are the question of faith rather than the question of proofs or reasoning.

Miskolc, Hungary
27 February, 2015

Peter Galbács

Acknowledgments

I am deeply indebted to my former supervisor, Prof. András Vigvári (1955–2014). I was a 2-year student when I became keenly interested in macro-financial theories. I think, he did the most for me by humbling my pride in the early times. It was he who revealed, with unrelenting honesty, all the shortcomings of my first writing which burst with conceit and which was, however, almost completely ungrounded and empty. Nonetheless, the lesson I had to learn was not mediocrity but prudence and judiciousness, and not the respect of authority but the commitment and humility to science and curiosity. He advised me to be patient and, instead of allowing me to be greedy of knowledge only superficially, gave me books in order to show me: there are no superfluous items in learning. Sooner or later, every bit of knowledge turns out to be a part of the big picture, fitting into items already known and supporting us in understanding the world, mostly by inducing more and more questions. He has taught me that ignorance is not to be hidden but to be admitted and that it is no problem if only question marks stand at the end of our thoughts. This is the only impulse of scientific investigations. All that I know about economics now I learnt directly from him or on his inspiration. During the years I spent by him as his student, I was forced to face serious trials again and again for which I was not prepared initially. However, they were the steps that gave me the most pieces of factual knowledge and necessary technical skills. But, without depreciating the things he did *as an economist*, it has to be admitted that I could learn the most from him *as a man*. Unfortunately, the painful limits of biological existence deprived both of us of rejoicing at the publication of this volume together. This book is dedicated to him, my beloved friend and master.

I have to express my gratitude to many people for supporting the creative process of this publication. First of all, Prof. László Vígh has to be mentioned. As one of the leading researchers (if not the utmost inquirer) of the theory of Keynes in Hungary, he has kept giving me hints of fundamental importance since the very beginning. He was the authority who gave the Hungarian edition of Weeks' important book into my hands which proved to be a determinative experience. All in all, his annoying questions and firm argumentations helped me the most. As far as the problems of fundamental importance are considered, he has an established and well-grounded point of view and insists on it to the utmost—however, his thinking never becomes dogmatic. By him, a young researcher is never taken as a subordinate counterpart to be forced to receive instant verdicts but as a discussion partner at his own level to get a share both in his grounded statements and bothering doubts. He is an independent philosopher of wide reading. I am sure, he could never guess the amount of questions and doubts that occurred to me not only against the big theories but also against my own precarious views during our conversations in his thick cigarette smoke at the university.

The researchers of the Department of Economics led by Prof. Dietmar Meyer at the Budapest University of Technology and Economics had held two discussion forums on the Ph.D. dissertation which, later, became the basis of this present publication. I owe a debt of gratitude to Prof. Meyer as the initiator of these meetings, since he was the first who was able to convince me that my investigations would deserve wider interest. As I had the opportunity to get to know him, he is deeply inspired with scientific curiosity. During these debates, some serious quarrels evolved with Dr. Zsombor Ligeti who, with his sharp remarks, called my attention to some major shortcomings of my manuscript. To tell the truth, it was a shocking experience then, since I did my best to believe: my ideas are perfect, complete, and well elaborated, so all these deficiencies should have remained completely hidden. He proved to have genuine human greatness when, after and in spite of the heavy debates, showed honest interest in the advancements of my research process. The encouraging and appreciating words of Dr. Ivan Major supported me more than he could ever think. During an informal conversation at the Institute of Economics, Hungarian Academy of Sciences, we discussed the most important and recent problems of theoretical economics that remained uncovered and undeservedly neglected in the dissertation. He mentioned the microfoundations of macroeconomics and urged me to immerse myself in the relating theoretical debates. This is the very reason why I have taken this problem as the most important (methodological) dilemma of neoclassical and mainstream economics.

Annamaria Parócza and Zsuzsanna Gerstmar, the staff at my department, proved to be substantial contributors to this book. I am deeply indebted to them for finding some publication that is very hard to hunt up and for providing me with a lot of items to read. As I spent the most of the research process simply reading, it was a great help that I could devote my free periods between the lectures to processing these texts. I am also very grateful to the Hungarian State Railways for the systematic delays, since, due to them, I could have a lot of time desired to be spent usefully and sensibly. Librarians at the College of International Management and Business of Budapest Business School, especially Edit Dutka and Andrea Lőrincz, must also be mentioned here with the greatest gratitude, since they supported my work far beyond their responsibilities. I am deeply indebted to Dr. Bence Marosán for the illuminating, very long, and sometimes even exhausting conversations about phenomenology and the philosophy of Edmund Husserl. I am sure, he is the most intelligent man I have ever met.

Last but not least, I should also like to extend my special thanks to Edit Koncz and Nikola Mihajlovic, who completed the task of proofreading the text at the price of great sacrifices, strenuously, and well beyond their power. Mrs. Koncz has taken the English manuscript in her hands in very hard times, when the possibility of failure was extraordinarily high. Her accuracy and self-sacrificing intention to help me proved to be an immensurable support. Thanks to her, the final stage of proofreading became a joint effort, which was gladsome creative work rather than stressful and strained slavery.

Contents

Chapter 1
Methodological Principles and an Epistemological Introduction

Truth emerges more readily from error than from confusion.
(Sir Francis Bacon)

There is something unchangeable in every man that [...] cannot be altered even by the milieu and the course of time. The idea that lives within, that makes them human beings and personalities and that never changes
(Sándor Márai)

From the intrinsic evidence of his creation, the Great Architect of the Universe now begins to appear as a pure mathematician.
(James Jeans: The Mysterious Universe)

1.1 Introduction

Economists study the ever-changing world that is quite complex to successfully analyze without making some selection. We search for general tendencies behind non-returning events that can be taken as proper grounds for scientific observation: purifying the events makes it possible to find and discover something that is persistent, recurring and substantial in the (perhaps seemingly) disordered and chaotic reality. Doing so, we are supported by economic theories that are, eventually, our analyzing devices: theories help us focus only on relevant factors, events and processes. We cannot say, therefore, that there exists only one relevant theory in economics, since there are various theories available as analyzing devices to study and describe certain phenomena. So, finding and choosing the right and adequate framework is always the mission of an economist. Knowing the history of economic thought is essential for even a further evolution of economics, because since economics as a science was born no significant authors have worked without comprehensive knowledge of the thoughts of his predecessors (Madarász 2000). However, this conclusion cannot be drawn by all the economists even today. Therefore, it is not evident for them that *a new monetary theory* on the one hand and *the history of monetary theories* on the other hand have the same root, and that something new can be achieved only starting from the ground of something old that should be known to us at least to stress what is to be exceeded by a new theory.

P. Galbács, *The Theory of New Classical Macroeconomics*, Contributions to Economics, DOI 10.1007/978-3-319-17578-2_1

All this had to be stressed to put it explicitly: the topic of this book is *not* a survey of a certain chapters of the history of our science, though, it is clear, it can also be interpreted in such a way. In our approach, the texts and models studied below are not relics from the history of economic thought, not show-pieces in a museum that describe the achievements of certain periods. The issues we study here are or can be relevant even today.

1.2 The Topic and Its Relevance

The theses of new classical macroeconomics were awarded the Nobel Prize in 1995. It is justified to think that it reached its peak by this event. However, monetarism was never able to have its views fully accepted. The honour of *Milton Friedman* and *Robert E. Lucas* did not bring a rest which would characterize circles of economics after a paradigm shift or which was present for a while after the appearance of Keynes. [→ **paradigm]** However, after the sweep of new classical macroeconomics, theorists and, perhaps, even the makers of operative economic policy, remained divided on the controversies over the scope of economic policy and its proper instruments, even if the possibility of a consensus was present.

Paradigm. There are a large number of comments about, reviews of and interpretations of the concept of sciences and scientific progress outlined by Thomas Kuhn—so, here, it is not needed to make a further attempt. It is more interesting to study the applicability of the concept of paradigm and Kuhn's model of scientific progress to economics.[1]

One extreme view is represented by Phyllis Deane (1978), a noted historian of economic thought with firm Keynesian attitude. In her interpretation, the history of economics is a history of successive paradigms. In the introduction to the Hungarian edition of her book mentioned above, the well-known Hungarian historian of economic science, Antal Mátyás, quoting Deane, distinguished five separate paradigms—that may be six, if monetarism is regarded as an individual and new paradigm. According to another potential view, the only paradigm shift in economics hitherto was the achievement of Keynes, since all the earlier theoretical systems were pointing towards the development of neoclassical paradigm. In her publication, Deane thoroughly studied the similarities between neoclassical theory and its predecessors, and, eventually, one could infer from her arguments that the main achievement of Smith had been to lay the theoretical foundations for neoclassical economics. However, it is unambiguous, on the one hand, that only the theory of Keynes can be compared with neoclassical economics in terms of significance and importance, and, on the other hand, that Keynes could attack his predecessors successfully at certain points. Therefore, it is justified to talk about a Keynesian paradigm shift which was not followed by

[1] Moreover, in the course of an attempt to interpret this account, we could not avoid facing the fact that Kuhn's model was not able to overwhelm the rivalling theories either in Kuhn's own times or later. Relativism of Feyerabend, logical positivism of the Vienna Circle, falsificationism of Popper or the theory of Lakatos (who opposed everyone and seemed to be a Popperian, while often corrected him) make together an intellectual environment in which Kuhn could hardly find supporters. For example, Bird (2009) highlights that there is and there was no characteristically Kuhnian school in scientific theory—mainly for Kuhn rejected the often radical theoretical developments joined his system.

further shifts, since monetarism returned quite directly to neoclassical economics that preceded Keynes.

Describing the history of economic science as a series of paradigm shifts is problematic not only because of the uncertainty as to the exact number of paradigm shifts, but the application of the notion of paradigm to the history of economic thought is dubious as well. It is a fact that social sciences warmly welcomed Kuhn's model of scientific progress, since disciplines that had not been allowed earlier to define themselves as sciences could step in a new phase of self-identity and self-awareness. Economics, sociology or psychoanalysis could not be regarded as sciences on the basis of either positivism or the criteria suggested by Karl Popper. Positivism stressed that the crucial requirement to be a science is the verifiability by reference to its predictive power, while, according to Popper, a science *as such* should be potentially falsifiable by its predictions. As far as any of these requirements are concerned, regarding economics as a science may be questioned. Kuhn suggested a more permissive and liberal criterion than these requirements on the grounds of which even disciplines charged with loud debates and dominated by theories that have multiple theoretical and methodological foundations could become sciences—though developing sciences, being in their pre-paradigm phase (Bird 2009).

Meanwhile, for Kuhn it was doubtful whether his theory (Kuhn 1962) is valid for social and human sciences and, moreover, he showed little interest in them. According to him, natural and human sciences are entirely different. Scientists use the prevailing paradigm, but social sciences appear to be hermeneutic and interpretative. Their aim is to understand individual or social behaviour and not to discover the fundamental and ultimate laws of physical-biological-chemical world around us. Within the realm of natural sciences, research produces new paradigms, and new ways of understanding nature from time to time, however, this endeavour is not conscious. Human sciences themselves are hermeneutic enterprises. A scientist does not reinterpret the analyzing apparatus available again and again, but tries to apply it to the most possible phenomena of the natural world. In human and social sciences, however, it is deliberately intended to reinterpret the studied processes, i.e. research always looks for new analyzing frameworks.[2] Moreover, Kuhn was totally uncertain whether some specialities of human and social sciences once may be able to find paradigms that can support normal, puzzle-solving research activities. His impression was that economics and psychology could be the first among current specialities at most to achieve this phase. So, Kuhn could hardly agree with the description of the history of economic thought as a history of paradigm shifts. The efforts of social sciences to find paradigms are discouraged by the continuous and never-ending change of social-cultural-economic-political reality. Finding paradigms in social sciences is not only impossible (or difficult, at least), but, perhaps, is an unnecessary and senseless attempt. Of course, questioning and analyzing the unchanged and uniform natural reality can be conducted on the basis of stable paradigms, but in social and human sciences, with ever-changing units under study, such paradigms may not be needed. Therefore, hermeneutic reinterpretation may be invariably required (Kuhn 1991).

---

[2] Broadly speaking, it well explains, firstly, why economics had been always constituted by a lot of competing theories and, secondly, why the emergence of a new theoretical account does not completely discredit the former systems. For example, economic developments in Japan during the 1990s could be clearly explained by the theory of Keynes (cf. e.g. Ozsvald and Pete 2003; Krugman 1998), while, at the same time, this theory and the economic policy recommendations made by Keynes somewhat faded away. It is also interesting that some authors wanted to base the after-2008 crisis management on the neoclassical paradigm founded by free market mechanisms (e.g. Erdős 2010), while international organizations urged a basically Keynesian solution (e.g. Blanchard 2008; Freedman et al. 2009; IMF 2009) and this reaction really seemed to be defendable on theoretical grounds (for further details see Galbács 2011a).

We are also in a precarious situation if we want to count the paradigm shifts in the history of economic thought occurred hitherto. To do so, firstly, we have to make it clear what a paradigm shift means (in other words, what the difference is between two paradigms). Without it, we can hardly answer the question, say, whether the replacement of neoclassical economics is evidently followed from the strengthening of the theory of Keynes. A paradigm shift, according to Kuhn, is *not* a synonym for a change in a prevailing (to use Kuhn's term the *current*) theory or in the way a science organizes its area of research (e.g. what is to observed, the questions to be asked, how these questions to be structured, or how the results should be interpreted, etc.). A paradigm shift is a more elementary achievement that can only occur after scientific crises (according to some counterarguments, a revision of the paradigm is not far from normal science either, and, moreover, some new and thoroughgoing results emerged not after such crises, but after revisions of the paradigm). A normal science is characterized by an insistence on its universally accepted paradigm, so anomalies are not necessarily enough to fundamentally discredit theories in a science—as opposed to Popper's view. Only extremely problematic observations accumulating in large numbers that are not properly explained by the accepted paradigm can weaken the power of the prevailing paradigm—these developments undermine the usual, puzzle-solving routine of a normal science, resulting in a scientific crisis. A new paradigm can only form after this stage, in the course and as a result of a scientific crisis.[3] Although, in the history of economic thought, the areas of research and the views formed on the fundamental questions and on their answers changed many times, an event that could throw economics into a state of crisis did not happen until the Great Depression of 1929–1933. Before this time, economics was in the phase of finding its way (or, in other words, in a pre-paradigm phase), since the changes in the answers to its fundamental questions (e.g. on the origin of value) were not triggered by crises but by thoughts being deadlocked. Although, for Kuhn, the revolutionary hunt for a new paradigm is driven by the anomalies that cannot be brushed aside or simply ignored, it is not obvious whether Keynes' answer really formed a new paradigm. It is true, of course, that the neoclassical theory starting from (and *presupposing*) the notion and the idea of a general equilibrium was not able to interpret the crisis of overproduction (according to Say's law, a crisis of overproduction and an involuntary unemployment could not emerge at all), and that Keynes efficiently stressed that a general macroeconomic equilibrium is an exception rather than a rule—however, this is a *complementing* and not a *denying* relation. While neoclassical economics is an axiomatic-deductive abstract theory presupposing that actors are under the conditions of perfect market competition and formal rationality, the theory of Keynes describes the functioning of actual macroeconomic systems. One could hardly speak of denial, since Keynes (1936) built even the most basic form of the quantity theory of money in his own theoretical system and, moreover, he regarded some of the neoclassical doctrines as special cases of his own system. It is out of the question that Keynes would *not* deny the equilibrium tendencies built in macroeconomic systems at all—denying the existence of

[3] At a point, Max Planck describes the paradigm shift as a special process (Planck 1933). According to him, a new paradigm does not gain ground so that the followers of the old framework gradually switch over—rather, a paradigm shift involves the exchange of the scientific elite; a new paradigm can only start reigning when believers of the prior framework gradually die off, in the literal sense. The emerging new generation is evidently open to a new conceptual system, to new theories and models. It has to be taken into account that mentioning Planck and the concept of paradigm side by side is anachronistic. Kuhn's seminal work was published in 1962 for the first time and if the one and a half decades that were taken by the development of the Kuhnian conception are also considered, we are still a couple of years after Planck. Kuhn was only 11 years old when Planck talked about the scheme of succeeding scientific thoughts in his cited lecture.

these equilibrium tendencies on theoretical grounds would result in a paradox conclusion according to which, eventually, involuntary unemployment is voluntary. Keynes "only" stressed the fact that the equilibrium presupposed by neoclassical theorists can be disturbed by numerous factors in reality.[4] The paradigm did not change—it did not to such an extent that some neoclassical axioms (e.g. the wage is equal to the marginal product of labour), or *postulates*, using Keynes' term, were not questioned even by him. The manifestation of neoclassical axioms in reality is disturbed and even hindered by the effects of actual economic and social environment. Broadly speaking, the same is true for the results of Vernon Smith's experiments.[5] The system of neoclassical axioms and *homo oeconomicus*

[4] As Keynes (1936) stresses it at the head of his *General theory*, the failure of neoclassical economics was caused by the fact that its postulates do not hold in reality, so economists cannot realize those socio-economic problems (involuntary unemployment, partial equilibriums) that are ruled out of their theory on conceptual grounds. Keynes, as it seems, was aware of that neoclassical economics is a *pure theory* (the term "pure theory" was actually used by Keynes himself in the *General theory*), that is to say, its postulates are not false *in themselves*, but *may not be relevant* occasionally as far as economic reality is concerned. Keynes, I think, neglected to highlight it for he wanted to deal neoclassical theory an annihilating blow *at a verbal level and in terms of operative economic policy*, and any efforts to find excuses would have implied arguments for underpinning neoclassical economics. Of course, it does not invalidate the details that will be said in Chap. 7 about the *complementing* relation between Keynes and neoclassical economics.

[5] The list of the theories aimed at confronting neoclassical economics is far from being complete. In his specification, Prof. Móczár (2010) mentions a number of related developments–however, these are experimental achievements in part, so the *complementing* relation mentioned with regard to experimental economics above applies to them as well. Only DSM-theorem (referred to as SMD-theory sometimes) can be regarded as really disturbing, but it rather concentrates on the contradicting relationship between micro- and macroeconomics and not on the (micro-level) postulates of pure neoclassical economics to question them. However, it would be a huge mistake to play down the significance of the DSM-theorem (named after the elaborators, Gérard Debreu, Rolf Ricardo Mantel, and Hugo Freund Sonnenschein), since, eventually, it questions the justification of the microfoundations of macroeconomics. The fundamental problem was posed by the relationship between individual and aggregate excess demand functions. As aggregate-level excess demand functions are important for macroeconomics, huge efforts were made to ground these functions on a micro-level. The seminal papers on this topic (Sonnenschein 1972, 1973; Mantel 1973; Debreu 1974) called attention to the fact that however precise we are in specifying the micro-level characteristics, the excess demand function for an economy is not restricted at all by these features. A consequence of the mathematical demonstration of the DSM-theorem was that the general equilibrium of a multi-market macro-system is not necessarily unique (the lack of uniqueness), i.e., we can find multiple price vectors that guarantee equilibrium and the emerging equilibrium states may be unstable in dynamic terms (Giraud 2009). The critique formulized by the DSM-theorem did not lack precursors: for instance, Scarf (1960) reported prices oscillating without any convergence towards the equilibrating vector under various conditions in simple general equilibrium models. According to Móczár (2006), it seems as if DSM-theorem could have dealt mainstream economic theory a final and destroying blow. The relation of mainstream economics to DSM-theorem is portrayed well by the *Handbook of Monetary Economics* (B. Friedman and Hahn 1990), in which the achievements underlying the theorem were not mentioned even in the References. Mainstream economics disregarded DSM-theorem (Nachane 2010; Rizvi 1994) the same as the inconsistency from the quantity theory and Walras' law (for further details see Chap. 3 below; cf. Patinkin 1989). Frank Hahn (1975) was one of the few who regarded DSM-theorem as the most dangerous critique of mainstream economics. It is understandable somehow, if we consider that the object of the critique, viz. mainstream economics, is an *economic* theory—while DSM-theorem (however impressive it is) is a mathematical

constructed, interpreted and analyzed on these grounds show what *would* be a man like if his behaviour was governed by only the rules of formal rationality and if his own job was to make economic decisions, while experimental economics shows how men behave under precisely specified laboratory conditions (that are, of course, not to be confused with the real world). This is the only possible relationship, since neoclassical theory created its own idea of man deducing it from the axioms, while experimental and behavioural economics formed its own concepts on empirical grounds, following an inductive logic (cf. Tversky and Kahneman 1974, 1986; Mérő 2004). Therefore, neoclassical axioms (as cornerstones of a pure theory) were not questioned either at this time, since these axioms were not intended to apprehend the real, existing men; so, they are un-falsifiable by experimental results.[6] If ever there was a paradigm in economics, only neoclassical economics can be regarded as such. [→ **homo oeconomicus, phenomenology**]

The major barrier to this consensus might be the radicalism of new classical macroeconomics. Lucas and his group showed reluctance to turn back to the Keynesian foundations (this demarcation also characterized the monetarism[7] of Friedman), but considerable opposition mounted to the theories of Friedman. The new classical system profoundly changed the views on the possibilities of economic policy, and offered novel recommendations on the policies to be followed. Aggregate demand management was replaced by the stimulation of the supply side, and activist economic policies should have been increasingly regarded as games played *against* market agents. So, it seemed that economic policy intervening and controlling deliberately on the one hand and actors on the other hand are not on the same side anymore. After all this, however, the question remained: *what is the actual goal of economic policy?* If monetary and fiscal policy can succeed only at the expense of outplaying (or deceiving) market participants, can we say that this goal is in accordance with the interests of the actors at all? It looks like economic policy knows better than people what they really need. [→ **paternalism**] I could not

demonstration. Even if we admit that the mathematical grounds of mainstream economic theory are wrong in some respects, the theory was still successful in resolving a lot of economic problems–while DSM-theorem could hardly support us in such a context. The prevailing paradigm is not required to be perfect at all but only to give solutions to several scientific problems. After all, mainstream theory performed well in these terms, and it is also sure that DSM-theorem is never going to help us (say) establish the theoretical grounds for the best operative monetary policy or finding the ultimate causes of cyclical fluctuations to found countercyclical economic policies. Moreover, it is not the possibility of multiple equilibriums that is highly problematic among the consequences of the DSM-theorem but the instability of these equilibrium states, since the latter can ruin the conviction of mainstream economics as to stable general equilibriums. As far as the possibility of multiple equilibriums is considered, it is worth remembering the mechanisms of ecological systems that can resist shocks within certain limits (equilibrium is stable), while, in case of a significant external impact, the ecosystem moves into another stable equilibrium with some changes in the main system-level characteristics (e.g. species composition, abundance—cf. e.g. Békés 2002; Bartha 2008; Mäler 2008). This feature is described by *resilience*, i.e. the capacity of flexible adaptation, extended recently to cover the functioning of socio-economic systems as well (e.g. Adger 2000; Noya and Clarence 2009). In the case of mainstream economics, the key doctrine is not the uniqueness but the stability of equilibrium.

[6] At a place, the complementing relation between experimental economics and mainstream theory is underlined by Vernon Smith (1991) as well.

[7] The Friedmanian wing of monetarism shall be referred to as orthodox monetarism below.

engage in considering questions of social philosophy in detail, though such an analysis could promise new aspects and results as well. Which is the point beyond which an economic policy following monetarist and new classical principles has to cope with the burden and the blame of paternalism?[8] Who empowers the government to play its dynamic plays against the mass of market agents?

Paternalism. Paternalism (a paternalistic action) is the case when an individual or a state (a government) interferes with another person against his will (i.e., whether he likes it or not) and is justified by the claim that the person interfered with will be better off or, at least, protected from possible harm which can be eliminated (or reduced) in the course of this action.

Although, at the theoretical level, conceptualization of paternalism does not seem to be complicated, considering certain cases may be fraught with difficulties. For example, we can say that John Smith acts paternalistically when he does something without the consent of Jack Duane, where something interferes with (i.e. limits) the liberty or autonomy of Jack Duane in order to improve his welfare by this thing. It must be noted that there are three necessary, but respectively not sufficient conditions here: a thing that restricts a given individual's liberty, the lack of approval, and, last but not least, an improvement in the individual's welfare (or, at least, an intention of this improvement). Condition one is quite easy to exemplify: the cases of physical force can naturally fall into this category. However, considering some special cases, these three crucial conditions together may raise doubts. If John Smith, being sceptical about the sense of responsibility of his younger son, instead of leaving him all his fortune, gives it to his older son with concrete instructions how to use it in the best interest of his younger son, it is unambiguous that John Smith acts in a paternalistic way. Notwithstanding, we cannot say that John Smith limits the younger child's liberty or autonomy through doing so (for a more profound conceptual introduction see Dworkin 2010). However, the question about the nature of paternalism differs from the questions as to the grounds on which and as to the cases in which one can paternalistically act.[9]

Although the problem of paternalism is an important field of moral and political philosophy and philosophy of law, it cannot be stated that it is unfamiliar to economics. Some of the arguments against paternalism can be directly applied in the debates on

[8] This issue shall get special importance apropos of the orthodox monetarist and new classical interpretations of the Phillips curve, since, as we will see, economic policy can mainly decrease unemployment by deceiving employees. It is doubtful that if *this* is the only way of reducing unemployment then this economic policy action really serves the interest of market agents, and, moreover, there can be found a social group in society that is better off by this intervention. It can raise one's doubts that an actual fall in real wages temporarily unknown to the employees is rather beneficial to the employers.

[9] For instance, Feinberg (1973), following in the Millian tradition, argues that paternalism is only justified when an actor not knowing anything about the consequences of his action would do harm to himself without a paternalist intervention–or the voluntariness of action should be questioned for some other reasons. A consequence almost naturally arises: an individual having a passion for hedonistic pleasures could hardly be prevented from gratifying these desires through legal measures if his decision is deliberate and voluntary and if his amusements do not concern others. However, Bretter (2004) demonstrates this seemingly loose interpretation to be underlain by a (perhaps too optimistic) view about individual's awareness and responsibility. Mill dares leave men to free will for they internalized such social values through the socialisation process that protect them from extreme and self-destructing instincts. So, any paternalist tendencies in law should be highly restricted, since they are completely redundant in addition to individuals' wisdom and awareness.

economic theory and on economic policy. For example, Feinberg (1973) argues against rough paternalistic acts saying paternalism turns adults into children not allowing their clear judgement to develop and to act. If we classify not only the cases of interferences with one's freedom and autonomy done in his best interests into the category of paternalism, but also certain manifestations of state provision, we can get a fair range of arguments *for* and as well as *against* paternalism. Feinberg's arguments are echoed in the words of Ludwig Erhard (1957) when the latter stood out against the allegedly too generous unemployment compensations saying that these transfers do nothing more than discourage the will to work and the efforts to find a new job.

The attitude towards paternalism is directly linked to the way we think about human beings. Paternalistic actions are only justified if we regard an individual as incapable of recognizing his own interest and welfare or the means necessary to serve it. However, if man is described as a rational creature serving his own interests and if market is regarded as a mechanism allocating resources and incomes in the most efficient way,[10] then paternalism may never be needed. For economic theory, welfare is not a certain kind of pleasure, but, in formal terms, satisfying the preferences, or the conditional optimization of the utility function (cf. Sydsæter and Hammond 1995). In the frameworks of mainstream economics, we should not think anything about the *factual* sources of individual's pleasure (happiness, satisfaction, etc.)—since individuals are supposed to be well informed, to have well-behaved indifference curves and to serve their best interests, no institution standing beyond markets is needed to promote individuals' welfare and to protect them. It directly follows from the circumstances that John Smith prefers option X to Y if and only if X is better (for John Smith, of course) than Y. Paternalism, i.e., serving and protecting individuals' welfare *under these conditions* would be a meaningless effort (Hausman 2008).[11]

In spite of the extremities, new classical macroeconomics has improved our knowledge of economics by raising numerous new aspects. Some of them had been already mentioned above—however, rational expectations hypothesis (REH) is the most salient. Rational expectations hypothesis was one of the cornerstones of this new theory, without which we could not speak of even inflation targeting today. However, when judging this school of economics, it should not be forgotten that this populous group itself was also divided, and, moreover, the vulgarization of their theory was accompanied by a considerable distortion of these ideas. All this significantly contributed to the fact that there is no solid, clear-cut and reliable view on the new classicals: the literature often reviews and summarizes the allegedly 'new classical' theses in an incorrect and incomplete manner. So, as long as there is no consensus on the nature of the theses, we can hardly expect a professional and rational debate on the scope and possibilities of economic policy that has the chance of harmony with these operational questions.

This volume tries to answer this challenge. Its major goal is to form a coherent picture of new classical macroeconomics while avoiding offering a simple collection of extracts. Our focus is on the original texts, but remarkable attention is paid to the fermentation process that was generated in the secondary literature by these texts. Isolated scientific developments are never awarded a Nobel Prize. It is hard

[10] Hayek (1988) discusses this realization in a very special way, viz. as the *fact* of the natural selection of societies relying on free market mechanisms.

[11] This finding almost directly echoes Mill's standpoint as to the relation between individual rationality and paternalism.

not to realize that how much time passes by from the emergence of a new and significant theory to the time it gets awarded. In Lucas' case, this amount of time was quarter a century. So much time was needed to get economics penetrated by the new theory and to get all the thoughts developed that are regarded as new classical macroeconomics. It seems that the new classical school is much more than the theses elaborated by Lucas, Sargent, Wallace and Barro. Therefore, attention is also paid to the theorists following the leading economists—all is done to paint a complete and sound portrait. However, while reviewing the theory and outlining its evolution, we aimed at presenting a critical analysis as well, so summarizing its content carefully and defining its exact scope were also regarded as express undertakings. As I believe, these goals are neither too modest nor too ambitious.

1.3 Our Guiding Principle: Active Control Or the Nihilism of Economic Policy

In this volume, there are quite few concerns devoted to questions of the history of economic thoughts. At a most, we might have made an attempt to locate the system of theories of new classical macroeconomics in the evolution of the economic thought. However, the range studied here raises some questions that made it a must to review the evolution of economics (at least, its certain chapters) in brief. It may seem (and the title of this section suggests it as well) that the competing theories of macroeconomics debated the role of the state in economies all along. So, this title suggests, according to this *subjective history*, that the fundamental question of the (macro) economic thought of the twentieth century (but, more or less, of the preceding periods as well) was the problem raised in the title.

Before justifying the adequacy of our idea, first, the duality mentioned in the title should be considered as the one that offers a choice for the authors of high theory[12]

[12] The notion of "high theory" is mentioned often in this work. We could attempt to give a direct definition, however, it seems better to put light on the very nature of this concept through an analogy. As an example from the literary history, the opposition of populist trash and high literature may be remembered. The only goal of trash going back even to verse-chronicles was always simple entertainment, so it was marked by a systematic and deliberate shallowing process—however, this cheapness and simplicity was not a result of an inability to create something better. Trash *voluntarily* does not want to reach higher regions (that is, to be of a higher level, more tasteful, complex and sophisticated). In this case, populism becomes a value in itself (cf. Gergő 1930). The relationship of high theory to its opponent can be characterized similarly–that can be labelled, say, as *vulgarized theory*. On this occasion, we must be satisfied only with a few characteristic features suggested. It seems that the goal of vulgarized theory is to be simple, popular and plain, while crowding out every other possible purposes (for example, the desire to get acquainted with reality). The textbook models can be mentioned as parts of a vulgarized theory that dominate higher education—however, nobody believes (nowadays) that they can give an accurate description of processes observed in reality. Their success should be explained on didactical grounds. Moreover, the *ad hoc* models and theories based on mere empirical

on the one hand and for the makers of operative economic policy on the other. If we talk about *active* control, it is a necessity that there should be a *passive* control as well. A rigid inflation targeting regime (that is built on and stuck to a non-flexible Taylor rule) being a recurrent theme in our analysis is an outstanding example for this mechanism. Similar rules are in operation within the scope of fiscal policy as well (Benczes and Kutasi 2010), though these cases should be considered as constraints on economic policy rather than as linking interventions to automatisms. Among the fiscal policy instruments, automatic fiscal stabilizers studied in Chap. 5 are the nearest to the Taylor rules of monetary policy. So, in the course of *passive control*, whether fiscal or monetary policy is talked about, economic policy intervenes in a given macroeconomic system *on the basis* of a given automatism (following a properly defined and declared rule), observing some crucial variables. *Discretionality is completely ruled out.* Deviating from the rigid rule is only possible at the expense of breaking (i.e. giving up) the rule. In our interpretation, the economic policy recommendations of Keynes can be mentioned as outstanding examples of *active control*. In this case as well, necessary actions are decided relying on some macroeconomic indicators, but using automatisms is far from it. The difference can be adequately described by saying that, in the case of active control, only the goal of economic policy is defined, but under passive (or, in other word, *vegetative*) control, an algorithm (a behavioural form or equation) is defined that makes intervention automatic on the basis of the received information set. So, in the latter case, no decision is part of the mechanism (Ligeti and Sivák 1978). Of course, there is only a soft line between active and passive control. Perhaps, we are not far from the truth if we stress the size of the information set that is utilized by each system. Under active control, such a huge amount of data may be analyzed and considered that cannot be compressed into a single reaction function (at least, we should not chase rainbows that we have identified all the relevant variables of the model that describes a given macroeconomic system) and there is no need to regard the effects of our interventions as functional consequences.

A third option is a view regarded here as the nihilism of economic policy. According to this, both the *necessity* and the *possibility* of active and passive control are questioned. As we shall see it soon, doubting the potential of an economic policy intended to exert control (either actively or passively) or its possibility makes two fundamentally different cases. In the former instance, this intention to control is rational, though success must not be expected. Roughly speaking, this was the standpoint of Milton Friedman. The latter case is ascribed to Lucas and his group. It is argued here: *this thesis of the new classical macroeconomics is conditional*, i.e. *it is valid only under certain restrictive assumptions*. Economic

experiences and tendencies are standing on the same side as well: researchers try to find a theoretical foundation for their equations specified in some way or other (perhaps, falsely). So, confusing the causal relations is an existing danger, and catching the right theoretical explanation can be completely missing. On this issue, there are views according to which applied economics remains intact by high theory researches (Horváth and Szilágyi 2004). The things said above cannot be regarded as an accurate definition–we just surveyed this issue in a hurry.

policy is naturally doomed to failure if there is no possibility of systematic interventions because of the unpredictability of the economic environment or the expectations of the actors. Precisely, the faith in *it* is regarded here as the nihilism of economic policy. The debate on the scope of economic policy and the evolution of the view on this scope are almost directly reflected in the changes in the concept of the Phillips curve, and, just because of this, it is not accidental that considerable attention was paid to this analytical device in this publication. Assuming a stable trade-off between unemployment and inflation and, then, broadening the scope of expectational errors reshaped Phillips curve again and again. It is hard to overvalue its significance in economics. Even if the theories formulated within its framework may be debated, we have to recognize that all chapters of the mainstream are inextricably linked to the Phillips curve. Perhaps, only the significance of the quantity theory of money or IS-LM model could be compared with it: all these three theories were the elemental parts of the language of high theory (as an interesting episode, it can be recalled when James Tobin asked Friedman to explicate his thoughts within the framework of the IS-LM model to help critics—cf. Tobin 1972).

One may ask why we only referred to duality in the head, while we were talking about three possible options hitherto (*active control* vs. *passive control* vs. *the nihilism of economic policy*). As an analogy, cherishing the hope of justification, the threefold structure of political ideologies may be remembered. Since these three ideas focussed on and stressed different elements of the achievements of political thoughts (cf. Bayer 1998), none of these ideologies can be placed at any point of the continuum linking the others—for instance, liberalism is not a "mixture" of conservatism and socialism, therefore it cannot and should not be regarded as an "in-between" case of other ideologies (Fig. 1.1).

It seems that such an approach to clarify the relationship of the economic policy alternatives above would be mistaken, since passive control, in our interpretation, is between the intention of an active control and the nihilist view. A passive control tries to exert controlling power but in a rigid way, ruling out any kind of flexibility. It has both the faith in the possibility of control and the scepticism of the nihilist views, since it does not believe *itself* to be meant for (or capable of?) intervening in the functioning of macroeconomic systems through using continuously the value judgements of the perceiving subject.

So, it is argued here that *choosing between active control and the nihilistic view* (or taking a stand between these two extremes) *was the fundamental problem of macroeconomic theories*—especially as the standpoint of passive control penetrated into high theory rather from somewhere outside theoretical literature, from

Fig. 1.1 A schematic description of political ideologies and economic policy alternatives

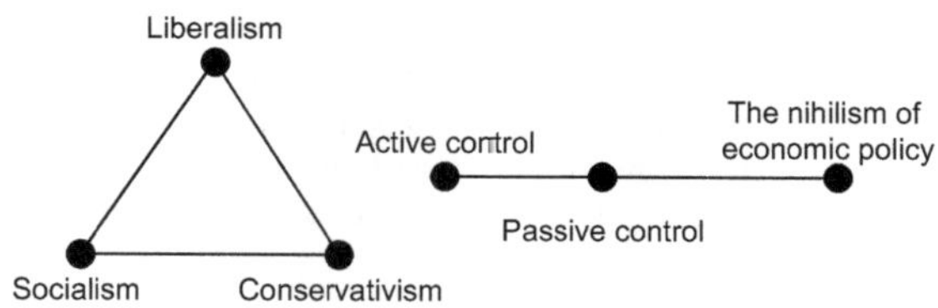

the realm of the operative economic policy. In the successive theoretical systems, the emphasis was shifted from time to time, and even if debates within high theory touched upon the sphere of operative economic policy (one can easily find such instances), this meant that macroeconomic policies also looked for new faiths and new instruments to describe and achieve their goals. It is not necessary to turn back to Say's law (cf. e.g. Mátyás 2002) or to the picture portrayed by Adam Smith (cf. Horváth and Szilágyi 2004) to get a nice instance of the sceptical-nihilist view. According to these notions, since individual optimization works fine, a socially efficient (general) equilibrium gets restored on its own (in other word, *automatically*) after external shocks, so there is no need for state interventions that would be harmful otherwise. During the interwar period, Wicksell elaborated the view most completely, according to which it is harmful if the monetary sector (and central bank policy) allows or facilitates the real sector to deviate from its natural state. For example, a credit expansion out of savings may trigger aggregate demand to exceed supply generating an inflationary pressure throughout the whole economy and even changing the vertical or equilibrium (that is, full employment) structure of production (Mátyás 1979). The resulting cumulative process is not the natural state of an economy, even if Wicksell did not find a reason that is sufficient on its own to divert monetary sector or the central bank[13] from credit expansion. In Wicksell's theory, although this process is limited in time, this period of overheating (or, in other words, economic expansion) does not result in stagnation or recession. When the market rate of interest equals the natural rate, it stops prices rising and restores the vertical structure of production. The economy makes a soft landing in this new equilibrium state. However, Wicksell's message is unambiguous: a control can be exerted from monetary side, but its application should be avoided and it should be made possible for the development path of the real economy to coincide with the natural way.

It is unnecessary to explicate the theory of Keynes in detail. For Keynes, an equilibrium state did not (necessarily) mean full employment, so the need for an activist economic policy became justifiable. This possibility of justification meant for him the urge to (over)stress the significance and effectiveness of fiscal policy. At the end of 1920s nobody could believe that full employment and maximal capacity utilization were the natural state of national economies. Keynes' theory, by announcing the programme for activist policies, aimed at answering this challenge. However, the drawback of his conception was revealed soon: though Keynes argued that under full employment, an increase in the quantity of effective demand

---

[13] This is the very reason why Wicksell's theory is hot even today. For example, recent bank privatization makes a central bank control over the monetary base almost impossible. This development had serious consequences on the economic growth of Hungary and the other new EU member states of 2004. The lesson is not more than commercial banks in two-tier banking systems, getting out of the tight control of the central bank, can generate serious pressures by giving excessive amounts of credit to market agents. It seems that Wicksell was among the first economists warning us of its dangerous effects. On the distorted patterns of the growth of the new member states see Csaba (2008).

has only limited inflationary effects (it is a complete misinterpretation of Keynes to say that there is no inflation before full employment has been reached or that inflation may remain stable), fine-tuning can be only managed by occasional overshoots indeed—so, inflation remains a real danger. So the fact that the public and political elite have started considering inflation rather than unemployment in the following decades (broadly speaking, from the middle of the twentieth century) contributed to the decline of the economic policy guidelines recommended by Keynes. Nevertheless, Keynes did not take the slightest notice of the welfare-decreasing effects of inflation, seeking and offering solutions mostly for unemployment. At most, the economic policy of Keynes stressed (carefully, dropping hints) only that fighting unemployment has inflationary costs.

Monetarism that emerged as the antithesis of the theory of Keynes denied activism. According to the monetarist theses, the effort and the intention to control in an active way is useless and harmful, therefore the famous constant money growth rule got into the centre of the new economic policy recommendations supported even by Robert Lucas. Fiscal policy cherished by Keynes was degraded to be a potential source of inflation, so these new suggestions urged to give up budget policy intentions (the time lags with which economic policy can exert its influence make it even impossible to effectively apply fiscal policy instruments). Incorporating adaptive expectations, theory could prove that maintaining unemployment under its natural level/rate is only possible at the expense of an accelerating monetary expansion.[14] New classical macroeconomics and the theory of real business cycles did not fundamentally alter this view, while the emphasises got relocated—however, the new classical idea of fiscal policy basically differed from the approach represented by orthodox monetarism, even if the spirit of activism suggested by Keynes did not haunt anymore. Special attention shall be paid to the notion according to which activist monetary policy creates only unwanted noises in the system, therefore its application should be avoided here again according to the suggestions phrased by Lucas' group. Emphasises got placed on supply side control (stimuli). Although supply side economics questioned the appropriateness and efficacy of countercyclical demand-side policies, economic policy was not condemned to inactivity anymore—notwithstanding, the spirit of activism did not emerge from these policy suggestions. For instance, by studying Ricardian equivalence and non-Keynesian effects, it will be demonstrated that the presuppositions of the theory should be considered very carefully when drawing economic policy consequences—and that *contesting the effectiveness of activist economic policy is possible only under some restricting and tenuous assumptions.* We shall be led to the same conclusions (true, by stressing logical inconsistency rather than realigning the assumptions) when it gets realized that after incorporating cyclical disturbances

[14] It should be mentioned here that it is quite difficult to distinguish between the mechanisms of fiscal and monetary policy in the theory of Friedman. These problems become extremely concrete when monetary expansion is realized by creating money.

into models, it was not possible to uphold the thesis about the ineffectiveness of anticipated monetary policy actions anymore.

1.4 Methodological Principles

Views on this alleged ineffectiveness of mainstream economics have emerged recently even in the Hungarian literature; in the English literature, they have a greater tradition (some of them will be mentioned in the text below). On the basis of this, it may seem that mainstream economics is in crisis, because it has lost its relevance from the point of view of real life problems. Even if it is true that the relationship between real life circumstances and economics has become imperfect, this discrepancy may be still attributed to various reasons. Among them, it is only *one* cause that is suggested by the views mentioned above, according to which (if it is correctly interpreted) mainstream economics is not able to offer adequate solutions to real life economic troubles. Doubtless, it *would be* a serious problem. In other words, it means that the theory somehow lagged behind reality, that is, it offers answers not to problems directly experienced by us or not to the problems we are concerned with *today*. However, this discrepancy emerges as well if operative economic policy cannot recognize the relevance of the theoretical answers. In this case, economic theory suggests adequate answers to significant and real issues that, however, remain unutilized. It *would be* a communication problem of economics, since it *would mean* that our science turns out to be incapable of formulating its answers as clear messages—*this case*, however, does not imply the irrelevance of the theory in any way.

Answering this critique should not be restricted to listing examples from economic policy and history. Moreover, as we shall see, these examples can be completely eliminated from the arguments. Accordingly, an attempt is made in this methodological introduction to identify and reproduce the way mainstream economics thinks (or, its world view) and, on the basis of these findings, the *relevant* way of criticizing mainstream economics is established. So, the primary goal of this introduction, beyond explicating the principles applied here, is to stress that mainstream economics can hardly be criticized for the lack of the adequate economic policy recommendations. The only message, the only "economic policy" recommendation of neoclassical economics emerged as a *pure theory* was the urge to establish markets that ensure completely free competition. Actually it was not, as we shall see soon, an economic policy strategy but an intention to merge an abstract, scientific-theoretical environment and the directly experienced reality into each other, the necessity of which was stressed by the results of the analyses based on the *pure* theory. Even if monetarism made attempts to phrase such suggestions, these proposals were broadly of similar nature, though, in some cases, the discussion was not about the *imperative* to merge, *but about the merger itself*. Meanwhile, mainstream economics remained a pure theory despite its conceit and a critic should always keep this fact in mind.

A terminological issue should also be clarified here, just at the beginning of this introduction. It is intended in this work to distinguish between *mainstream* and *neoclassical* economics. This distinction is deliberate, of course and consistent, but ponderous sometimes, if not unnatural. From among these two concepts it is easier to define the second one that is identified *here* as the neoclassical economic theory of the nineteenth century laying the foundations for today's standard microeconomics. There are multiple and competing definitions of mainstream economics of mainly sociological character that do not try to define mainstream by its theoretical content. According to a possible approach, mainstream economics consists of theories that are accepted by the highest academic circles at a given time (Colander et al. 2004). Citing another definition rooted in sociology, the mainstream is the theory that constitutes the curricula at the most prestigious universities and is awarded the most illustrious scientific prizes (Dequech 2007). The foundations of both definitions are formed by categories (acceptance, prestige, etc.) that can be only dubiously filled with purport and used for practical classifications. However, the theoretical heterogeneity and temporal variability of the content of the mainstream is still bothering, since even if a consensus on the definition of the academic elite was reached, we could still talk about the mainstream in the context of certain periods. While the effects of the neoclassical school on mainstream economics were remarkable from the very beginning,[15] for example, experimental and behavioural economics should be also regarded as parts of the mainstream according to the above-mentioned criteria, the results of which were often intended to directly confute the theorems thought to be traditionally the elements of mainstream economics. To avoid all these complications, mainstream economics is identified *here* as the neoclassical synthesis models also including monetarism (from the theory of Milton Friedman to the ones of new classical macroeconomics) (for further details see Weeks 1989). Of course, it can be debated how uniformly this branch of theories is rooted in neoclassical economics—for instance, Hoover (1990) argues that new classical macroeconomics rather goes back to the doctrines of Walras, while Friedmanian monetarism to that of Marshall; however, it is clear that these theories are actual parts of mainstream economics, so they are worthily assessed in terms of practical applicability. So, this definition perfectly serves our purposes (cf. e.g. Samuelson and Nordhaus 1998, where functioning of macroeconomic systems was tried to be explained using these successive theories).

1.4.1 A Diagnosis

According to the major findings of the above-mentioned critique, mainstream economics did not support *effective* economic policies of any countries in the

[15] Frank Hahn (1984) makes it unambiguous by dropping a few short hints that a strong relationship exists between neoclassical and mainstream economics.

world (Veress 2007), and, *for instance*, considering theoretically the experiences of Eastern European transition economies did not happen at all. Moreover, even if there were attempts to apply the theses of mainstream economics in practice, these efforts remained fruitless (Csaba 2009).

Phenomenology. Regarding its topic, the critique questioning the relevance of mainstream economics considerably draws from the crisis effects described by Husserl in his Crisis essay[16] (Husserl 1976), since it can be gathered from reading these critical remarks that mainstream economics does *not* apply to directly experienced life (the directly experienced reality), to real life problems, or, more specifically: it is not the directly experienced life (directly experienced world) that the mainstream is concerned with.

According to Husserl phenomenological critique, sciences participating in the crisis formulate their theses *not* in relation to the directly experienced reality, rather to an artificially constructed, abstract world (or sphere). So, modern sciences describe their notions and theorems on the basis of their special world view. According to this reasoning, modern sciences are guilty of trying to substitute their own notions for the knowledge gained from naive/everyday experiences. The answer to this crisis is not the restoration of a naive cosmology, of course—it is only the critique according to which observations of modern sciences want to replace everyday experiences that are, otherwise, irreplaceable (for example, Einstein's physics invalidated the Newtonian concepts of absolute space and time, while they were still demanded by everyday experience and the natural attitude, both built on a natural assumption of their existence—cf. e.g. Hawking and Mlodinow 2005). In other words, modern sciences have existed in an abstract environment differing from the (directly experienced) reality all of us live in.

Phenomenology was intended to be a new idea of sciences. The monumental philosophical movement and epistemological criticism launched by Husserl aims at studying "phenomena", that is, appearances of things, or things as they appear in our experience. Husserl readily used Kantian terminology (e.g. Husserl 1931), and, actually, the problem he raised is strongly linked to the thoughts of Kant (1783, 1787). (Through experience, we cannot reach the Kantian things-in-themselves, since they are given to us not in their real nature, independent of the perceiving subject, though everyday attitude does not even question our ability to know sensory qualities—cf. Biemel (1936). Husserl did not deal with this relationship, and he was satisfied with his notion that there are no things-in-themselves, that objects exist for us, and are for us what they are, only as objects of actual and possible consciousness; cf. Husserl (1931) Experience, perception and consciousness were in the centre of interest of phenomenology (cf. DW Smith 2009). Evidently, neither those sciences that act in their abstract environment nor laboratory human sciences (experimental psychology and social psychology,[17] experimental economics) meet the requirements of the scientific ideal established by phenomenology. "Back to the things themselves!"—it became the watchword of phenomenology, and this ideal inspired several

[16] The two-volumed Hungarian edition of Husserl's crisis essay was used when writing this methodological introduction. The first volume (Husserl 1998a) contains the main text, while the second (Husserl 1998b) includes the comments published in the first German edition (1962) plus an editorial epilogue to support the reader in interpreting the Crises and in reviewing the life-work of Husserl (Mezei 1998).

[17] For example, Bungard (1997) stresses that, during laboratory experiments, we can observe at most how individuals behave in an artificial environment; moreover, it also has to be taken into account that individuals make efforts to find out the purpose of the experiment and to behave accordingly (or even contrary, on purpose).

disciplines.[18] Sociology, for example, was especially responsive to the methodological recommendations of phenomenological criticism (cf. Hernádi 1984), but there were ground-breaking efforts in literary theory and science (Ingarden 1931; Maclean 1986). In social psychology, attempts by Erving Goffman were particularly interesting in which he did not take interpersonal actions out of their actual environment. So, in this view, social psychology is not a laboratory science but the science of social behaviour studied as it happens in reality. Economics, however, has not shown any particular interest in the methods of phenomenology so far, only the first few questions have been raised, extending the crisis problem to economics (for further details see Galbács 2011b).

Husserl's view seems to be especially interesting if one considers the fact that, nowadays, experimental methods gain ground against pure theory. Husserl (1910) sharply criticised experimental psychology for its belief that laws of thought can be obtained by using experimental methods of psychology (practically, all this was about the psychological foundations of logic). However, the relationship of logic and psychology is completely different: logical truths do not depend on whether someone accepts them or not—i.e. what is logically true does not depend on any act of evaluation or judgement.[19] Acts of judging are empirical acts done in space and time—however, what is logically true (a logically true thing) is not tied to any particular place or time. So, we have to distinguish things that are true in themselves from things that are regarded by someone as true. Psychology can only study the latter, while the former is (or should be) the subject of an inquiry of pure logic (cf. Pivčević 1970). So, the efforts of psychology should not aim at disproving any arguments belonging to the scope of pure logic. The theses of mainstream economics that are based on abstraction may be *logically* true in spite of the fact that (their logical truth, of course, is not an a priori assumption but should be a question to investigate) economic actors, studied under laboratory conditions, do not confirm them to be the basic principles of their actions. Refuting an axiomatic-deductive theory cannot be psychologically initiated.[20] [→ **paradigm**]

Phenomenology is an esoteric philosophy to a considerable extent and, moreover, Husserl and most of the members of the phenomenological movement are very hard to read. Husserl tended to briefly and obscurely express his thoughts, so every effort to interpret him requires meticulous reading. Phenomenology exerted its influence, first of all, on sociology. Phenomenological sociology has grown to be a highly fruitful discipline overseas. Although mainstream economics has been moving in completely different directions, economists should not turn deaf ears to new aspects and findings that can enhance our knowledge by complementing mainstream theses. Phenomenological sociology elaborated a theory and a methodology that hold out a promise of applications within economics. By models inheriting the concept of homo oeconomicus (that is, because the mainstream organized itself as a pure theory) we cannot learn anything how people make economic

---

[18] Phenomenology as a theory of science should not be discussed in such an oversimplifying and brief way at all–this short and striking allusion is not intended to substitute for a systematic introduction.

[19] Using Husserl's terms: the law of gravity had already been true when Newton had not discovered it yet, therefore neither he nor others could regard it as either true or false (Husserl 1900; cf. Kusch 2009).

[20] In this context, two circumstances need to be suggested. Firstly, Husserl did not object abstract sciences *in themselves*, only when they replaced the directly experienced reality by the world they created. An axiomatic-deductive pure economic theory grounded in abstraction cannot be condemned even by Husserl himself as long as it clarifies its relation to directly experienced reality and refrains from merging these spheres into each other. Secondly, for Husserl, the right and effective answer to the objections against these trying-to-merge abstract sciences is *not* made by experimental disciplines (the laboratory environment of which is abstract and far from reality again) but by phenomenological sciences.

decisions in everyday life, since the man of mainstream economics (whatever degree of complexity his behaviour shows) is eventually a machine, defined as a system of behavioural equations.

Phenomenology, and, hence, social sciences with phenomenological interest started paying attention to everyday life and attitude, and this effort, partly, goes back to Edmund Husserl's scientific views. The arguments are clear. The rigour of natural sciences had become a common ideal, and non-natural (i.e. social) sciences tried to meet its requirements under the burden of this ideal that seemed to be so appealing. Phenomena were regarded by them as manifestations of natural laws identifying their job as to find and describe the laws that govern the things studied. Conscious mind, together with the physical world, was naturalized, since acts in the conscious mind were assumed to be guided by stable laws. Although the critique by Husserl (1910) was directed towards experimental psychology that had the ambitions to reform sciences and that regarded exact mechanics as its model, his arguments can be extended to all other social sciences. As it has to be stressed, there is a fundamental difference between *nature* and *mind* as objects of scientific research. Phenomenology and disciplines applying phenomenological methods aim at performing a pure, substantial and direct analysis of conscious mind that is also aspired by psychology (and even by sociology and experimental economics), which can only *indirectly* study consciousness. Responses of data suppliers or results from observations of experimentees under laboratory circumstances ensure access to only exterior manifestations of the conscious mind. Consciousness and physical-natural world cannot be studied through the same method of direct observation, since consciousness is not experienced as appearing similarly to factual things. A psychical being appears as itself through itself. The latter is a crucial argument for separating physical and psychical world—and for the view according to which psychological being cannot be studied through methods and on grounds that regard natural sciences as models.

Husserl suggested studying how the world comes to have its straightforward existence for us, how there arises in us the real world and the constant consciousness of the universal existence, so Husserl (1976) wanted to consider the how of the pregivenness of the world. Life-world is one of the corner-stones of phenomenological sociology: it is the world (in everyday sense) *in* which we live in our natural attitude and at which our conscious life is directed. Phenomenological sociology studies the consciousness of individuals living in life-world and creating societies. So, phenomenological sociology describes how individuals living in life-world with natural attitude generally think: it tries to clarify how life-world (as a general and unexamined ground for the natural world view) supports our experience and knowledge (Berger and Luckmann 1966).

The life-world incorporates not only the surrounding material objects and our social environment but a culturally determined frame of reference (pregiven knowledge) that helps us interpret our experiences. This unexamined ground, by its pregivenness, supports us to classify our perceptions and experiences into reference schemata available and to conceptually identify the objects of perceptions. The stress was on the subjective characteristics of experience and knowledge. Our experiences and acts of experiences are supported by the *types*: we classify new experiences into types and thus they fit into relevant reference schemas, or, if it is not possible because of a bothering contradiction, a problematic type gets modified. This knowledge is *par excellence* social knowledge, since it is not only constructed from an individual's own (previous) experiences but from experiences that were transmitted to him from his fellow-men (say, from his parents and teachers during the socialization process). The man of the natural attitude accepts certain definitions, assumptions, prescriptions and types the totality of which, i.e. his everyday knowledge, makes it possible to act in the life-world (in everyday life) following routines and to interpret this world. Acts of an individual are not isolated actions, but parts of a universe created by him together with other individuals assumed and experienced similar to him, in which universe any individual acts are committed taking others' actions into consideration and using specific interpretative strategies.

Studying everyday knowledge and consciousness characterizing individuals in natural attitude eventually helps us understand how people (even as economic actors) consider and solve everyday problems, how they call and apply their cognitive schemes, the types, the behavioural routines and how they deal with confusing situations—if, say, a situation turns out to be unsolvable through usual routines (e.g. Garfinkel 1957), or an inconsistency of individual's stock of knowledge, that was hidden hitherto, becomes evident (Schütz and Luckmann 1975).

These highlighted and brief examples can clarify the fact how far this theoretical enquiry is from the way, for example, mainstream economics tries to discover the *functional* (sic!) attributes of the economic man. By regarding man as a social being and, particularly, questioning the manner of experiencing and interpreting, we get far from the aspect of mainstream economics that is basically mechanical and non-inquisitive about consciousness. We should not identify the existing, actual individuals with robots mimicking mathematical functions and giving well-defined answers to well-defined stimuli if we want to understand and not to oversimplify and abstract what happens in the conscious mind. Stressing *this* can be the program of a phenomenological economics.

Modern sciences (modern, in Husserl's sense) break away from real life through abstraction, and this poses the danger of diminishing their relevance as well as attention to and interest in real life problems. Of course, it is not a speciality of economics. Galileo's abstract physics built on mathematics played a key role in the emergence and spreading of abstract sciences. Galileo created his system deductively, broken away from experiments with actual bodies. Aristotelian physics (physics as natural philosophy, in a pure qualitative sense) aimed at studying actual (i.e. non-abstract) bodies and relations yet (Bréda 1983; Maróth 2002). Galileo's veer was somehow a consequence of reasoning followed from a view of Aristotle, according to which there is no void as an ideal an abstract medium, as *geometric* space (Physics 214^{b}12–216^{a}26; cf. Ross 2004). And if there is no void, no *actual* shapes can be placed in it to be objects of scientific investigation. This conflict was solved by Galileo, who placed geometric (that is, *ideal* and, it must be stressed, *not real*) figures in the geometric space choosing their dynamics to be the object of his inquiry. Through this, physics advanced to be an abstract-ideal science as well that was organized by the axiomatic-deductive scheme (for an example see Kovács 2004). The ultimate goal was to mathematize (geometrize) nature, since, according to the reasoning, theses consistent and pure to the utmost can be gained by mathematics (Galileo 1632). So, Galileo had recognized the features and potentials in mathematically supported reasoning that, later, made mathematics become the model and ideal of sciences, and, moreover, the form and common language of reasoning.

The deliberate goal of mathematizing, the intent to mathematically reconstruct the empirically observed processes is to *avoid* experiments based on empirical observation. Creating (deducing) theses became a speculative act facilitated by the apodictic nature of formalized science. According to Husserl's commentaries, Galileo's step was, eventually, a necessity, since we should not believe that there are multiple worlds existing simultaneously, though reality differently shows itself in perception for all of us. The goal and *sense* of mathematizing is to discover and describe the *real* (unchanging-unchangeable and eternal) nature that hides behind phenomena being only relatively and confusedly perceivable. For this purpose, mathematics and geometry seemed to be the right instruments, since these sciences created an objective world starting from ideal objects—geometry displays something that *objectively* exists (Fink 1936). This concept offered the possibility to grab the very thing from reality flowing before our eyes as a boundlessly complex run (using Husserl's words, from the "Heracletian flux") that (hiding behind phenomena) can be attributed to the real nature. Supported by this, the new idea (i.e. model) of sciences emerged that emphasised the possibility of describing the world in its totality (Husserl 1976).

A thesis of Galileo was in direct parallel with thoughts on real nature in which Galileo stressed that perceived phenomena consist of only *subjective* qualities we attribute to the world as it is in itself that can be understood only in geometric terms, since nature-in-itself is mathematical (as Galileo said, the philosophy of nature was written in the language of mathematics; cf. Jesseph 2004). Questioning the possibility of sensory experience was further deepened by René Descartes, who stated, following Galileo, that world as it is in itself cannot be perceived at all, saying our sensory experiences do not necessarily represent the world. Being in the prison of perception bearing no resemblance to the external world, one does not have any knowledge of the things of reality, so, on the grounds of observation of reality (that is, based on experience), we cannot draw universal conclusions (Descartes 1637; cf. Tuck 1989). *Motion* became the central concept of the new physics still functioning as natural philosophy as well. The physical world got interpreted as a system that follows its own laws, and, simultaneously, the faith became common and usual according to which mapping and describing reality standing behind sensory experiences is the ultimate goal. So, if the world is governed by this mechanical pattern resulting from the motion of minute particles, then genuine understanding is only possible if it is supported by mathematics. And, of course, as another statement (equivalent in its meaning, but more weighty in its phrasing), it was also believed to be true that the world can be understood *indeed* by using mathematical apparatus. We have not spoken of a general causal order of the world, according to which every event goes go back to reasons. In this concept, nature and the world can be discovered in their totality through mathematics, so *causality* and *exact causality* had the same meaning (cf. Biemel 1935). The burgeoning of mechanical philosophy (as a world view) was also undeniably strengthened by medical science that underwent quick development. Human organism was regarded as a complex, but *coherent* system, eventually, as machinery the operation of which can be comprehended by laws governing physiological operation (Boros 2007). It needed only a small step to extend this mechanical philosophy to cover moral actions. According to this view, everything, including human thoughts and acts as well as motivations (Hobbes 1640) can be described and explicated by mathematically specified laws of motion[21] (Jesseph 2004).

[21] Of course, this concept of motion is more abstract than motion in its mechanical sense. The philosophical thesis that motion is an inalienable feature of matter (that features all kinds of

Extending the principle of and the searching for exact causality to consciousness was, eventually, inevitable. If the physical world, the causal relations of nature can be described *without* taking mental phenomena into consideration, then there remains no room for some specifically non-physical factors of human nature in this system. Spiritual-moral-psychical acts *must be* actions that follow physical laws, since, *if* those acts were *not* so, there would exist a further system of causality simultaneously with the causal relations entirely determined by physical laws (Biemel 1935). So, these spiritual-moral-psychical actions can (and *should*) be described through the same physical-chemical formulas as physical nature (cf. Husserl 1910). In this view, there is nothing specifically human in human beings, since individuals behave under the rules of exact causality. *They are mathematical functions eventually*. It is no wonder that Milton Friedman, in relation to human behaviour, talked about a money demand *function* (Friedman 1966) when giving the methodological foundations for mainstream economics—or that the theory of Keynes (1936) definitely denying the existence of such exact causal relations could be part of mainstream economics only "neoclassified", i.e. through the neoclassical synthesis. **[→ liquidity trap]**

Physics is inspired even today by the ideal of describing consciousness on a physical basis. These efforts exert serious influence on economics as well, since, as we have seen, economics tries to do the same when we construct mathematical models to describe individuals' (economic) decisions. Today's physics also believes that, since physical laws giving the background for biological processes are all known now, the lack of knowledge of material processes does not hinder biology anymore in becoming a sub-discipline of mathematical physics. Although determinism is not taken as valid today because of Heisenberg's indeterminacy principle, it is not a serious obstacle, since estimates from physical calculi would be considerably accurate because of the low energy levels of human body. So, physics seriously aspires to establish a complete (or, in other words, a final, ultimate or master) theory that, as a coherent theoretical framework, fully describes all physical aspects of the universe (cf. Weinberg 1993). This state of knowledge seems to be unreachable for the only fact that human brain and body consist of too many particles, hence there are so many equations to solve to predict human behaviour that is far beyond our abilities (cf. Hawking 1993). At least, it can be obvious that physics does believe that consciousness and human behaviour can be reduced to physical processes and that a physically grounded (exact) model of them is unrealizable only because of capacity limits. This is a serious undertaking!

Somehow it is beyond our task to react to this idea, since this is not the task to provide a thorough (or just a superficial) critique of the man concept of mainstream economics rooted in physics or even that of modern physics. However, it seems that there is a clear difference between a mathematical model that approaches behaviour as a pure theory and the concrete desire to entirely describe this behaviour in

manifestations of matter) applies to both quantitative and qualitative changes of matter. All this does not alter the view that quality changes go back to quantitative modifications.

physical terms. It is argued here that economics shares the faith according to which the essence of things can be well grabbed mathematically which faith is acceptable as a scientific principle. However, saying that the world *in its totality* can be described also well (even perfectly) in mathematical terms is much more serious.

Before taking a step forward, let's make a short detour at the problem of naturalizing of consciousness (cf. Husserl 1910). The heart of the problem lies in cognition. Philosophy of mind stresses that (the act of) perception *cannot* be empirically approached—if one tries, he surely misses the point. The classic example is seeing. Let's imagine that we see a brown cow. The brownness of the cow, that is, the brown colour that we experience when we see the cow is *not* an objective feature of the cow, being independent of the perceiving subject. It is evident that there occurs some physical-chemical processes in the course of seeing (that can be well grabbed empirically), but these processes do not give an account of seeing itself, the act of seeing in the conscious mind. They only show what biochemical processes frame the act of seeing—i.e. what kind of biochemical processes serve as a necessary physiological background for seeing, since without them seeing as an act of the conscious mind cannot be even imagined (we say an act of the *conscious mind* and not a process of the *nervous system*). However, seeing (and, indeed, all kinds of experience) is something different from that is empirically described by these processes. Seeing as an experience occurs in the conscious mind. When we see a brown cow, this experience does not follow from the way in which seeing the cow occurs within the framework of certain physiological processes (physical impulses received by biological sensors) going through our brain and the entire nervous system. To take a rare, but outstanding example: in the case of patients suffering from Cotard's syndrome (or, in other words, walking corpse syndrome), the metabolic activity of the brain cells informs well of the processes in the conscious mind—or, more precisely, of the lack of these processes (Charland-Verville et al. 2013; Joseph and O'Leary 1986). Patients lose the functioning of the conscious mind: they have no smell, taste, their thoughts disappear and all kinds of connections to the external world cease to exist. If the brain regions supporting our self-awareness do not function physiologically, it displays some external effects and manifestations of the processes in the conscious mind—however, the exams used to map metabolic activity do not reproduce what is going on in patients' consciousness (only externally informing us of its physical-physiological aspects).

So, what is going on in consciousness, and what is it that cannot be grabbed by naturalistic approaches grounded in physics? Eventually, all the processes in the conscious mind can be listed here. For example, the brownness of the cow definitely joins in the sight in our consciousness, and the emergence of this colour experience cannot be explained by the biochemical processes in the background anyway. It is out of the question that our experiences occur in the framework of certain physiological processes, as it has been mentioned above, since our sensors connect in this way to the external world. However, the real question is that how sensory stimuli from the external world (or the things from which these stimuli come from) and the picture emerging in and for our consciousness (if we talk about seeing—generally speaking: *our experience*) connect to each other, and how the latter follows from the former. The core of the problem is that things appear as phenomena, as beings-in-our-consciousness, but the *correlata* (correlatum in plural) of these phenomena, i.e. the external material objects (i.e. things-in-themselves) are not *immediately* or *directly* given to us. We cannot know what the objects are like in the outside world in their objective existence, independently of our conscious mind, since we cannot catch them without references to our consciousness (of course, there are schools of epistemology that believe in direct realism, that the experienced external material objects are immediately given to us—this extreme view, however, is hard to prove—cf. BonJour and Lyons 2013). The picture emerging in our consciousness cannot be compared to anything else, since there is no picture existing independently of our conscious mind. Who could ever perceive a picture that is independent of any conscious minds? How can a thing existing in consciousness be independent of consciousness? There is (perhaps) no colours being independent of our conscious mind (at least, we have no available support to analyze this issue), since a colour is nothing else but a sensory experience triggered by rays of light coming to our eyes on certain wave-lengths. But colours should not be identified with either these rays of light or the chemical reactions triggered by their interaction with sensors. Colour is rather the way in which these rays emerge in and for our consciousness mind. So, according to this view, colour *cannot* be an objective feature of external material objects. Well, we can say that any knowledge we acquire of the world, this knowledge is always a combination of a sensory and a subjective element: an external and an internal (coming from the conscious mind) element coincide with each other here (Boros 2009). This is the essence of the aprioristic knowledge raised by Kant. *In experience, our consciousness somehow influences that which is experienced—at least the way in which it emerges in and for consciousness*. The mere existence of the things-in-themselves cannot trigger any knowledge without the perceiving subject. Thing-in-itself— conscious mind—interaction: any of our knowledge relating to the world originates in this threefold system, and, therefore, internal and external world cannot be even separated. The point is that the perceiving subject is always implied in all kinds of knowledge: knowledge is always possessed by the perceiving subject. And, as János Boros (2009) phrased the problem: we do not even know which part of knowledge is external and which part comes from the perceiving subject.

Although economics takes human beings into consideration, it is never interested in the individual actions and processes going through in them but, firstly, tendencies of behaviour labelled as laws and, secondly, the outcomes following from the interactions of numerous, fictive and completely similar economic actors. The individual is a black box for economics, containing processes about which economics obviously knows, but it does not take its significance into consideration (because of the approach and the scientific strategy economics has). Of course, these inner processes have outer representations: an individual's acts, and behaviour studied by psychology and the biochemical processes that can be monitored well by instruments. Perhaps, it can be grabbed most effectively here that how the approaches of psychology, psychophysics and neurology and philosophy of mind differ. The former analyze what the black box does—the latter questions what is going on in this black box. Of course, this particularity of psychology and psycho physics was inherited by even mainstream economics, determining fundamentally the approach our science has.

1.4.2 Economics Is a Rationalizing Science

Before we devoted some paragraphs to the problem that was raised by getting into consciousness, we went too far ahead. A statement was phrased according to which mainstream economics rooted in neoclassical views shared in the efforts to mathematize human behaviour and relationships, reducing these interdependencies to mechanical, exact causal relations. Our first task is to demonstrate that neoclassical economics with its strong tendencies towards mathematization also joined the traditions of modern natural sciences. Here, it means that neoclassical economics has deduced its theses on the functioning (behaviour) of abstract-ideal entities from a few underlying principles and postulates using mathematical apparatus, and, doing so, it has followed axiomatic-deductive methods and, hence, rigorous methodological rules. This feature of synthesis models is an inherent part of the neoclassical heritage—for instance, Friedman, in his Nobel lecture, stressed that it is no use distinguishing between natural and social sciences (cf. Friedman 1977). Eventually, it was the intention and the program of transforming economics (as a social science) into a natural science.

Neoclassical texts directly reveal this intention. Being inspired with the conviction about the existence of causal relations and laws (Menger 1871; or Marshall 1890) economics started idolizing mathematics and physics in terms of both self-organization and methodology (Jevons 1888). Neoclassicals examined the functioning of abstract entities under hypothetical, i.e. non-actual conditions[22] (Walras 1874). The role that abstract geometric space had played in the physics of Galileo was taken by equally abstract environment of the absolutely free competition in neoclassical economics that was filled with rational actors making only economic decisions as analogies with ideal geometrical shapes. Actions of economic actors are not simply determined. Although, for example, it is presumable that even the weather may be considered as a determinant of human behaviour (cf. e.g. Durkheim 1897) and one can easily produce models to take its effects into consideration (for a witty example of the relationship between the weather and economic actions see Mérő 2004), the functioning of economic actors was believed to be determined by only a *few*, mathematically formulized laws reminding of the laws of motion in mechanics. It is not an overstatement to say that the efforts to investigate market equilibrium emerging as a resultant of opposing forces (market *forces*—sic!), to

[22] At the quoted place Walras defines pure economics as "the theory of the determination of prices under a hypothetical régime of perfectly free competition."

describe market as an automatic equilibrium system, or to theorize market systems as having tendencies towards general equilibrium—these are all the heritage of physics[23] (cf. e.g. Szegedi 2003; Newton 1687; this conception emerges at various places, see e.g. Kopányi 2003; Varian 1999; Samuelson and Nordhaus 1998). So, economics has become a *pure science*. It declared *not* to investigate *life* that would allow the presence of the (perceiving) subject as well as a research programme (it is not stated that economics did not make effort to collect pieces of knowledge of reality: the ultimate goal of all theories is to describe reality—however, it is not indifferent *how* and *where* a theory gains the ammunition for this description and explication), but to *deductively* consider the consequences of presupposed absolutely free competition[24] (Walras 1874; Marshall 1890). Economy has become an abstract mathematical system (Zalai 2000), that can be well described in terms of, firstly, the relation of elements (i.e. actors), the environment, the inner structures and the mathematically interpreted processes and, secondly, the presupposed features of these processes. On the basis of the absolutized axiom of utility maximization, economic decisions can be eventually interpreted as optimization problems. Actors (consumers and producers) possessing *all* the relevant and necessary information and considering the utilizing alternatives of the scarce resources available try to take the optimum decision regarding their preferences (their objective function)—so, since all its necessary circumstances are given, an economic decision becomes a rational action *within the theory*. Homo oeconomicus as presupposed by neoclassical economists is an idealized, abstract creature that can be characterized by an intention to exchange (when he acts in order to maximize his utility coming from either consumption or production) and whose *only* task is to take economic decisions (Lehmann 1968). For homo oeconomicus, there is no time, no social and natural environment, he is ageless, he has no whims, and his decisions are not biased by occasional external effects. So, his behaviour only reflects the *objective* and consistently prevailing economic laws established by formal

[23] The following rows are especially typical and squealing: "The concept of general equilibrium refers not to the existence of an unstable and instantaneous equilibrium, but to the abilities of an economy that plays of forces point towards an equilibrium state (as a resultant of individual actions, on parts of the market and in the entire economy) due to self-regulating mechanisms of market itself. Market relations are characterized by stability." (Kopányi 2003)

[24] According to Walras "We have, perhaps, at last reached the place where we can see the importance of a scientific formulation of pure economics. From the viewpoint of pure science, all that we needed to do, and all that we actually have done up to the present, was to treat free competition as a datum, or rather as an hypothesis, for it did not matter whether or not we observed it in the real world, since, strictly speaking, it was sufficient that we should be able to form a conception of it. It was in this light that we studied the nature, causes and consequences of free competition." Marshall highlights that "the laws of economics [...] are statements as to the effects which will be produced by certain causes [...] subject to the condition that *other things are equal*, and that the causes are able to work out their effects undisturbed. On this account it has been called a hypothetical science [...]. [...] Almost every scientific doctrine [...] will be found to contain some proviso to the effect that other things are equal [...]. [...] These conditioning clauses are not continually repeated, but [...] in economics it is necessary to repeat them oftener than elsewhere [...]." (italics in original)

rationality.[25] [→ **homo oeconomicus**] Keeping Husserl's interpretation in mind, we can say that human (and social) sciences, similarly to natural sciences, i.e. abstracting from everything subjective, constrained themselves to phrase only objective truths (Husserl 1976).

Homo oeconomicus. Social sciences have definite man concepts that highlight those aspects of man and his behaviour by idealization that are relevant for the given human science. According to these concepts, man is solely of *this* or *that* sort, but not *different* sort. *Homo sociologicus* is the man as sociology idealizes it—depicting man as a social being. Establishing its own man concept, sociology stressed that man living in society adapts to norms and values. So, man is a social being, and it is impossible to define and characterize him scientifically as a creature existing outside of society. Eventually, it is society that endows man with the essence of a human being—and it can deprive him of this essence as punishment outlawing him with the intention of improvement (Foucault 1975). Ergo, sociology interprets the functioning of man and society as an interaction between individuals and society—however, psychology puts the emphasis on individual's mental factors. So, *homo psychologicus* is a being whose behaviour is determined by individual mental circumstances more strongly than by values, norms and social environment.[26] When political science created its own *homo politicus*, it placed man's efforts to assume the political power[27] in the centre (cf. Bihari and Pokol 1998). Social psychology also has its own interpretation of man, establishing its approach and research strategy (Csepeli 1997), even if this concept does not have a distinctive label, according to which human behaviour is biased by the effects of groups *of certain kind* and *of certain size* interacting with him. Social psychology also talks about mental effects, but the sources of them are not to be found in individual's personality development[28] (cf. e.g. Mead 1970). The man of economics, *homo oeconomicus*, who continuously takes rational considerations and economic decisions as a non-social idealized being is not alone—even if he is regarded and presupposed by economics as lonely.

Although the man concepts is synthesizable and although methodologies of disciplines can be extended and enriched by considering the aspects and approaches of other sciences

[25] As an outlying parallel, we can remember Italo Calvino's allegory of the nonexistent knight (Calvino 1962), whose strange behaviour governed by ice-cold rationality is infinitely far from the contingencies of everyday life.

[26] Although sociology and psychology have different man concepts, they are not radically different after all. Sociology is ready to rely on the explanatory power of both social and individual psychological factors. It is highly characteristic for theories of deviance (cf. e.g. Andorka 1997).

[27] It is interesting that homo politicus can be found in the literature as well as in a completely different interpretation. According to this notion also having roots in political science and political philosophy, *homo politicus* concentrates on justice and social well-being (Faber et al. 2002). Man, being rational as a consumer, has a further utility function, that is determined by the state and conditions of society–accordingly, the utility realized by *homo oeconomicus* is dependent on his *own*, while that of *homo politicus* turns on *social* well-being (cf. Nyborg 2000). However, this reasoning definitely clarifies the fact that man concepts of different social sciences always try to grasp only certain aspects and not the totality of human behaviour.

[28] Simmel (1900) keeps listing scientific man concepts. Of course, these conceptions should not be confused with those trendy concepts proliferating in enormous quantities that, instead of idealization, summarize briefly some characteristics of human behaviour and translate them into Latin in order to appear as highly scientific. Recently, we could talk about *homo cyber sapiens* (that is, the man who can extend his biologically determined intelligence thanks to new technologies; cf. Steels 1995), or *homo creativus* (who is simply creative; cf. Inkinen 2009).

(cf. Andorka 1997), it must be noted that these man concepts, in their clear forms, are quite far from each other (this is the reason why synthesizing may be particularly exciting). However, it is interesting that, while economics presupposes its man unchangeable and timeless, sociology puts emphasis also on situations in which homo sociologicus steps out of the world closely defined by norms and values. So, compared to economics, sociology pays greater attention to study settings when man leaves the environment assumed in terms of science and when the set of determinants of behaviour changes. For example, Robert Merton (1968) readily examined the cases in which society prescribes goals and ways to achieve them that are (partly or entirely) unacceptable for the individual. Hypothesizing the denial of values, he could understand certain deviant (criminality, drug abuse, alcoholism) and extreme forms of behaviour (rebellion). For Émile Durkheim (1897), the breakdown of norms and values endanger the existence of society by disintegration (these were the very arguments on which the attempts to apply criminal law to enforce moral norms were based on—for further details see Bretter 2004 and Hart 1989) and this process can be blamed for the proliferating of certain forms of suicide. [→ **paternalism**]

So, the ultimate goal of neoclassical economics was *not* to describe *life*, the directly experienced reality, but to generate hypothetical outcomes comparing to which actually observed processes (may) differ. Of course, these differences may constitute the subject of research—however, in terms of *economic policy* and definitely *not* in terms of the pure theory (Heller 1927). Neoclassical economics is, therefore, a science operating with ideal (ideal-typical) concepts. Hence, the reasoning is totally wrong (however, respectable, because of their vehemence) in any case that criticizes neoclassical economics on the basis of the cardinal utility theory claiming that *actual* costumers do not surely assign their psychological satisfaction to the set of real numbers. Neoclassical customer is not more than an abstract model that renders the mathematized/ideal-typical formulization of a psychophysical law by its utility function—but there is not any (implicit) statement that phrases the *sameness* of the two entities (it must be stressed that it is *not* argued here that neoclassical economic theory is free from any error—only that it cannot be rationally criticized starting from the foundations mentioned above).

As we have seen, the relation of economics and physics emerges as a separate problem. It is a commonplace to highlight the connections between physics and economics within the circles of historians of science—however, it really matters *which* e physics and *what kind of* physics we refer to. Although the main characters stressed the connections to Newtonian physics, Philip Mirowski (1984) put emphasis on effects of the nineteenth century physics. His arguments are weakened by the fact that he regarded Neoclassicals as quite inexperienced in that physics to which, according to him, we should attribute the major influence—so, the relationship seems to be rather far-fetched. It is out of the question that certain neoclassical concepts came from the 19th century physics—however, firstly, the methodological foundations of economics actually rooting in the heritage of Newton seem to be more important in terms of the case in point; and, secondly, we should not forget about the platonic tradition in which economics also had a share by following Newton. So, there are two (and not independent) kinds of influence here that affected economics rooted in neoclassical views more strongly compared to the nineteenth century impacts.

Newton determined the methodological principles of mathematical physics based on empirical foundation up to modern times and these principles also directly affected economics. Newton developed the model of how to simultaneously apply inductive and deductive logic that complement each other to create a scientific system. Eventually, Newton's physics consists of theorems that *deductively* follow from the principles because of the apodicticity of mathematics. This is the same methodology that was applied in Euclid's geometry—with the crucial difference that Newton referred to reality when he tried to establish his principles. The essence of this method is that we *empirically* gain pieces of knowledge that will be generalized through induction and the entire structure of a discipline will be built on the principles obtained this way. In his famous Principles (Philosophiæ Naturalis Principia Mathematica—Mathematical Principles of Natural Philosophy, 3rd edition, 1726) Newton wrote the following: "The qualities of bodies, which admit neither intensification nor remission of degrees, and which are found to belong to all bodies within the reach of our experiments, are to be esteemed the universal qualities of all bodies whatsoever." When economists try to establish the concept of *home oeconomicus* through either observation or introspection, they act in the same spirit. They observe how individuals behave in situations of economic decisions and the common elements of these observations (e.g. the law of diminishing returns or the utility maximization axiom) will be chosen for principles of economics by following the same inductive logic. Doing so, economists assume that these are the common features of human beings they grasped. It is *abstraction* we meet in the case of both Newton's physics and the neoclassical theory, since the goal was to grasp the *unchangeable and timeless* essence of phenomena. When Newton talked about axioms obtained (empirically) from observed phenomena, he refers to phenomena as ultimate sources. For example, Newton created the concept of the material point by following this abstraction method so that he abstracted from the dimension and shape of any perceptible object, preserving only inertia and translational motion. Material point is the ultimate and common feature of all bodies. Neoclassical economists created the indefinitely abstract notion of *homo oeconomicus* by following the same procedure. Economists abstract from all individual and personal qualities in order to get to those characteristics that embody the essence of economic activity.[29] Eventually, it is the substance of the economic

[29] Although it may seem that abstraction is a simple and obvious action, it is not true anyway. When a researcher looks for the essence of something and abstracts from all the inessential elements in order to find it, there exists a danger that he abstracts from something relevant as well (he throws out the baby with the bathwater). It looks that neoclassical economics also committed the same error sometimes. For example, Say's law is based on a special concept of money, since it is needed to abstract from using money as a store of wealth in order to hold the law. This is one of the main weaknesses of the neoclassical theory that were highlighted by Keynes. If we start from money that can be stored as well and used not only for the transaction of current business, then the possibility of overproduction crises immediately emerges. An abstraction resulting in such a concept of money is questionable at least, since it is at stake that whether crises and considerable recessions should be regarded as minor and dysfunctional episodes, as occasional fluctuations or as fundamental characteristics of the capitalist socio-economic system.

man that they try to grasp. Any characteristic beyond it only disturbs the functioning of this essential core.[30] Axioms (fundamental laws—in economics, we can label them as *behavioural laws*) are rules commonly appearing in phenomena and the truth of the deduced theorems follows from the truth of the axioms (laws) because of the fact that we are constrained during the act of deduction by the apodicticity of mathematics. *A right axiom embodies a universal and uniform experience*. Newton's methodological principles help us, economists, understand what a right axiom is like and that when economics talks about the ultimate foundations of its own theoretical system, it does not refer to arbitrary axioms, but to axioms that are in accordance with observed phenomena and that can be obtained from observations through abstraction. And Newton also helps us understand that induction as a theorizing logic and strategy is a necessary instrument for laying the foundations for even an axiomatic-deductive science (cf. Shapere 1988).

Economics also follows Newton when it forms notions beyond axioms to explain observed (and even non-observed) phenomena. This is a much more difficult task and economics, like physics, loses the ground from under its feet. Economists (and physicists as well, following Newton—cf. Feynman 1964) make *certain* assumptions (hypotheses) about behaviour (functioning), and, since their theoretical system is built on mathematical foundations, the consequences followed from the axioms in accordance with those assumptions can be compared to reality (under some restrictions), and therefore, it can be judged whether these notions can be accepted as explanations for behaviour. Economists, like Newton (but, actually, all physicists since Newton), form explanative conceptions even on things that cannot be directly observed in a way that they try to deduce these notions from observed phenomena (i.e. experience) as much as possible. Astronomical research into the Oort cloud, to mention a simple instance, reveals this logic quite well. The Oort cloud is the homeland of comets situated roughly at a distance of 10,000–100,000 astronomical units (AU) from the Sun (an astronomical unit is defined as

Grabbing the essence is not a simple and evident cognitive act at all and results in faiths that, though referring to reality, cannot be rationally confirmed (Polanyi 1964; 1966). This is the same controversy that has been held by equilibrium economists against their disequilibrium counterparts (and *vice versa*) on the elements and consequences of their theories for generations. Since their approaches as a whole are based on faiths in equilibrium-disequilibrium, this controversy remains desperately unwinnable. Probably this unwinnableness explains how these two schools (i.e. equilibrium and disequilibrium economics) could survive and even get closer to each other despite their antagonism between fundamental assumptions. So, it is a fairly ambitious and, perhaps, an evidently hopeless undertaking to find the objective substance. For example, Max Weber could totally evade the problem of the objectivity of abstraction, since, though he also applied this cognitive technique when creating his ideal-types, he did not attribute any real-objective existence to these theoretical constructs (we will return to this issue below).

[30] In connection with Lucas' island models and agent-based models, we shall see that this abstraction technique is fraught with danger in the case of, say, creating a society from abstract agents presupposed to be infinitely far from reality. In such a case, economists examine the alleged social actions of dissocialized beings, so they may gain definitely distorted results and conclusions. Analyzing these results requires special caution and self-discipline from the researcher (at a place, it was also suggested by Galbraith 1973 in a quite implicit manner).

the mean Earth–Sun distance). In this enormous distance the gravitational effect of the Sun is already very weak. Starting from here, comet nuclei move towards the inner Solar System. We cannot *directly observe* (to put it simply: *see*) the hundred billions of comet nuclei orbiting in this wide region, but we still know a lot of them—mainly based on the information from calculations of orbits of those comets that become directly observable (visible) roving into the inner regions. These orbits show that all these objects come from the same area of the Solar System (i.e. from the Oort cloud): although we cannot see them, by experience, we still know they exist. This scientific logic helps researchers make further steps. For instance, Murray (1999) details some arguments for the presence of a distant large undiscovered planet in the outer Solar System by observing and numerically analyzing the orbits of long-period comets coming from the Oort cloud, denying a former conception according to which these comets start off towards the inner Solar System following a random pattern (e.g. Weissman 1996). This well illustrates that science can study things and can formulate theories about things that cannot be *directly* observed at all. Economics has witnessed similar cases. Milton Friedman's famous permanent income hypothesis (a central topic of Chap. 5) refers to an income concept (i.e. the permanent income) that cannot be observed and numerically measured. We still know of its existence, mainly through the structure and the time pattern of consumption suggesting that consumers *do not* decide on expenditures and savings considering their current incomes. What we are certain of is only the fact that the scheme *supposed* by Keynes (being a hypothesis similar to the permanent income theory, in which consumer expenditures are determined as a function of current incomes) does not work, since data (observations) contradict it (in other words, falsify it, in the Popperian sense). Economists form an assumption on the basis of which data (observed facts) can be harmonized and which corresponds to the fundamental experience of the mechanisms of economies (i.e. there *must* be a relation between income and consumption). If we proceed rightly (i.e. our axioms are obtained from experience through abstraction and no logical errors occur during numerical deductions), such an explanative conception that was tested in reality can help us understand how things work. *This* knowledge goes far beyond the simple mathematical formulization of behavioural laws which is, otherwise, a mere mathematical problem. Economics always deals with these two issues separately, though simultaneously. The Phillips curve, as a mere mathematical law, can be reduced to a simple regression problem. Understanding things going beyond experience (observations) needs theorizing, since, in this case, we try to clarify the causes that explain the empirical Phillips curve (its emergence and mechanisms). This is the same logic as the one that stands behind Newton's widely debated hypotheses.

In connection with these considerations, we take a short hint about Milton Friedman's famous and ill-famed F-twist (we shall return to it in the following chapters). In his Principles (Principia Philosophiæ, 1644), Descartes stated that "I wish what I shall write later to be taken only as an hypothesis which is perhaps very far from the truth. But, even though these things may be thought to be false, I shall consider that I have achieved a great deal if all the things which are deduced from them are entirely in conformity with the phenomena. [...] their falseness [i.e. falseness of his hypotheses] does not prevent what will be deduced from them from being true and certain." This can be almost literally related to Milton Friedman's positivist program (Friedman 1953), in which all hypotheses and axioms are judged by whether the theory based on them generates predictions consistent with reality (i.e. with observations). It is argued here that, by this principle, Friedman drastically deviated from the methodological ideas characterizing mainstream economics before him (and, actually, after him as well). It is definitely the most important part of Friedman's positivist program when he writes about the issue of the density of leaves around a tree. If, as he said, we want to model the positioning of leaves, and if a model gives the *best* predictions (best, i.e. these predictions are most consistent with observed data) the (obviously absurd) hypothesis of which assumes that leaves are rational, utility-maximizing actors, then this model is acceptable and highly useful. If a model generates predictions that are consistent with reality, this fact does not imply that its axioms (and premises) are right. However, this is definitely neither Newton's way (cf. Fehér 1977) nor that of economics of neoclassical roots, since mainstream theory, as it is argued throughout this volume, tried to map and *truly* understand how things work. Economists, like Newton, insist (as firmly as possible) on making assumptions that originate in experience (i.e. that are obtained from experience by abstraction), i.e. they refrain from designing foundations (using imagination) that cannot be empirically judged (cf. Janiak 2009; Feynman 1964) and the application of which is only justified by the consistency of their derived consequences with observed facts. So, if the mere empirical consistency of the consequences means the *only* criterion for choosing our presumptions, axioms and explanations, then we do not posses any kinds of knowledge of reality—or, if still, this knowledge is gained only accidentally (that is, we blundered upon the structure of reality). Doing so, it can happen that starting from wrong assumptions, we arrive at right consequences (i.e. consistent-with-reality). And doing so, we cannot actually know how long the applied conception will prove to be useful. This is the very reason why abstraction (that starts from observing reality, as Newton stressed as well) played a crucial role in the self-organization of mainstream economics. In later chapters, we shall see that it was rarely fruitful when economists did not insist on this requirement and when abstraction based on experience was replaced by alchemy considering mere empirical performance. When neoclassical economics was born, the founding masters created its system considering *not* the empirical consistency of the consequences derived from the axioms. Moreover, as we have seen, they highlighted the pureness and abstractness of their theory, also implying that the predictions generated are allowed to be far from the empirically detected outcomes.

As far as our second statement, i.e. the platonic foundations of mathematical economics is concerned, Newton's influence is equally elementary in this regard. Following Newton, economics got a share in an intellectual heritage that can be traced back to Plato and even beyond. Of course, it is extraordinarily interesting to look for the ultimate roots of a thought, however, this pursuit must be given up beyond a certain stage. Ancient Greek philosophy is so rich that we can hardly find a modern western thought that does not have an archetype in this sphere—and, moreover, Greek philosophy drew considerably from eastern thoughts. Looking back in time, traces become more and more ambiguous, since parallels do not necessarily indicate connections to the East (cf. Wallis 1972). Plato was the most paramount giant of this tradition, in fact the entire western philosophy should be considered as a series of footnotes to Plato (cf. Kristeller 1955). As far as the case in point is concerned, here, we have to highlight *the efforts to grasp the substance* in the platonic heritage. The thought that there exists a timeless and unchanging-unchangeable sphere above (or beyond) sensible-material world was a commonly held belief even in the times before Plato and Newton himself, as a religious man and experienced thinker in platonic philosophy (e.g. Slowik 2013), wished to get to know just this hidden and substantive core of the world. If one tries to calculate the trajectory of a launched arrow under Newton's laws, then he only grasps a pure *form* of this motion that is not disturbed by contingencies. His laws, *the laws of physics*, are elementary laws. An actual arrow is affected by air resistance, its flight gets perturbed by the wind and is influenced by the nature of the material it was made of, etc. To incorporate these effects, it is needed to make the theory modified and (more and more) elaborated. However, the laws written in the language of mathematics describe the *essential core* in comparison to which the experienced contingencies can be registered and interpreted.

Without the intention of a systematic enquiry to be devoted to Newton's physics and to the platonic foundations of mainstream theory of neoclassical roots, some short hints should be dropped here. Heraclitus (c. 535–c. 475 BCE) had already taught, before Plato, that the sensible world surrounding us permanently changes, therefore we cannot rely on our sensory experiences, but the truth (i.e. the ultimate reality) can be found under the covering surface (i.e. the proper, genuine and ultimate order of things, to put it in other words), and this is to be studied and investigated by "scientists". So, the real nature of things is hidden from our senses. Pythagoras (582–496 BCE) had another thought that turned out to be equally important. According to it, truth lies hidden in numbers and in their ratios—these two traditions unified soon, yet in the ancient times (cf. Kristeller 1955), exerting their influence together. Numbers, by their nature, are also the instruments of cognition. *This idea still determines the way natural sciences try to interpret and describe reality*. If that is the case, fundamental faith in the nature of reality spreads from Pythagoras through Plato to Newton and even beyond him: numbers do reveal the ultimate reality being concealed beneath the surface and inaccessible by our senses. So, the ultimate reality can be grasped and described through *numbers* (in modern terms: *equations*): reality reveals itself to our knowledge. Parmenides (c. 540 or 515–510–c. 460 or 435–430 BCE) taught, similarly to the views above, that truth unfolds itself only to reason (logos) but not to our sensory faculties. Parmenides already stressed that, being the first, that there are two worlds around us: one of them is the world of appearances, filled with things open to sensory experience, while the other is the world of ultimate reality that can *only* be unveiled by reason (but, by reason, it can *really* be unveiled) and does not show itself in sensory experience. So, it was Parmenides who originally phrased the significant view that separates a physical and a metaphysical world and, through it, fundamentally influenced the way in which physics and economics developed later. It is not an overstatement to say that it has been the most determinant thought in the history of western philosophy and sciences: Parmenides, Plato starting from him, and the entire movement of Platonism spanning centuries have essentially affected the way we think today.

By the time of Plato (427–347 BCE), these fundamental thoughts had been available. Plato elaborated them to be a special system undoubtedly characterizing him. Plato was a genuine philosopher, though he did not step forward from nowhere. He also believed in the Pythagorean thought (Maróth 2002) that reality can be described by numbers and ratios (e.g. Timaeus 32 A–B) and the special idea of two worlds (Timaeus 49 A). The unchanging, perennial world (the realm of Forms) that gives the objective blueprint of perfection for all the continuously occurring-changing-decaying objects in our physical-sensible world never reveals in sensory experience, it can only be grasped by reason. Forms (or Ideas) are the actually existing, abstract, non-physical things, being the *essences* of the various objects on Earth. These chaotically circling objects are only shadows mimicking the Ideas so that these objects can only imperfectly represent them. Forms are the ultimate causes of their plural representations in particular objects around us. In the world of appearances flowing before our eyes, in spite of the chaotic and continuous change and contingencies, there is something timeless, perennial and never-changing in our objects: this is believed to be the Ideas. Something that is the essence of being a chair, *chairness itself*, in the countless chairs; something that is the essence of love in the countless appearances of love; and something that is the trajectory of *the* arrow unaffected by any contingencies (and easily formalizable in mathematical terms) in the countless arrows launched. The realm of the Platonic Ideas, the intelligible world offers the possibility of true knowledge: this knowledge is only achievable by contemplating and grasping the world of Forms. Moreover, we get to the notion following from the views above that there is a world existing independently of the experiencing subjects that can be known rationally, so the objective essence of the world can be grasped and revealed. These views together are among Plato's most successful thoughts with a brilliant carrier (cf. Stead 1998). Above or behind the contingencies there is something the knowledge of which means the only and ultimate, reliable and timeless knowledge. If we review Plato's cosmology, the idea deserves special attention through which Plato described the system of the Universe (we should say today: the Solar System) in his terms. Of course, this system was formalized in the geocentric framework, but (and this is the most important for us now, in respect of the later development of natural sciences) he specified the distances of the orbits of the celestial bodies (including the Sun) from the Earth (i.e. the radii) by numbers (ratios), in Pythagorean manner (Timaeus 36 D; cf. Heath 1921). So, the Universe was structured following an order that can be grasped in numerical terms.

Newton, by his Platonism, laid foundations that still influence the approach of physics. This statement remains true even if one considers the fact that the Platonizing tendencies of physics (in today's sense) were not started by Newton.

Aristotle's qualitative physics was replaced by the quantitative (i.e. mathematical, the recent form of) physics thanks to Platonic efforts. This direction determined the scientific attitude and endeavour of Kepler, Galileo and Newton. Thanks to the influences of Platonism, modern sciences interpret the natural world in quantitative terms, i.e. in a mathematical framework (Kristeller 1955). We cannot overstate the fact that Platonism strongly determined the development of physics and the other natural sciences, since even those times were inspired by this Platonizing idealism in which physics has already deviated far from Newton's mechanics. Einstein, in his essay on Maxwell, writes that "sense perception only gives information of this external world or of »physical reality« indirectly" (Einstein 1934), that is, the essence of the physical world cannot be grasped through sense perception. However, Einstein represented modern physics, in which scientists, forced to deviate from certain principles of Newton, create the elements of their theories increasingly in a pure speculative way,[31] and not deducing from observations. The world, the system of laws of physics is of mathematical character, laws of physics speak to us and express themselves in the language of mathematics—well, it has remained a never-changing principle of physics (cf. Feynman 1964). In his philosophy, Michael Polanyi, being a scientist too, expresses this view in a very interesting way. Polanyi (1964, 1966) often stressed that scientists, by discovery, reveal a latent manifestation of reality (*hidden reality*, in his own terms) that is not open for everyday perception, forming a sphere that is somehow *above* the sphere we can directly experience (*beyond* physical world—as Plato said, the actually existing things are *aspatial* and *atemporal*, giving us not too much help). The thing which scientists surmise and which they try to grasp and describe explicitly, is a level of reality existing above physical objects—*above*, since none of us can reach it by directly experiencing the sensible world. This is quite a direct expression of Platonic idealism that determined, as we have seen, the approach of physics. Scientists observe the physical world (e.g. measure it), but the things that they have been searching for and that is taking *form* by their experience—these are the actually existing things of Plato, the Ideas or Forms, the hidden order and system of reality. Polanyi (1964) supports this notion by a quotation from Dirac: "Nature's fundamental laws do not govern the world as it appears in our mental picture in any very direct way, but instead they control a substratum of which we cannot form a mental picture without introducing irrelevancies." To put it in other words, laws of physics cannot be reached in direct experience. This physical world only refers to the hidden and for most of us not discernible laws governing how this world works—this is the very essence of how Polanyi described scientists' intuition and of the platonic ideas.

[31] Today, the consistency of assumptions with reality is not a predominant scientific requirement (cf. Hawking 1988). Moreover, one can read it at numerous places in the literature that dissecting this consistency is even a nonsense problem (Hawking 1993). Scientists often use the heuristic techniques (see Kahneman 2011) that the most simple assumption of the possible postulates is always the nearest to reality.

Nothing has exerted greater and more elementary influence on mainstream economics than these platonic tendencies of which it has had a share, imitating physics. There is no denying that 19th economists were affected by physics of their own time through, say, certain concepts (as Mirowski states), but, as we have seen, there are stronger links between economics and Newtonian classical physics. For instance, the approach and the strategy of abstraction is such a link, since the nineteenth century was the very period in which physics started abandoning the principles of Newton of the method of creating basic assumptions. However, it is almost completely indifferent what roads physics took in modern times, since by the time these changes of physics developed, economics had already formed its own methodological profile. Recently, we have seen that economics starts showing methodological interest in the (statistical) methods of modern physics, with the explicit intention to exceed mainstream theory (for further details see Mellár 2015). However, none of us knows whether these attempts shall turn out to be fruitful and imperishable. So, if we want to identify roots of economics in physics, it seems that the common methodological principles of economics and Newtonian physics have greater weight. It is only an addition to these that economics, through physics, is definitely rooted in the platonic heritage—and it must be stressed that these platonic tendencies were due to Newton's interconnecting role. The goal of both mathematical natural and social sciences is to grasp the perennial and unchanging sphere of reality—to reduce the infinitely diversified natural and social phenomena to only a few general rules (Feynman et al. 1964). The physicist searching for the thoughts of God, as Stephen Hawking (1993) phrased, looks for the Platonic Ideas. Polanyi (1964) refers to the Pythagorean tradition to which physics owes its attitude valid even today that natural laws can be formulized in mathematical terms. This quite directly offers the foundations for the world view of mathematical physics: there exists a reality of nature behind/above/beyond phenomena that can be only intellectually grasped and that can be (and should be) formulized well in mathematical terms, since the world of our everyday experience is so complex and confusing as these timeless laws can appear only faded, covered and disturbed by innumerable incidents (e.g. Feynman 1964). These laws are only visible to an expert eye—to the eye of physicists and economists. The development in physics from Kepler through Galileo to Newton was highly determinant in terms of the history of the economic thought, since the concept of man to be described by functions (behavioural equations) is rooted just here.

Before continuing, it has to be noted that searching for the Platonist origins of physics and, through it, of economics needs careful attention (we return to this problem in Chap. 3 for further details). Although it is a commonplace in the literature to refer to the Platonic tradition in terms of the birth and development of mathematical physics, we should not forget about the influence of Aristotle (and others, of course—cf. Brague 1993; see Randall 1940). For example, both the desire for an empirical founding as a central principle in Newton's methodology and the ideal of a deductive science that starts from axioms show the influence of Aristotle still prevailing until Newton's time (cf. Ross 2004). In some cases, even Platonic influence is questioned saying, for example, that, though the Renaissance showed a special interest in the philosophy of Plato (indeed, this interest never ceased to exist), Renaissance Platonists were not engaged in mathematics—or that scientists of the Renaissance (e.g. Galileo) were not Platonist at all. In the Renaissance a totally complicated situation arose, since, at that time, the Platonic movement declared an explicit intention to harmonize Plato and Aristotle (e.g. Trinkaus 1983; Copenhaver 2008) and, moreover, the Aristotelian tradition is not denied in the literature even by those who highlight the role of Plato in terms of the (later) development of physics (e.g. Kristeller 1944). A definite and satisfying answer cannot be suggested here (but it is not even needed—let historians of philosophy keep working). We could just enumerate some arguments that are thought to support the conception of platonic roots. We can ironically say that if a mathematized discipline such as econometrics is in a state that it can *support* and *falsify* any thought *at the same time* (Seater 1993), it is perfectly true for the history of thoughts. So, definite answers are still to be awaited. But, for now, let us return to our genuine questions.

So, homo oeconomicus is the result of a consistent abstraction-idealization process. The fundamental axiom of neoclassical theory, the law of diminishing marginal utility, followed from the highlighting of Weber-Fechner's law in psychophysics (Mátyás 1969), according to which the growth of subjectively perceived intensity of recurrent stimuli with the same physical intensity is decreasing. The same law emerges in the law of diminishing marginal returns, since the returns of factors of production are deduced from the utility of consumer goods through derived factor demand (Lehmann 1968). In order to express these laws in analytical terms, differential calculus of mathematical analysis (derivatives) got applied through assuming such partial utility and production functions as mere *models* the functional-analytical properties of which manifested well the meaning of axiomatic assumptions. Mainstream economics itself, because of its neoclassical roots at least, organized itself as standing on axiomatic-deductive foundations, and this feature created the possibility for further and later modifications. If one alters the axioms (in part or even on the whole) or their number, he gets a new system, a new theory and new models (cf. e.g. Galbács 2008; or Galbács 2009; Galbraith 1973 details how one can approximate models to reality incorporating more and more aspects; on the axiomatic discussion of relations of Euclidean, absolute and affine geometry see Coxeter 1961). The concept of man that stands in the background of these economic models manifests a small degree of variability (rather a high degree of stability indeed), and, at most, the assumed degree and nature of rationality makes the difference—and the number of variables that are supposed to take effects on the behaviour of homo oeconomicus in terms of economic performance. Within the realm of mainstream economics, it is literally a question of quantity. According to an implicit view of mainstream economics, the increasing of the number of the relevant variables approximates models to reality, over and above and, of course, at the expense of intensifying complexity. If homo oeconomicus appearing in textbooks in microeconomics is a pure ideal-type existing infinitely far from reality (Weber 1904), then broadening the circle of laws of behaviour that are taken into consideration in these models eventually supports an approximation to the directly experienced reality. However, the scientific strategy remains the same: the man of these models is structured through an idealization process based on abstraction whose behaviour is dominated by some hard rules analogous to natural laws that can be well expressed by equations. To put it more precisely, it is the direct extension of laws of nature to human behaviour.

This technique is Newtonian inheritance, since classical mechanics was designed by Newton in a way that one can make the predictions of the theory increasingly approximate to the observed facts by modifying the initial assumptions (i.e. reducing the level of abstraction and taking more and more factors into consideration). For example, the orthodox monetarist interpretation of the Phillips curve could emerge by exactly applying this technique, in which the axiom of full information was replaced by another one that postulated the mechanism of adaptive expectations (Friedman 1957); or the Phillips curve of new classical macroeconomics in which the axiom of rational expectations got introduced through another switch (Lucas 1972, 1973; Sargent and Wallace 1975).

It reveals a great deal about the approach of sociology that also embarked on interesting methodological adventures around the end of the nineteenth century—and it remains the case even if one considers that these two conceptualizing attempts (i.e. of economics and sociology) implied dramatically different strategies of facing reality. While neoclassical economics investigated the laws governing market, consumer and economy as a whole, the approach of Max Weber *did not concentrate on this*, but on arranging the chaotically and inscrutably complex phenomena of the social reality by ideal-types through which an *interpretative understanding* of this reality can be possible. *According to him*, a *complex* description of social processes or an all-embracing causal explanation is an incomprehensible, hence an impossible mission. Even if these laws were available, they would not prove to be too helpful in studying cultural and social reality, since, according to this approach, only unique constellations matter in doing so—but these special patterns always follow from the former states of the social world (so their formation is not governed by such laws). Laws are only responsible for the transformation (from one state into another), and can only offer the rules of transformation, but an interpretative understanding of constellations cannot be only carried out by them (Weber 1904). Because of this uniqueness, laws of social and cultural phenomena cannot be found anyway, since no one can generalize such laws from the special and perhaps unprecedented causal relations. Therefore, in Weber's sociology, creating ideal-typical concepts is the only way of understanding and interpreting reality. Doing so, a researcher, in accordance with his own purposes (i.e. by a value-driven action), selects and highlights, or, *extremizes* (accentuates) certain elements of reality then arranges these elements into a logically consistent image free of contradictions. The concepts created in this way cannot be found in reality *in their pure form* and cannot be either verified or falsified empirically (so, an ideal-abstract concept does *not* exist *as it is in itself*, but shows itself in concrete-actual manifestations more or less, if abstraction is *right*). However, conceptualizing routine of neoclassical economics differs from this approach in that Neoclassicals focused exactly on finding and deducing *laws* (in accordance with the efforts of natural sciences), while ideal-types of Weberian sociology only supported the interpretative understanding of past events *with no* references to causal laws (Molnár 1999). Even if the method and strategy of creating ideal-typical concepts (i.e. abstraction) are common, these are two opposing scientific programs eventually. Weber offers an excellent description and a user's guide to the technique of abstraction that also directly applies to the conceptualizing strategy of mainstream economics that is on a completely different track. The relevant difference is that Weber did not attribute any real and objective existence to his pure concepts regarding them as only theoretical constructs (this is the reason why he got involved in long-lasting debates with Marxist social theory—cf. Peschka 1975). But anyway, the technique of abstraction and accentuation elaborated by Weber turned out to be the most important theorizing action of mainstream economists.[32] Thus, Weber was not an "idealist" sociologist; he did not search for a level of reality above physical objects. His ideal-types did not serve the purpose of grabbing an objectively existing substance.[33] So, while Weber concentrated on the interpretative understanding of unique constellations and not on searching for objectively existing (and, especially, mathematically graspable) substance and laws, neoclassical economics made conscious efforts to formulize laws. However, this is *not* a methodological mistake, since neoclassical economics *as a pure theory* did not show any interest in concrete and individual social-economic phenomena. *Neoclassical economics is a pure technical discipline*. The recognition of this is necessary so as to avoid calling it to account for either moral maxims (cf. e.g. Barr 2004) or investigating *concrete* reality (i.e. the unique constellations) and the poor strategy of the investigation into *this* reality.

The assumptions that stand behind the models operating ideal-typical concepts should *not* be arbitrary. Idealizing means highlighting something that really *exists*. We can get to these ideal-types through abstraction disregarding those features that do not have primary importance relating to our case in point. It is known that these

[32] Weber made an express criticism against neoclassical economics for its epistemological specialities (for a brilliant summary see Csontos 1993). However, this critique was ambiguous. While Weber could superbly grasp the main features (the *essence*) of the methodology of economic theory, his commentaries were confined to calling economics to account for the methodological aspirations of another discipline, i.e. his own sociology. Indeed, this is not a sharp critique as all similar critiques are so, since in such cases one confronts a discipline with the methodology and strategy of another science (that is preferred by him). Moreover, this demarcation can explain well why economics tends to forget about Weber who was regarded previously as an insider (cf. Tribe 2006). Probably, neither of the sides (i.e. Weber himself and mainstream economics then taking its form) realized that, in spite of evident and fundamental discrepancies, there are a lot of epistemological resemblances.

[33] Even if some authors comment his conceptualizing efforts ambiguously. At a place, Attila Molnár (1999), one of the leading researchers of Weber in Hungary, illustrates Weber's ideal-types through the parable of the wheel (as a physical object) and the circle (as an Idea). This is not a wise choice at all, since the circle as an Idea is an objectively existing thing, and one can easily draw the conclusion from this (even if Molnár finds an excuse for his decision) that Weber tried to create such concepts through his methodology.

momentarily non-relevant elements probably exert effect on *actual* manifestations of the concept to be idealized, but now our goal is to grasp the *only* presently relevant aspect of the functioning (e.g. behaviour) of the phenomena in its pure form. Any further effects beyond this function (which is *marked* as central) should be regarded as influence that diverts actual processes from allowing the *marked-as-central* function to be exclusively manifested. Since actual and infinitely complex functioning is the resultant of countless factors, it may seem evident that an ideal-type shall be far from actual instances from which we create this concept through abstraction. Homo oeconomicus, *in its pure form*, cannot be found in reality, since this theoretical concept only shows how a man *would* work if the only function of his life *were* utility maximization through rational economic decisions. Even if this concept is far from the image of individuals (people of flesh and blood), i.e. from the picture we think of men *in general*, the assumptions behind homo oeconomicus are *not* arbitrary. On the contrary, it is the so-called *limit-concept* (referring to Kant's *Grenzbegriff*) of the features disturbed by real relations deprived of these actual influences—exactly in the same way as a spatial point (in geometry) is the limit-concept of actually drawn and drawable points, if we take the disturbing fact into consideration that none can draw a geometrically pure point under actual circumstances. Actual points can only more or less *approximate* to the point taken in terms of geometry. We have to say that idealized concepts *cannot be* arbitrary in their nature, since there is *nothing* added during idealization to the category to be idealized (e.g. in the case of homo oeconomicus to the category of human being in general), but, on the contrary, it is abstracted from features regarded as his own characteristics under actual conditions but as *non-relevant in terms of the current investigation*. So, creating homo oeconomicus, we suppose nothing about man that he would not be able to perform if he were not hindered by features and circumstances, peeled off during abstraction, in a pre-idealized (i.e. in an actual) state. And *vice versa*, if a concept *also* contains features that are *not immanent* characteristics of the actual category to be idealized, then it is *not* an ideal-typical (non-idealized) concept.[34] These concepts (at least partially) are based on arbitrary assumptions. However, the highlighting of non-actual features is *not* idealization but arbitrary alchemy. The problem of contradicting reality can be caught in the very act only here. Ideal-typical concepts are neither real *by nature*, but, of course, *nor* unreal, that is, they do not oppose or contradict reality (to describe this relationship, we need adjectives similar to the trichotom of "moral-immoral-amoral", since the antonym of "real-unreal" is not satisfactory here: ideal-types are *not* real in their pureness, since they are results of abstraction started from reality, but *not* unreal either, since we try to accentuate actual features in abstraction—one should say that these ideal-types do not emerge in reality in their pure

[34] Referring back to the critique of cardinal utility theory: actual people *have the ability* to link real numbers to consumption bundles in terms of the utility realized, they just do not necessarily do it (since, say, it is exhausting or boring). However, the law of diminishing marginal utility can be still explicated quite well by the abstraction based on it–so, cardinal utility theory is not more than an abstraction-based model.

form, but this does not follow from a contradiction to reality). It is similarly true that the behavioural laws gained in this way may be at a great distance from reality, *since the most general* (that is, the most comprehensive) *laws are farthest from reality*, i.e. from the completeness of reality. To make a law embrace what is common in the most possible phenomena it is needed to be infinitely abstract, that is, to be meagre in its content (Weber 1904). The more actual behavioural effects are traced back to a general law, the less it can explain the individual and concrete emergences (and, following from this, the more behavioural effects are covered by a law, the more general and universal this law itself will be), *since the more these behaviours are different, the less they resemble each other and, hence, the less similarities they share*—the similarities that are exactly manifested by the law in point. Remaining in this context, one can mention the efforts that try to extend the law of diminishing returns (as well as the differential calculus) to suicide, sexual behaviour, prejudice or alcoholism and further deviancies (and perhaps tooth brushing) stating that, for example, rational cardio- and nephropathic patients must necessarily be alcoholics, since, otherwise, they would step in the other-world with non-similarly exhausted organs (for a summary see Becker 1993; Cserne 2000). So, no one should call neoclassical economics to account for the *experiencable* actuality (their possibly great distance from reality) of either concepts (and axioms) grounded by idealization or laws derived from them or postulated. The only relevant aspect is that of arbitrariness: our ideal-typical concepts *must* follow from acts of *abstraction* (as we have seen it above in the case of the neoclassical economics or Weber's methodology).

1.5 Conclusions

In Husserl's view, the crisis of modern sciences was caused by their efforts to mathematize our life-world, i.e. the world we live in directly and to state that their idealized constructions are identical with reality (Mezei 1998). It has been argued hitherto that neoclassical economics did *not* try to apply this aspectual misinterpretation. Although its methodology became similar to Galileo's guilty physics and shared the faith in mechanical philosophy, we have to realize that, on the basis of texts by the early masters, they did *not* substitute the abstract world to be scientifically investigated for directly experienced reality. According to these texts, early Neoclassicals pursued their studies deliberately assuming abstract environment and elements differing from reality. It may seem as if the mainstream theory based on the traditions of neoclassical economics is free from the vice of the modern sciences.

Separating directly experienced reality from idealistic-utopian model environments was given even in the case of orthodox monetarism. To put it differently, orthodox monetarist economists stressed that their models and theories did *not* offer a complete (realistic) description of this directly experienced reality. It could not be otherwise, since Milton Friedman, clarifying the methodological approach of mainstream economics in his seminal paper (Friedman 1953), argued for the accidentality of the connection between assumptions behind models and the reality being indirectly investigated. In other words: if the estimates generated by a theory and the observed facts are near enough, then the model in question performs well in terms of predictions, and, *according to this view*, this is the *only* measure to judge theories.[35] Reality of assumptions, calling to account for this reality, as we have seen, is *not* a proper ground for criticizing mainstream economics, indeed (in the sense that one should not criticize these models if their assumptions do not emerge in directly experienced reality in their pureness; only the *rightness of the abstractions* has special importance and relevance in this context). This is out of the question that ideal-typical concepts may be far from directly experienced reality (that is, they cannot be directly experienced in their pure forms). Similarly, pure geometry is based on unreal assumptions as well and could be criticized on this basis. However, the distinction between features *not emerging* in reality and *not existing* must be stressed. The (say, momentarily) non-emerging (but existing) features can emerge by abstracting from the further circumstances that are hindrances to this emergence. As we have seen, *this* is the essence of abstraction. So, the critique against the methodology of mainstream economics is semantically grounded in part, since Friedman ignored this discrepancy embracing both ideal-typical (i.e. *not* resting on an unreal foundation) and arbitrary conceptualization methods under the label of "unreality". Without this distinction, stressing unreality is not sharp criticism against sciences and theoretical models operating with ideal-typical concepts. Ideal-types are by definition far from reality, but they are still not arbitrary concepts, however, Friedman did not exclude this latter case (i.e. the arbitrariness) from the possible methods of building axiomatic systems.

But regarding the outcomes of ideal-typically grounded investigations as goals to be accomplished makes the success of mainstream economics doubtful. An ideal-typical concept does not have any moral content. A spatial point, a Euclidean plane cannot be *good* or *bad*. Similarly, the idea(l-type) of the actor capable of forming unbiased expectations is not good or bad, either, in *moral* sense. For Neoclassicals, using ideal-types meant nothing more than creating ideas above reality that were known to be only more or less approximated by actual actors. This is an acceptable theorizing view, since comparing reality to theories built on ideal-types can help us understand, say, *how* and *why* actual processes deviate from the idealized functioning of idealized actors. The use of ideal-typical concepts (but only them!) is not a valid reason for criticism made against mainstream economics, even if these ideal-types are infinitely far from their pure manifestations under actual conditions—so, even if it is known that one or other idealized concepts of mainstream economics will not emerge in reality *in their pureness*. Idealized concepts of pure geometry are none the further from our reality. No one can ever see actually (sensorially)

[35] Although neoclassical synthesis itself was not homogenous on the acceptance and truth of this view (cf. Samuelson 1963), *this* is to be regarded as the methodological foundation of mainstream economics (cf. Weeks 1989).

experiencable spatial points, Euclidean planes or lines running to infinity. If mainstream economics was blamed for using similar, ideal-typical concepts, then the scientific quality of mathematics or geometry should be questioned as well. A problem arises if (1) one attributes moral contents (an imperative meaning) to these ideal-types. Orthodox monetarism can be just blamed for this mistake (e.g. Friedman 1962); however, the error of merging directly experienced reality and mathematized scientific-theoretical environment into each other was *indirectly* made by even neoclassical economics (apropos of an imperative connotation), since Neoclassicals urged to achieve market efficiency in reality—an outcome that was deduced in a pure theory under assumed conditions such as absolutely perfect competition or atomistic and omniscient economic actors.[36] Or, if (2), as a direct manifestation of the Husserlian crisis, one confuses these ideal-types and reality, as it happened in the case of new classical macroeconomics. If our axioms, in part or in toto, are *not* the results of idealization, then regarding the conclusions drawn on these grounds as goals to be accomplished is totally nonsense indeed (for example, according to the Friedmanian methodology, replacing the geocentric model by the heliocentric conception would not be needed, since the process under investigation, i.e. the Sun rises in the east and sets in the west, can be accurately predicted by both models). If the purpose of pure economics is to generate hypothetical outcomes to which actual conditions (of reality) can be compared and the possible differences can be scientifically investigated (in economic, sociological, or institutional terms), then a comparison between predictions of a theory based on an *non*-abstraction axiomatic system and actual circumstances is not a scientific problem, since the causes of the discrepancy are to be found *in theory itself*. In this case, our concepts are *not* the limit-concepts of directly experiencable elements, processes, structures and relations. And also in this case, actual conditions cannot be regarded as the ones that hinder the emergence of our conclusions deduced from ideal-typical concepts. All the coincidences (and discrepancies) between theory and reality are mere contingencies, since if our axioms (partly or wholly) do not contain anything that can be attributed to actual elements and processes (as characteristics that do not emerge in their pureness), then the outcomes (laws) derived from them cannot be the border-line cases of actual conditions either. Because of the accidentality of its assumptions, orthodox monetarism is not quite capable of giving either a realistic or an abstraction-based description of reality. On these grounds, of course, adequate economic policy recommendations cannot be made either. Orthodox monetarism could not go further than a mere repetition of the advantages of absolutely free competition (with constant money growth rate offered by Friedman is a great example for it—Friedman 1960; new classical macroeconomics was not too consistent in suggesting this rule—cf. Sargent and Wallace 1974).

---

[36] In Walras' terms, continuing the former part: "Hence free competition becomes a principle or a rule of practical significance, so that it only remains to extend the detailed application of this rule to agriculture, industry and trade. Thus the conclusions of pure science bring us to the very threshold of applied science."

The methodological error arising from the *direct* merging of actual circumstances and abstract theory also occurred in new classical macroeconomics, since this theory, still organized by following the axiomatic-deductive scheme, regarded its conclusions based on the axioms of a presupposed general macroeconomic equilibrium or the Walrasian auctioneer revivified in the notion of the labour unions as reality itself. According to Weeks' (1989) crucial statement, "REH [and, consequently, new classical macroeconomics] purports to specify the actual behaviour of agents with regard to real world influences. [...] In the case of the REH its own assumptions demand it to stand or fall on the test of realism."[37] So, *the relation of mainstream economics to reality underwent a special restructuring process*. While Neoclassical economics deliberately applied the distinction of the directly experienced reality and the idealistic-utopian theory and models (with merging happened through an imperative only), this relation became accidental in the case of orthodox monetarism (with holding the imperative mentioned above), and, finally, new classical macroeconomics eliminated the border-line once and for all. Up to the new classical turn, mainstream theory did *not* mainly state that directly experienced reality is analogous to their utopian models. However, they definitely raised the issue of the *imperative* (and, as it follows: the *possibility*) of merging: neoclassical economics and orthodox monetarism permanently argued for the normative content of their findings. (That is not a less exciting problem of how the conclusions drawn from the behaviour of atomistic actors under utopian conditions may be manifested in reality. Results fundamentally have to change if also homo oeconomicus is regarded as a social being—and these two, contradicting assumptions cannot be held simultaneously. If the axiom of atomistic nature of actors as one of the pillars of the theory is changed, we must necessarily arrive at *different* conclusions.) By analyzing the effects of monopolies and oligopolies that mean our socio-economic reality (Galbraith 1967), neoclassical economics explicitly argued for the need of calling absolutely perfect competition into existence. Through the modification of *actual* conditions, Neoclassicals tried to replace *them* again and again by their abstractly grounded inferences, and, citing efficiency criteria, urged to manifest their ideal-typical concepts under actual circumstances (cf. e.g. Walras 1874).

[37] This statement, however, seems to be dubious when one considers Lucas' (1980) words, according to which "a 'theory' is not a collection of assertion about the behaviour of the actual economy but rather an explicit set of instructions for building a parallel or analogue system – a mechanical, imitation economy. A 'good' model, from this point of view, will not be exactly more 'real' than a poor one, but will provide better imitations." As Morgan and Knuuttila (2012) emphasize, artificial world models of Lucas are constructed to represent the outputs and *not* the behaviour of a system. Although the aspect mentioned by Weeks (i.e. comparing new classical theory with reality) remains a relevant principle throughout this analysis, it is argued below that new classical models and REH are pure, idealization-based theoretical constructions which cannot be expected to give a comprehensive description of reality. Of course, it has to be noted here that new classical macroeconomics and REH are far more than the theory of Lucas (and Sargent, Wallace or Barro). For instance, the intention to specify the actual behaviour of market agents is attributed to Shaw (1984) by Weeks, rather then to Lucas. New classical macroeconomics is a populous and heterogeneous camp.

1.5.1 A Summary: The Ways of Critique

Choosing the way of critique fundamentally depends on whether we have an intention to save (or, at least, to *find excuses* for) mainstream economics. It was argued above that, under the evolution of mainstream theory, its relation to reality underwent a peculiar rearrangement in which the total and direct merging of directly experienced reality (i.e. our life-world) and mathematical models into each other emerged only as a relatively late development, but the imperative of it was present all along. *This* was the way how Husserlian scientific crisis emerged in economics. So, a twofold critique can be justified. Firstly, since mainstream economics organized itself as an axiomatic-deductive science, a critique may be directed at (1) the rightness of axioms (i.e. whether axioms underlying a given theory follow from an idealizing-abstracting process), or (2) the logical order (logical consistency) of the theoretical system founded on these axioms. In this latter case, economic theories are regarded as systems that gain their axioms through idealization starting from the ground of reality, but that draw their conclusions through applying only logical actions in a closed system—a system that is *not* linked to our life-world directly, but only through abstraction (for the sake of good order, it still holds that such a theory investigates reality, but through applying abstraction). Secondly, we can check (3) whether a theory proved to be successful in describing reality realistically, assuming it had such purposes. A pure (i.e. idealization-based) theory cannot be directly relevant in terms of life-world problems. Of course, new classical macroeconomics can be criticized while holding this, since they promised to give a realistic portray of reality, as Weeks (1989) highlighted it. In this case, a pure theory is *not* regarded as a pure theory, i.e. it is not treated in accordance with its abilities, though, it is completely admitted by the ambitions of the theory itself. So, within mainstream economics, only new classicals are justified to be called to account for making concrete and successful economic policy recommendations. But judging the lack of success is controversial at least (though listing examples from economic history is not undertaken here), if one takes the fact into consideration that suggestions (e.g. Lucas 1990; Feldstein 1986) of supply side economics growing from new classical theory became common-places (e.g. Hall and Taylor 1991).

Since, before the new classical turn, relating mainstream theory to reality took place in two steps (first, there was an intention to deliberately apply a construct created as a pure theory to actual conditions, and not a pure theory was considered and stated to give a *realistic*—and not an abstract—description of actual conditions; excluding orthodox monetarism for its accidental relation to reality), the theoretical-logical system created in the first step can be analyzed and judged on its own. So, we have the opportunity to study and analyze mainstream economics regarded as a *pure science*, as a pure logical system. If we do not take this opportunity, we will also commit the same methodological error that was made by mainstream economics. In this case, criticism should *not* be based on a comparison of theory and actual conditions, i.e. directly experienced reality, since mainstream economics did not mainly try to give realistic descriptions of actual real-life conditions (with new classical macroeconomics as the only exception because of its explicit purpose to portray reality). Therefore, it is not surprising that national economic policies *cannot* be established on the basis of mainstream views, since the recommendations were formulated assuming an idealized-abstract environment. Theory exactly suggested making our world similar to the artificial environment of their models, and, if we succeed in it, the successful economic policy will be immediately available on the basis of the theoretical recommendations. Certainly, it is not stated here that mainstream economics should not be criticized—however, it really matters what criticism is directed at. *Arbitrary* and *wrong* (i.e. non-abstracted) *assumptions* or *confusions in the logical order* are such features in the case of which criticism is justified. In spite of the conceited new classical revolution, mainstream economics remained a pure science all along, and any criticism always *has* to take it into consideration. *This pureness and all its consequences have to be excluded from the scope of criticism.* A comparison to directly experienced reality, to our life-world or objections to the strategies of operative economic policies can only be rational in the case of new classical macroeconomics.

By now, our view on the scope of mainstream economics should be clear. Neoclassical theory and mainstream economics based on it remained a pure theory all the while, in spite of Friedman's methodological blunder, and as such, it did not even characteristically attempt to give a realistic description of reality. As a pure theory, it gives a framework to the researcher's hand from the hypothetical outcomes of which directly experiencable processes shall deviate, and further investigations can be devoted to identifying the causes and determinants of these discrepancies. Say, the results of experimental and behavioural economics founded on empirical observations and on an inductive logic cannot be interpreted as refutations of the underlying axioms of synthesis models. Neoclassical economics generates such predictions that show how human beings *would* behave if they were professional decision making machines following only the principles of formal rationality; on the contrary, experimental results describe what we are actually like. The inferences of these two theories perfectly *complement* each other. [→ **paradigm**] The only exception is new classical macroeconomics: though this chapter of the neoclassical synthesis was also organized as an axiomatic-deductive theory, similarly to the preceding sections, its purpose was to paint a *realistic* portray of our life-world. And, accordingly, findings of experimental and behavioural economics offer powerful refutations of, say, rational expectations taken as actual expectations formation mechanism of actual individuals (however, rational expectations hypothesis is still valid as an abstraction, that is, it is based on idealization).

The excessive ambitions of new classical macroeconomics can be easily labelled as self-esteem problems. Judging economic policy success is a separate problem. The normative quality of results of the pure theory emerged through the imperative connotation as a similar self-esteem problem in the case of neoclassical and mainstream economics. If this was regarded as a mistake, and one can hardly find other options, the same *aspectual* misstep would be taken when sounding the economic policy failures of mainstream economics. Although new classical macroeconomics as a pure theory could hardly have undertaken to give a realistic description of reality, in its case, talking about economic policy failures would be thoughtless taking into consideration how the recommendations of supply side economics became a common-place.

References

Adger WN (2000) Social and ecological resilience: are they related? Prog Hum Geog 24 (3):347–364

Andorka R (1997) Bevezetés a szociológiába. Osiris, Budapest

Aristotle (350 BCE) Physica. English edition: Aristotle (2008) Physics (trans: Waterfield R). Oxford University Press, Oxford

Barr N (2004) Economics of the welfare state, 4th edn. Oxford University Press, Oxford

Bartha S (2008) A vegetáció viselkedésökológiájáról, avagy milyen hosszú is legyen egy hosszú távú ökológiai vizsgálat. In: Kröel-Dulay G, Kalapos T, Mojzes A (eds) Talaj-vegetáció-klíma kölcsönhatások. Köszöntjük a 70 éves Láng Editet. MTA ÖBKI, Vácrátót, pp 73–86

Bayer J (1998) A politikai gondolkodás története. Osiris, Budapest

Becker GS (1993) The economic way of looking at behavior. J Polit Econ 101(3):385–409

Békés V (2002) A reziliencia-jelenség, avagy az ökologizálódó tudományok tanulságai egy ökologizált eposztemológia számára. In: Forrai G, Margitay T (eds) Tudomány és történet – tanulmánykötet Fehér Márta tiszteletére. Typotex, Budapest, pp 215–228

Benczes I, Kutasi G (2010) Költségvetési pénzügyek–hiány, államadósság és fenntarthatóság. Akadémiai, Budapest

Berger PL, Luckmann T (1966) The social construction of reality – a treatise in the sociology of knowledge. Anchor Books, New York, Hungarian edition: Berger PL, Luckmann T (1998) A valóság társadalmi felépítése – Tudásszociológiai értekezés (trans: Tomka M). Jószöveg, Budapest

Biemel W (1935) Beilage I, zu § 9. In: Husserl E (ed) (1976) Die Krisis der Europäischen Wissenschaften und die Transzendentale Phänomenologie – Ein Einleitung in die Phänomenologische Philosophie. Martinus Nijhoff, Hague, pp 349–356

Biemel W (1936) Beilage II, zu § 9a. In: Husserl E (ed) (1976) Die Krisis der Europäischen Wissenschaften und die Transzendentale Phänomenologie – Ein Einleitung in die Phänomenologische Philosophie. Martinus Nijhoff, Hague, pp 357–364

Bihari M, Pokol B (1998) Politológia. Nemzeti Tankönyvkiadó, Budapest

Bird A (2009) Thomas Kuhn. In: Zalta EN (ed) The Stanford encyclopedia of philosophy, Fall 2009 edition. http://plato.stanford.edu/entries/thomas-kuhn. Accessed 8 Oct 2010

Blanchard OJ (2008) Cracks in the system–repairing the damaged global economy. Finance Dev 45(4):8–10

BonJour L, Lyons J (2013) Epistemological problems of perception. In: Zalta EN (ed) The Stanford encyclopedia of philosophy, Spring 2013 edition. http://plato.stanford.edu/entries/perception-episprob. Accessed 9 Nov 2013

Boros G (ed) (2007) Filozófia. Akadémiai, Budapest

Boros J (2009) A megismerés talánya. Áron, Budapest

Brague R (1993) Europe, la voie romaine. Criterion, Paris, Hungarian edition: Brague R (1994) Európa, a római modell (trans: Somorjai G). PPKE BTK, Piliscsaba

Bréda F (1983) Egy világrendszer keletkezéséről–adalékok Galilei Dialogójához. In: Galilei (ed) Párbeszédek a két legnagyobb világrendszerről, a ptolemaiosziról és a kopernikusziról (trans: Zemplén J). Kriterion Könyvkiadó, Bukarest

Bretter Z (2004) Politika a határon: a Devlin–Hart vita. Kalligram, Pozsony

Bungard W (1997) A proszociális magatartás empirikus kutatásának módszertani problémái. In: Lengyel Z (ed) Szociálpszichológia. Osiris, Budapest, pp 23–35
Calvino I (1962) The nonexistent knight and the cloven viscount. Random House, New York
Charland-Verville V, Bruno MA, Bahri MA et al (2013) Brain dead yet mind alive–a positron emission tomography case study of brain metabolism in Cotard's syndrome. Cortex 49 (7):1997–1999
Colander D, Holt RPF, Rosser B Jr (2004) The changing face of mainstream economics. The Long Term View 7(1):31–42
Copenhaver B (2008) Giovanni Pico della Mirandola. In: Zalta EN (ed) The Stanford encyclopedia of philosophy, Winter 2008 edition. http://plato.stanford.edu/entries/pico-della-mirandola. Accessed 15 May 2012
Coxeter HSM (1961) Introduction to geometry, 1st edn. Wiley, New York, Hungarian edition: Coxeter HSM (1973) A geometriák alapjai (trans: Sztrókay K). Műszaki, Budapest
Csaba L (2008) The new kind of macroeconomic populism. Public Finance Q 53(4):601–616
Csaba L (2009) A szovjetológiától az új intézményi közgazdaságtanig–töprengések két évtized távlatából. Közgazdasági Szemle 56(9):749–768
Csepeli G (1997) Szociálpszichológia. Osiris, Budapest
Cserne P (2000) Gary Stanley Becker, a preferenciák és a közgazdaságtan imperializmusa. Szociológiai Szemle 10(1):113–127
Csontos L (1993) Max Weber on the methodological foundations of economic theory. Working paper 93–992. Department of Economics, University of Connecticut, Connecticut
Deane P (1978) The evolution of economic ideas. Cambridge University Press, London, Hungarian edition: Deane P (1997) A közgazdasági gondolatok fejlődése (trans: Szepesi G). Aula, Budapest
Debreu G (1974) Excess demand functions. J Math Econ 1(1):15–21
Dequech D (2007) Neoclassical, mainstream, orthodox, and heterodox economics. J Post Keynesian Ec 30(2):279–302
Descartes R (1637) Discours de la méthode. Ian Maire, Leiden. Hungarian edition: Descartes R (1992) Értekezés a módszerről (trans: Szemere S). Ikon, Budapest
Descartes R (1644): Principia Philosophiæ. Apud Ludovicum Elzevirium, Amstelodami. English edition: Descartes R (1982) Principles of philosophy (trans: Miller VR, Miller RP). Kluwer, Dordrecht
Durkheim E (1897) Le suicide: Étude de sociologie. Alcan, Paris, Hungarian edition: Durkheim E (2003) Az öngyilkosság: szociológiai tanulmány (trans: Józsa P). Osiris, Budapest
Dworkin G (2010) Paternalism. In: Zalta EN (ed) The Stanford encyclopedia of philosophy, Summer 2010 edition. http://plato.stanford.edu/entries/paternalism. Accessed 14 Dec 2010
Einstein A (1934) Mein weltbild. Querido, Amsterdam, Hungarian edition: Einstein A (2000) Hogyan látom a világot? (trans: Szécsi F). Gladiátor, Budapest
Erdős T (2010) Economic recovery–a few topical theoretical problems. Public Finance Q 55 (2):251–277
Erhard L (1957) Wohlstand für Alle. Econ-Verlag, Düsseldorf, Hungarian edition: Erhard L (1993) Jólétet mindenkinek (trans: Zalán P). Konrad Adenauer Alapítvány, Budapest
Faber M, Petersen T, Schiller J (2002) Homo oeconomicus and homo politicus in ecological economics. Ecol Econ 40(3):323–333
Fehér M (1977) Utószó. In: Newton I (ed) A világ rendszeréről és egyéb írások. Magyar Helikon, Budapest, pp 391–426
Feinberg J (1973) Social philosophy. Prentice Hall, Englewood Cliffs, Hungarian edition: Feinberg J (1999) Társadalomfilozófia (trans: Krokovay Z). Osiris, Budapest
Feldstein M (1986) Supply side economics–old truths and new claims. Am Econ Rev 76(2):26–30
Feynman RP (1964) The character of physical law. MIT, Cambridge, MA, Hungarian edition: Feynman RP (2005) A fizikai törvények jellege (trans: Gajzágó É). Akkord, Budapest

Feynman RP, Leighton RB, Sands ML (1964) The Feynman lectures on physics. Addison–Wesley, Reading, MA, Hungarian edition: Feynman RP, Leighton RB, Sands ML (1969) Mai fizika (trans: Bozóki G, Gombosi E, Nagy E et al). Műszaki, Budapest

Fink E (1936) Beilage III, zu § 9a [Die Fragenach dem Ursprung der Geometrie als intentional-historisches Problem]. In: Husserl E (ed) (1976) Die Krisis der Europäischen Wissenschaften und die Transzendentale Phänomenologie–Ein Einleitung in die Phänomenologische Philosophie. Martinus Nijhoff, Hague, pp 365–386

Foucault M (1975) Surveiller et punir–naissance de la prison. Gallimard, Paris, Hungarian edition: Foucault M (1990) Felügyelet és büntetés–a börtön története (trans: Fázsy A, Csűrös K). Gondolat, Budapest

Freedman C, Kumhof M, Laxton D, Lee J (2009) The case for global fiscal stimulus. IMF, Washington

Friedman BM, Hahn FH (1990) Handbook of monetary economics. North-Holland, Amsterdam

Friedman M (1953) The methodology of positive economics. In: Friedman M (ed) Essays in positive economics. University of Chicago Press, Chicago, IL, pp 3–43

Friedman M (1957) A theory of the consumption function. Princeton University Press, Princeton

Friedman M (1960) A program for monetary stability. Fordham University Press, New York

Friedman M (1962) Capitalism and freedom. University of Chicago Press, Chicago, Hungarian edition: Friedman M (1996) Kapitalizmus és szabadság (trans: Bősze SM). Akadémiai, Budapest

Friedman M (1966) Interest rates and the demand for money. In: Friedman M (ed) (1970) The optimum quantity of money and other essays. Aldine Publishing Company, Chicago, IL, pp 141–155

Friedman M (1977) Nobel lecture: Inflation and unemployment. J Polit Econ 85(3):451–472

Galbács P (2008) A racionális várakozások elméletének fogalmi inkonzisztenciájáról. Competitio 7(1):103–125

Galbács P (2009) Estimation of the structural balance based on the OECD approach. Public Finance Q 54(4):540–555

Galbács P (2011a) A globális pénzügyi és reálgazdasági válság kezelésének hatása a lisszaboni folyamatra. Külgazdaság 55(1–2):46–64

Galbács P (2011b) A mainstream közgazdaságtani elmélet irrelevanciájáról–egy metodológiai megközelítés. Köz-gazdaság 6(1):5–26

Galbraith JK (1967) The new industrial state. Houghton Mifflin, Boston, Hungarian edition: Galbraith JK (1970) Az új ipari állam (trans: Hantos É). KJK, Budapest

Galbraith JK (1973) Economics and the public purpose. Houghton Mifflin, Boston, Hungarian edition: Galbraith JK (1978) Közgazdaságtan és a társadalmi cél (trans: Szalai S). Kossuth, Budapest

Galileo (1632) Dialogo sopra i due massimi sistemi del mondo. Giovanni Batista Landini, Florence. Hungarian edition: Galilei (1983) Párbeszédek a két legnagyobb világrendszerről, a ptolemaiosziról és a kopernikusziról (trans: Zemplén J). Kriterion Könyvkiadó, Bukarest

Garfinkel H (1967) Studies in ethnometodology. Prentice-Hall, New Jersey

Gergő E (1930) A ponyvairodalom kritikájához. Nyugat 23(17):364–367

Giraud G (2009) From non-tâtonnement to monetary dynamics within general equilibrium theory–the limit-price exchange process. CNRS–Paris School of Economics, Paris

Hahn F (1975) Revival of political economy–the wrong issues and the wrong argument. Econ Rec 51(135):360–364

Hahn F (1984) Equilibrium and macroeconomics. Basil Blackwell, Oxford

Hall RE, Taylor JB (1991) Macroeconomics–theory, performance and policy. W.W. Norton, New York/London, Hungarian edition: Hall RE, Taylor JB (2003) Makroökonómia–elmélet, gyakorlat, gazdaságpolitika (trans: Eső P, Lóránt G). KJK–Kerszöv, Budapest

Hart HLA (1989) Law, liberty and morality. Oxford University Press, Oxford, Hungarian edition: Hart HLA (1999) Jog, szabadság, erkölcs (trans: Krokovay Z). Osiris, Budapest

Hausman DM (2008) Philosophy of economics. In: Zalta EN (ed) The Stanford encyclopedia of philosophy, Fall 2008 edition. http://plato.stanford.edu/entries/economics. Accessed 13 Apr 2012

Hawking SW (1988) A brief history of time from the big bang to black holes. Bantam, New York, Hungarian edition: Hawking SW (2003) Az idő rövid története (trans: Molnár I, Egri G). Akkord, Budapest

Hawking SW (1993) Black holes and baby universes and other essays. Bantam, New York, Hungarian edition: Hawking SW (1999) Einstein álma és egyéb írások (trans: Ungvárai Nagy Z, Ungvárai J). Vincze, Budapest

Hawking SW, Mlodinow L (2005) A briefer history of time. Bantam, New York, Hungarian edition: Hawking SW, Mlodinow L (2006) Az idő még rövidebb története (trans: Both E). Budapest: Akkord Kiadó

Hayek FA (1988) The fatal conceit–the errors of socialism. University of Chicago Press, Chicago, Hungarian edition: Hayek FA (1992) A végzetes önhittség–a szocializmus tévedései (trans: Pásztor E). Tankönyvkiadó, Budapest

Heath T (1921) A history of Greek mathematics. Clarendon, Oxford

Heller F (1927) Etikai tudomány-e a közgazdaságtan? In: Heller F (ed) (2006) Etikai tudomány-e a közgazdaságtan? Aula, Budapest, pp 141–152

Hernádi M (ed) (1984) Fenomenológia a társadalomtudományban. Gondolat, Budapest

Hobbes T (1640) The elements of law. In: Gaskin JCA (ed) (1990) The elements of law, natural and politic. Oxford University Press, Oxford

Hoover KD (1990) The new classical macroeconomics: a sceptical inquiry. Blackwell, Oxford

Horváth ÁB, Szilágyi K (2004) Konszenzusból nyugvópontra–elmélettörténeti áttekintés a makroökonómia viharos évtizedeiről. Külgazdaság 48(11):4–19

Husserl E (1900) Logische Untersuchungen. Niemeyer, Halle, English edition: Husserl E (2001) Logical investigations (trans: Findlay JN). Routledge, London/New York

Husserl E (1910) Philosophie als strenge Wissenschaft. Logos 1(1):289–341, Hungarian edition: Husserl E (1993) A filozófia mint szigorú tudomány (trans: Baránszky Jób L). Kossuth, Budapest

Husserl E (1931) Méditations cartésiennes–introduction à la phénoménologie. Armand Colin, Paris, English edition: Husserl E (1960) Cartesian meditations–an introduction to phenomenology (trans: Cairns D). Springer, Dordrecht

Husserl E (1976) Die Krisis der europäischen Wissenschaften und die transzendentale Phänomenologie. Martinus Nijhoff, The Hague, Hungarian edition: Husserl E (1998a–b) Az európai tudományok válsága I–II (trans: Berény G, Mezei B, Egyedi A, Ullmann T). Atlantisz, Budapest

IMF (2009) Fiscal implications of the global economic and financial crisis. IMF, Washington

Ingarden R (1931) Das literarische Kunstwerk. Max Niemeyer, Halle, Hungarian edition: Ingarden R (1977) Az irodalmi műalkotás (trans: Bonyhai G). Gondolat, Budapest

Inkinen S (2009) Homo creativus–creativity and serendipity management in third generation science and technology parks. Sci Public Policy 36(7):537–548

Janiak A (2009) Newton's philosophy. In: Zalta EN (ed) The Stanford encyclopedia of philosophy, Winter 2009 edition. http://plato.stanford.edu/entries/newton-philosophy. Accessed 11 Feb 2012

Jesseph DM (2004) Galileo, Hobbes, and the book of nature. Perspect Sci 12(2):191–211

Jevons WS (1888) The theory of political economy, 3rd edn. Macmillan, London

Joseph AB, O'Leary DH (1986) Brain atrophy and interhemispheric fissure enlargement in Cotard's Syndrome. J Clin Psychiatry 47(10):518–520

Kahneman D (2011) Thinking, fast and slow. Farrar, Straus and Giroux, New York

Kant I (1783) Prolegomena zu einer jeden künftigen Metaphysik, die als Wissenschaft wird auftreten können. Johann Friedrich Hartknoch, Riga. Hungarian edition: Kant I (1999) Prolegomena minden leendő metafizikához, amely tudományként léphet majd fel (trans: John É, Tengelyi L). Atlantisz, Budapest

Kant I (1787) Kritik der reinen Vernunft, 2nd edn. Johann Friedrich Hartknoch, Riga. Hungarian edition: Kant I (2009) A tiszta ész kritikája (trans: Kis J). Atlantisz, Budapest
Keynes JM (1936) The general theory of employment, interest and money. Palgrave Macmillan, London, Hungarian edition: Keynes JM (1965) A foglalkoztatás, a kamat és a pénz általános elmélete (trans: Erdős P). KJK, Budapest
Kopányi M (2003) Mikroökonómia. KJK–KERSZÖV, Budapest
Kovács Z (2004) Geometria–az euklideszi geometria metrikus megalapozása. Kossuth Egyetemi Kiadó, Debrecen
Kristeller PO (1944) Ficino and Pomponazzi on the place of man in the universe. J Hist Ideas 5 (2):220–226
Kristeller PO (1955) The classics and renaissance thought. Harvard University Press, Cambridge, MA, Hungarian edition: Kristeller PO (1979) Szellemi áramlatok a reneszánszban (trans: Takács F). Magvető, Budapest
Krugman P (1998) It's baaack! Japan's slump and the return of the liquidity trap. Brookings Pap Eco Ac 29(2):137–206
Kuhn TS (1962) The structure of scientific revolutions. University of Chicago Press, Chicago, Hungarian edition: Kuhn TS (2002) A tudományos forradalmak szerkezete (trans: Bíró D). Osiris, Budapest
Kuhn TS (1991) The natural and the human sciences. In: Hiley D, Bohman J, Shusterman R (eds) The interpretative turn–philosophy, science, culture. Cornell University Press, Ithaca, pp 17–24
Kusch M (2009) Psychologism. In: Zalta EN (ed) The Stanford encyclopedia of philosophy, Winter 2009 edition. http://plato.stanford.edu/entries/psychologism. Accessed 15 Nov 2010
Lehmann H (1968) Grenznutzentheorie. Dietz Verlag, Berlin, Hungarian edition: Lehmann H (1971) Határhaszon-elmélet (trans: Radnóti É). Kossuth, Budapest
Ligeti I, Sivák J (1978) Növekedés, szabályozás és stabilitás a gazdasági folyamatokban. KJK, Budapest
Lucas RE (1972) Expectations and the neutrality of money. J Econ Theory 4(2):103–124
Lucas RE (1973) Some international evidence on output-inflation tradeoffs. Am Econ Rev 63 (3):326–334
Lucas RE (1980) Methods and problems in business cycle theory. J Money Credit Bank 12 (4):696–715
Lucas RE (1990) Supply-side economics–an analytical review. Oxford Econ Pap 42(2):293–316
Maclean I (1986) Reading and interpretation. In: Jefferson A, Robey D (eds) Modern literary theory–a comparative introduction. Batsford, London, pp 122–144
Madarász A (ed) (2000) Közgazdaságtani eszmetörténet. Osiris, Budapest
Mäler KG (2008) Sustainable development and resilience in ecosystems. Environ Resour Econ 39 (1):17–24
Mantel RR (1973) On the characterization of aggregate excess demand. J Econ Theory 7 (3):348–353
Maróth M (2002) A görög filozófia története. PPKE, Budapest
Marshall A (1890) Principles of economics, 1st edn. Macmillan, London
Mátyás A (1969) Fejezetek a közgazdasági gondolkodás történetéből. Kossuth, Budapest
Mátyás A (1979) A polgári közgazdaságtan története az 1870-es évektől napjainkig. KJK, Budapest
Mátyás A (2002) A korai közgazdaságtan története. Aula, Budapest
Mead GH (1970) Mind, self & society. The University of Chicago Press, Chicago, Hungarian edition: Mead GH (1973) A pszichikum, az én és a társadalom–szociálbehaviorista nézőpontból (trans: Félix P). Gondolat, Budapest
Mellár T (2015) Szemben az árral–rendhagyó közgazdasági előadások. Akadémiai, Budapest
Menger C (1871) Grundsätze der Volkwirtschaftslehre. Wilhelm Braumüller, Wien, English edition: Menger C (2007) Principles of economics. Ludwig von Mises Institute, Auburn

Mérő L (2004) Az élő pénz–a gazdasági vállalkozások eredete és az evolúció logikája. Tericum, Budapest

Merton RK (1968) Social theory and social structure. Free Press, New York, Hungarian edition: Merton RK (1980) Társadalomelmélet és társadalmi struktúra (trans: Berényi G). Gondolat, Budapest

Mezei B (1998) A tiszta ész krízise–Edmund Husserl kései munkájáról. Utószó. In: Husserl E (ed) (1998b) Az európai tudományok válsága II. Atlantisz, Budapest, pp 263–329

Mirowski P (1984) Physics and the'marginalist revolution'. Camb J Econ 8(4):361–379

Móczár J (2006) [Az] Arrow–Debreu-modell és a Kornai-kritika harminc év után. Közgazdasági Szemle 53(2):175–194

Móczár J (2010) A közgazdaságtan válsága–Neoklasszikus versus keynesi közgazdaságtan. Magy Tud 171(3):318–330

Molnár AK (1999) Max Weber. Novissima, Miskolc

Morgan MS, Knuuttila T (2012) Models and modeling in economics. In: Mäki U (ed) Philosophy of economics. North Holland, Amsterdam, pp 49–87

Murray JB (1999) Arguments for the presence of a distant large undiscovered solar system planet. Mon Not R Astron Soc 309(1):31–34

Nachane D (2010) Liberalization, globalization and the dynamics of democracy in India. London School of Economics Asia Research Centre, London

Newton I (1687) Philosophiæ Naturalis Principia Mathematica. Jussu Societatis Regiae, London. English edition: Newton I (1934) Mathematical principles of natural philosophy. University of California Press, Berkeley/Los Angeles/London

Noya A, Clarence E (2009) Community capacity building–fostering economic and social resilience. Project outline and proposed methodology. OECD, Paris

Nyborg K (2000) Homo economicus and homo politicus–interpretation and aggregation of environmental values. J Econ Behav Organ 42(3):305–322

Ozsvald É, Pete P (2003) A japán gazdasági válság–likviditási csapda az új évezredben? Közgazdasági Szemle 50(7–8):571–589

Patinkin D (1989) Money, interest, and prices. MIT Press, Cambridge, MA

Peschka V (1975) Max Weber jogszociológiája. Akadémiai, Budapest

Pivčević E (1970) Husserl and phenomenology. Hutchinson University Library, London

Planck M (1933) Ursprung und Auswirkung wissenschaftlicher Ideen. In: Planck M (ed) (1944) Wege zur physikalischen Erkenntnis. S. Hirzel Verlag, Leipzig, pp 243–257

Plato (360 BCE) Timaeus. English edition: Plato (2000) Timaeus (trans: Zeyl DJ). Hackett, Indianaoplis/Cambridge, MA

Polanyi M (1964) Science, faith and society. The University of Chicago Press, Chicago

Polanyi M (1966) The tacit dimension. Doubleday, Garden City/New York

Randall JH (1940) The development of scientific method in the school of Padua. J Hist Ideas 1 (2):177–206

Rizvi SAT (1994) The microfoundations project in general equilibrium theory. Camb J Econ 18 (4):357–377

Ross D (2004) Aristotle. Routledge, London/New York

Samuelson PA (1963) Problems of methodology–discussion. Am Econ Rev 53(2):231–236

Samuelson PA, Nordhaus WD (1998) Economics, 16th edn. McGraw-Hill, New York

Sargent TJ, Wallace N (1974) Rational expectations and the theory of economic policy. The University of Minnesota, Minnesota

Sargent TJ, Wallace N (1975) "Rational" expectations, the optimal monetary instrument, and the optimal money supply rule. J Polit Econ 83(2):241–254

Scarf H (1960) Some examples of global instability of the competitive equilibrium. Int Econ Rev 1 (3):157–172

Schütz A, Luckmann T (1975) Strukturen der Lebenswelt. Luchterhand, Neuwied

Seater JJ (1993) Ricardian equivalence. J Econ Lit 31(1):142–190

Shapere D (1988) On the methods of science. In: Leplin J (ed) (1995) The creation of ideas in physics–studies for a methodology of theory construction. Springer, Dordrecht, p 13–27
Shaw GK (1984) Rational expectations – an elementary exposition. Harvester Press, Brighton
Simmel G (1900) Philosophie des Geldes. Duncker und Humblot Verlag, Berlin
Slowik E (2013) Newton's neo-platonic ontology of space. Found Sci 18(3):419–448
Smith DW (2009) Phenomenology. In: Zalta EN (ed) The Stanford encyclopedia of philosophy, Summer 2009 edition. http://plato.stanford.edu/entries/phenomenology. Accessed 25 Dec 2009
Smith VL (1991) Rational choice–the contrast between economics and psychology. J Polit Econ 99(4):877–897
Sonnenschein H (1972) Market excess demand functions. Econometrica 40(3):549–563
Sonnenschein H (1973) Do Walras' identity and continuity characterize the class of community excess demand functions. J Econ Theory 6(4):345–354
Stead C (1998) Philosophy in Christian antiquity. Cambridge University Press, Cambridge
Steels L (1995) The homo cyber sapiens, the robot homonidus intelligens, and the 'artificial life' approach to artificial intelligence. Vrije Universiteit, Brussel
Sydsæter K, Hammond P (1995) Mathematics for economic analysis. Prentice Hall, New Jersey
Szegedi P (2003) Előszó Newton természetfilozófiai írásaihoz. In: Szegedi P (ed) Newton válogatott írásai. Typotex, Budapest, pp 45–83
Tobin J (1972) Friedman's theoretical framework. J Polit Econ 80(5):852–863
Tribe K (2006) A lost connection–Max Weber and the economic sciences. In: Az KL, Borchardt K (eds) Das Faszinosum Max Weber. Die Geschichte seiner Geltung. UVK Verlagsgesellschaft, Konstanz, pp 313–329
Trinkaus C (1983) The scope of renaissance humanism. University of Michigan Press, Ann Arbor
Tuck R (1989) Hobbes. Oxford University Press, Oxford, Hungarian edition: Tuck R (1993) Hobbes (trans: Bánki D). Atlantisz, Budapest
Tversky A, Kahneman D (1974) Judgement under uncertainty–heuristics and biases. Science 185 (4157):1124–1131
Tversky A, Kahneman D (1986) Rational choice and the framing of decisions. J Bus 59(4):S251–S278
Varian HR (1999) Intermediate microeconomics. W.W. Norton, New York/London, Hungarian edition: Varian HR (2001) Mikroökonómia középfokon–egy modern megközelítés (trans: Bara Z, Temesi J). KJK–KERSZÖV, Budapest
Veress J (2007) Búcsú a "klasszikus" gazdaságpolitikától. Paper presented at the 6th international conference of the Faculty of Economics, University of Miskolc, Miskolc-Lillafüred, 10 Oct 2007
Wallis RT (1972) Neoplatonism. Duckworth, London
Walras L (1874) Éléments d'économie politique pure ou théorie de la richesse sociale. L. Corbaz & Cie, Lausanne, English edition: Walras L (1954) Elements of pure economics or the theory of social wealth (trans: Jaffé W). Allen & Unwin, London
Weber M (1904) Die Objektivität sozialwissenschaftlicher und sozialpolitischer Erkenntnis. Archiv für Sozialwissenschaft und Sozialpolitik 19(1):22–87, English edition: Weber M (1949) Objectivity of social science and social policy (trans: Shils EA, Finch HA). In: Shils EA, Finch HA (eds) The methodology of the social sciences. Free Press, New York, p 49–112
Weeks J (1989) A critique of neoclassical macroeconomics. Macmillan, London, Hungarian edition: Weeks J (1998) A neoklasszikus közgazdaságtan kritikája (trans: Szalai Z). Aula, Budapest
Weinberg S (1993) Dreams of a final theory–the search for the fundamental laws of nature. Hutchinson Radius, London
Weissman PR (1996) The Oort cloud. In: Rettig T, Hahn JM (eds) Completing the inventory of the solar system. Astronomical Society of the Pacific, San Francisco, pp 265–288
Zalai E (2000) Matematikai közgazdaságtan. KJK–KERSZÖV, Budapest

Chapter 2
The Rational Expectations Hypothesis as a Key Element of New Classical Macroeconomics

Now, what I want is, Facts. Teach these boys and girls nothing but Facts. Facts alone are wanted in life.
(Charles Dickens: Hard times)

2.1 Introduction

The contradictions and the reality of the underlying assumptions and conclusions of the rational expectations hypothesis (REH) will be reviewed in this chapter. It will be argued that crucial elements in new classical macroeconomics lead to confusion in the argument, therefore unbiasedness needs to be reconsidered. Our critique is based primarily on the fact that new classical macroeconomics is set to deliver a realistic description of our life-world and, therefore, their models can be compared directly to life-world conditions. However, it will be seen that the portrayal of a typical agent as the REH postulated it is valid *only as an ideal-type*, that is, we will have a problem unless this model is regarded as a pure theory. Although the degree of rationality postulated by the REH cannot be a requirement under life-world conditions, the rational expectations hypothesis is still not a futile theoretical construct, since by finding a proper scope for this theory we can offer marvellous theoretical support for the mechanisms of inflation targeting. If *exaggerated* assumptions are set aside (that is, we *ease* the presuppositions) by which a monetary authority capable of and destined for leading market expectations can be included, we end up having a moderate version of the hard definition of the REH given by Muth, while certain consequences of this concept will be consistent with the assumptions of the original (hard) definition—assumptions that otherwise seem indefensibly and groundlessly hard. [→ **experimental economics**]

The outcome of talking about the *easing* of assumptions of macro-models (and of all kinds of logical models actually) is dubious. Instead, one should, perhaps, stress the incorrectness of these presumptions by underlining that an assumption can be right or wrong (and nothing else) and that we can either maintain it or not. Emphasizing the possibility of eased assumptions implies that assumptions that are applied in various theories can be *more or less* correct. Is this not an absurd idea? It might be useful to build our arguments upon an analogy. The central problem of empirical studies investigating the validity of Barro-Ricardo equivalence was

P. Galbács, *The Theory of New Classical Macroeconomics*, Contributions to Economics, DOI 10.1007/978-3-319-17578-2_2

whether the fluctuations of consumption-related expenditures in the theory of Keynes are, and to what extent, in accordance with actual processes (see Chap. 5). A fairly flat pattern of consumption expenditures follows from the permanent income hypothesis; by contrast, a strong volatility follows from the Keynesian theory. Every change in disposable income leads to proportional fluctuations of consumption affected by marginal propensity to consume; while the permanent income hypothesis predicts a much smoother time pattern. Whether consumer expenditures fluctuate and to what extent is an empirical problem and, on the very basis of empirical tests, they can be sufficiently described in quantitative terms. Consumption expenditures *may not* fluctuate *at all*, they can be *excessively* volatile or may show a *considerable* steadiness. If such qualifiers are not satisfactory for a precise analysis (and they can hardly be so), we can say that the time series data of consumption expenditures are more or less scattered around their trend. Therefore, *it seems that the predominance of a certain theory can be measured in quantitative terms*. Observed facts may be closer to the predictions of the Keynesian or the monetarist theory; a clear case is expected to be only a theoretical possibility. Anyway, what would a clear and striking manifestation of the Keynesian consumption pattern stand for? Is there an economist who can substantiate that an empirically estimated numerical value of the marginal propensity to consume indicates the clear manifestation of the Keynesian theory? The Keynesian consumption model and the permanent income hypothesis differ in the implied numerical values of autonomous consumption and marginal propensity to consume, since a higher exogenous component and a lower marginal term characterize the latter. These two theories can be regarded as two extremities of possible consumption patterns. But which values of the estimated parameters should be labelled as clear manifestations of one theory rather than the other? It could be hard to find a reasoned answer.

Therefore, in this sense, we can talk about *the extent to which a theory predominates*. In this way, easing and modifying certain assumptions of a model on a continuum is not, therefore, possible? It is true, of course, that such easing leads to another model, although it would be a transformed (constrained? modified?) version of the original construct. In our view, assumptions can be eased and although their literal predominance may be implausible; it is still possible, however, to find an eased set that is (more) consistent with life-world conditions.

Assumptions of a model are not (or not necessarily) binomial variables. That is, setting aside an assumption does not necessarily imply the contrary of the original setting. Suppose that an assumption on the political environment is present in an economic model. In this case, setting aside the assumption of a democratic scenery does not imply a totalitarian dictatorship. Similarly, in the case of the market system, if one does not use a monopolistic system as a starting point, this does not result in the presumption of an absolutely free competition, since there are plenty (in fact, innumerable) of intermediate structures between the two extremities of free competition and monopoly. Therefore, it is not absolutely free competition that comes after easing the rigour of a monopolistic environment but duopoly, followed by multiple agent models of oligopolistic competition, all of which should be regarded as a shift toward free competition.

In his well-known methodological paper, Milton Friedman (1953) argues that if one ends up having a theory consistent with reality in its predictions through applying unrealistic assumptions, one should depart neither from our model nor those assumptions.[1] As it has already been mentioned above, Friedman's example is identifiable with the behaviour of leaves around a tree: by establishing a model in demonstrating their density and thus finding a successful construct through comparing the predictions of some alternative models in the assumption that leaves are rational agents capable of applying mathematic optimization methods—then this model cannot (and should not) be contested. As an analogy, the geocentric and heliocentric models of the Solar system come to mind. Both of these theoretical constructs were successful in predicting the observed trajectory of the Sun in the sky (cf. Weeks 1989). Therefore, in terms of the *correctness of a model*, it does not matter whether it is capable of delivering consistent estimations *and nothing else*. We do not (and *cannot*) discover the actual mechanism of reality, hence we cannot explain a change in the consistency of the predictions if it does occur. *As long as the model works properly, its outputs are consistent with reality but not for the reason the model itself suggests*. We, critics, should not be satisfied with consistent predictions, since doing this would be tantamount to putting a stop to the development of science, since if the only ground of our judgement is the *consistency of predictions* (and *not* the *correctness of the model*), to exceed the geocentric model would not have been necessary at all.

Considering the lessons of Chap. 1, the assumption included by Friedman's leaf model is *incorrect*, since it stems from an arbitrary alchemy, not from abstraction. However, it is excessive rather than *incorrect* to assume the existence of agents in the rational expectations hypothesis, capable of establishing expectations consistent with the objective probability distribution. The axiom is right: if, in actual fact, economic agents *had* the methodological experience and the information-based background (that is, if they were *homo oeconomici*) necessary for the establishment of such expectations, nothing *would* prevent a particular theory from presenting a realistic description of reality in synch with its underlying intentions.[2] However, human beings are not what the REH assumes them to be. By the way, this concept *may* be accepted in terms of a pure theory and it should not even be criticized for this: if Man was an agent created to be an analogy of the homo oeconomicus and an actor in applications running on computer screens, this portrayal would be acceptable. However, if the theory aims to describe reality *realistically*, this *excessive* assumption cannot be maintained since it implies the immanent failure of the theory itself (i.e. in its endeavour to offer a realistic description of reality). When referring to the easing of the REH assumptions, a concept of agents might be used as a point of departure (this, in fact, is the direction we have taken) who, rather than

[1] At some point Friedman said that empirical tendencies alone prove unsatisfactory to suggest a certain relationship between variables; clarification of the theoretical background is of paramount importance (Friedman 1970).

[2] However, it is always dubious whether theories should, in fact, paint a realistic and comprehensive picture of reality. Is drawing a map on the scale of 1:1 sensible? (cf. Varian 1999)

establishing their expectations independently, borrow them from a specialized institution, where these expectations, which are not unbiased, are set to approximate the objective probability distribution as much as possible. The distinction between *incorrect* and *excessive* assumptions applied throughout this publication is based on this very principle. The analysis is about a twofold exploration of assumptions: the first question is whether an assumption is based on abstraction; the second is whether the assumption itself is in accordance with the goals and ambitions of the theory which the assumption is applied by. It should be emphasized that assumptions are eased not for the purpose of turning erroneous presuppositions into right ones—a wrong assumption cannot be fixed, only replaced (and replaced it should be: the F-twist cannot be regarded as a legitimate scientific approach in economics). However, intensifying and easing correct abstractions lead to model families the analysis of which may enhance our knowledge.

The problem of whether the denial of new classical doctrines is based on the irrelevancy or on the excessiveness of their assumptions, is difficult to address. We will see in Chap. 3 that the general equilibrium analysis (general equilibrium considered as a reference case) relies on the presumption of equilibrium. Simply put, it is the axiomatic postulation of equilibrium. To claim that it is the most crucial element of neoclassical theory is not an exaggeration. It is argued here that *as an abstraction,* the assumption of tendencies pointing towards a state of equilibrium should not be labelled incorrect: if changes in prices are not hindered, the state of equilibrium will be established. *This* is the fundamental mechanism of market as such.[3] Nineteenth century neoclassical models—by distinguishing real-life circumstances from the model environment—applied the pure theory in a proper way (the only danger was the notion of *imperative* referred to above). However, the general equilibrium in the new classical theory as a presupposition is considered *excessive* since the promise here was to provide a realistic description of reality. In other words, there is a discrepancy between the purpose of the theory and the assumption applied. However, to presuppose an equilibrium and equilibrium mechanisms is still *not unreal*, since, after all, the abstraction that resulted in the axiomatic system is correct. Were the opportunity to present itself, a state of equilibrium would in fact emerge—the efforts to prove this are the greatest merits of mainstream economics.

2.2 Preliminary Remarks

It goes without saying that economists have been puzzled by the rational expectations hypothesis and the new classical macroeconomics by which it was created, and that the scope of the theory itself still lacks clarification. Despite the fact that

[3] Júlia Király (2000) claims that highlighting the equilibratory or non-equilibratory nature of macroeconomic systems is a matter of faith and attitude. We are of the view that there is more to it than that. General equilibrium models are abstraction-based constructions that grasp the fundamental mechanisms of a purely economic setting through their ideal-types postulated as axioms.

the REH, complete with the paradigm of continuous market clearing, underscores the predestined failure of the activist fiscal and monetary policy,[4] the theory has become an integral part of modern economic policy recommendations (the theses by Nobel awardees *Finn E. Kydland* and *Edward C. Prescott* should be mentioned here). In pure logical terms, it can be clearly proven that rational expectations are of crucial importance when explaining the mechanism of inflation targeting (cf. Harasztosi 2004); or at least, it can be realized that the criterion suggested by Muth holds, according to which the predictions of the relevant theory and expectations of agents coincide; even if the *functionality itself* (i.e. the fact that inflation targeting works effectively) as it will be seen later, does not follow from the Muthian definition. Still, huge gaps in our knowledge in economics are becoming visible. The REH in its rough and pure form does not seem to take production costs into account (cf. Erdős 1998); according to the classic texts, inflation depends only on expectations. Monetary policy also relies on the rationality of expectations—however, some central bank inflation models trace price dynamics back to costs alone (for one of the most important developments in this research area see Várpalotai 2003). If, in terms of econometrics, costs alone are enough to explain inflation (i.e. it is virtually unnecessary to exceed the Keynesian foundations, at least in inflation theory), what is the role of the (more and more) often cited expectations within the mechanism? Today, our knowledge is surely imperfect to provide satisfactory answers.

The role of expectations in price dynamics has been one of the most fundamental questions of high theory for decades—even if, from time to time, the importance attributed to it is seen fading away. For example, as early as 1912, Mises pointed out that price dynamics may occasionally deviate from the base provided by the quantity of money (during hyperinflation, for example), and that in such cases the value of money will be determined by expectations (cf. Mises 1981). As for the progress and improvement of theories that followed, this observation was perceived to be rather an annoying one, which economists did try to disregard, since what it implies is the invalidity of the quantity theory (since, as a consequence, it is not the actual quantity of money that will determine the price level after all). Weird as it sounds, orthodox monetarists and new classicals did not set aside the rough form of quantity theory even for a single moment while highlighting the role of expectations—the scope of expectations was constrained to expectation errors, and the consequent relation between price levels and quantities of money was, as it were, conceived of as the outcome of *deus ex machina*.

The functionality of inflation targeting and its characteristics highlight another important circumstance. Monetarism (which includes new classical macroeconomics as a radical wing[5]) established its theory while stipulating a very strong

[4] On the effects of rational expectations in a non-equilibrium environment and the efficiency of anticipated actions of countercyclical economic policies see e.g. Fischer (1977).

[5] This statement is not as evident as it is appealing and palatable. Certain authors stress that new classical macroeconomics may be classified as monetarist (e.g. Mátyás 1984), while others claim

limitation—and this feature significantly confined the relevance of the theory itself. It appears, therefore, that the monetarist theory can only be applied to describe macroeconomic systems where the intermediate target of the central bank is to control the quantity of money. However, to use the quantity of money as an operative target is a practice no longer paramount[6] (it never proved to be very successful as its applicability has raised problems as a result of a recent breakthrough in financial innovation[7]). The conclusion that can be drawn from all of this (and is, in fact, often drawn) is that the monetarist theory no longer deserves any attention.[8] However, this is not the case. Modern central bank policy, based on inflation targeting, is capable of using certain instruments much more efficiently now (as opposed to how this was done in the past) and, as a result, affecting the price level with considerable efficiency. Put quite simply, *the success of inflation targeting regimes today stands a good chance of making up for the failure in the control of the quantity of money*. Therefore, monetarism should not be forgotten. Instead, it should be put on the agenda again and one should consider which elements of the theory are relevant (in other words, which elements can be made relevant again) and the relevancy of the central bank's control of prices in the context of the REH should be looked into, since this control can be regarded as the predominance of a tendency resembling the quantity equation. Never should a theory be regarded as a monolithic entity; it is more like a mosaic whose individual segments are linked by assumptions and conclusions. This peculiarity has in fact been recognized by current theories in economics, for in the absence of this recognition the theory of rational expectations today would not exist without efforts being made for the entire monetarist theoretical system (whatever is it) to be rammed down our throats.

In this short chapter, an attempt is made to perform the easiest part of this job, i.e. the reality behind the underlying assumptions of the REH (the most viable and most debated component of monetarism) is investigated, given the possibility of a

---

that orthodox monetarism follows Marshall, while new classicals follow in the footsteps of Walras (Hoover 1990). Evaluation can be less of a challenge if we believe Milton Friedman, who cited the Walrasian equilibrium in the context of the natural rate of unemployment (Friedman 1968). Meanwhile, Frank Hahn does in fact write about the new classicals when describing monetarism (Hahn 1980).

[6] This is distinction is justified. M_1 aggregate is a possible intermediate target, while the control of the monetary base (M_0) is an accommodating operative target.

[7] On the special aspects and its implementation, along with the instruments involved in this financial innovation see Vigvári (2004). In this context, one need only think of widely (increasingly so) used e-money, that part of money, used in transactions, which is all but uncontrollable from the perspective of monetary authorities. Time and again, the development of financial instruments also underscores the necessity of the redefinition of monetary aggregates (Anderson and Kavajecz 1993).

[8] It must be stressed that the REH and the new classical macroeconomics are not synonyms. REH, having parted company with new classical macroeconomics, became an independent school of thought. By any measure, REH is interwoven via its links to monetarism (see the details regarding the roots of orthodox monetarism and new classical theory).

comparison to a directly experienced reality; cf. Weeks 1989. On the other hand, hidden logical inconsistencies are also highlighted. However, we do not turn a blind eye to those attractive characteristics of the theory which have led to its eventual success. First and foremost, our aim here is to paint a coherent picture of what is known as the rational expectations hypothesis. We do not need to strain ourselves to see the difficulty of the task at hand as followers of the hypothesis do not form a homogenous group in terms of the assumptions and conclusions.[9] Therefore, practical difficulties that put a question mark behind the plausibility of the hard definition of the REH are in the focus of our attention. Considering these problems will move us on to an alternative (moderated) version of this hard definition which, stemming from the asymptotic unbiasedness of estimations, accepts the *biasedness* of *individual* expectations and introduces a central institution (i.e. the central bank) to make the predictions of the relevant (quasi-relevant) model and the expectations of agents coincide. So, the basic requirement incorporated in the hard definition will eventually be met, albeit through an alternative and unusual mechanism.

2.3 The Concepts of Rationality

In the literature of the REH, in view not only of new classical macroeconomics but all other schools that have assumed (a certain degree of) rationality of expectations in their theory, not a single competent author has worked in the last few decades without referring to a seminal paper by Muth from 1961 (Muth 1961). Initially, the idea emerged in a context that differs fundamentally from the framework in which Robert E. Lucas applied it a decade later. Mention should also be made of the fact that the rational expectations hypothesis did not emerge in Muth's paper for the first time. Jan Tinbergen, in his 1932 paper entitled *Ein Problem der Dynamik*, specified a model that resembles Muth's later work in conceptual terms—with considerable differences, of course. Tinbergen was aware of the fact that due to the uncertainty that surrounds the future, agents form expectations (although in an attempt to simplify the model, he assumed away the difference of expectations). These expectations are rational (the term *vernünftig* is used in German) if we assume that they are consistent with economic relations. Moreover, expectations can, in specific cases, be replaced by the results of an economic theory—and this definition rhymes well with Muth's concept and with the concept established in the relevant economic theory (cf. Keuzenkamp 1991). Recently, the relevancy of the REH has been questioned on the ground that the future is mostly uncertain, therefore it cannot be scanned via rational expectations. Thus, imagination has to replace

[9] Equally important is the fact that developers and adopters of the REH have relied on the mathematical apparatus more than every previous school. However, it is also true that their theoretical inferences and theorems are literally *derived conclusions*, i.e. they come from the way equations are specified—to form generalizations is, sometimes, a slow-moving process characterised by constraints of its own.

reason. It is becoming more and more evident that there is a huge gap between economic models and the actual mechanisms of the market. Traditionally, these theoretical models apply static general equilibrium frameworks to predict market processes, assuming that agents optimize their behaviour while relying on rational expectations. However, markets are dynamic; moreover, they are characterized by a process of continuous innovation, self-enhancing, emotional outbursts and uncertainty. In this world of uncertainty, individuals are driven by feelings, emotions, intuitions and imagination rather than rational expectations or conjectural estimations of future utilities (Bronk 2009).

According to Muth's oft cited definition, expectations of firms[10] (or, to use accurate wording, the subjective probability distribution of outcomes) will be in the region specified by the prediction of the theory itself (i.e. the objective probability distribution of those outcomes) for the same information set. In other words, rational agents expect an outcome the emergence of which is the most probable—and this is the criterion by which the rationality of expectations can be assessed. Muth tries to offer an interpretation of his own theory; the most important element among the special features highlighted by him is by stressing that a publicly announced prediction will not exert any influence on the macroeconomic system under scrutiny, unless the prediction itself is based on insider information, i.e. on an information set unavailable to market agents.[11] Interpretations on Muth highlight that this particular description of the formation of expectations does not entail the assumption of perfect foresight by market participants or access to the widest spectrum of information available—what in fact this is all about is that agents try to use their information in the most efficient way, while their anticipation is directed towards the most probable outcomes,[12] and they rely greatly on the predictions of the relevant economic theory while forming their expectations (cf. e.g. Pete 2001). By the way, the relationship between the assumptions of the REH and those of the PFH (Perfect Foresight Hypothesis) is often unclarified; sometimes they are regarded (mistakenly) as synonyms (cf. e.g. Bryant 1983). A key precondition for a proper interpretation is the need to distinguish between the *deterministic* and *stochastic* models, since the former (and that alone) makes it possible for the REH to be identified with PFH (Barro and Fischer 1976). In stochastic models, assumptions do not determine unambiguously whether or not a certain event will occur—however, deterministic models are a different kettle of fish: results are evident here

[10] It is, perhaps, not an exaggeration to broaden the scope of this statement to describe all market participants.

[11] Rationally formed expectations are not sufficient for the ineffectiveness of discretional monetary policy—to eliminate stickiness is another requirement, while information sets related to monetary policy and other agents also need to be identical (cf. Dickinson et al. 1982). It will be argued here that this complementary part of Muth's thesis, which is ignored by high theory (i.e. only insider information can modify market expectations) is crucial for the validity of this theory.

[12] It may be confusing, since we shall talk about expectations here as if they were discrete points—whereas it is obvious that expectations tend to constitute probability distributions. These discrete points should be regarded as point estimations.

(*occurrence* vs. *non-occurrence*). This distinction is even clearer if we consider the fact that a deterministic model delivers a concrete numeric value by way of a prediction through the substitution of inputs (and under the influence of the parameters); by contrast, a stochastic model predicts a probability distribution because of the scope of contingent events. This is exactly what characterizes statistical and econometrical predictions, since confidence intervals indicating the probability distribution of a given point estimation can be provided along with estimations. Reading between the lines, John Weeks' (1989) analysis seems to convey the message that refuting the PFH is not that simple. Proponents of this argument often claim that the volume of information that needs to be gathered and processed for a perfect foresight far outstrips qualities still possible to be explained by the utility maximization axiom—according to this line of thinking, forming perfect foresights would no longer be economical, while its possible emergence cannot be ruled out entirely. Instead, Weeks voices his criticism on conceptual grounds making an attempt to deny the implicit presumption according to which knowing and predicting the future is an act that depends entirely on the quantity of information processed. It is obvious that it would only be possible to perfectly foresee the future under deterministic circumstances, where contingencies are ruled out. Kenneth Arrow himself, by using a suggestive parable, demonstrates that if contingencies do become part of the game, future outcomes cannot be foreseen with certainty, though it is clear that possessing and processing more and more information can reduce uncertainty (Arrow 1973).

Before moving on, we need to do some more thinking in connection with the concept of *rationality* and *rational expectations* for a while. First of all, mention must be made of a circumstance highlighted by Muth himself. While reading his paper, it becomes evident that a rationally formed expectation (or, in other words, rational formation of expectations) is a cognitive act that can hardly be compacted in a mathematically specified expectation operator. Whether it be a simple, extrapolative (or its purely algebraic derivative, i.e. naive, or regressive) or adaptive scheme, the process of expectation formation can be described without any problem by using difference equations. [→ **expectation operator**] Muth, on the other hand, stresses that rationality (of expectations) is a radically different quality, and this category does not refer to the formation method through which the numerical values of expectations are generated (i.e. the expectation methodology) but to the relationship to processed information (a prediction which is rational in a Muthian sense can be formed through various methodologies, justified and permitted in view of a situation). The weakest definition of the REH does not go beyond the claim that market participants, when forming their expectations, include in their consideration not only the past values of the given macroeconomic variable but all information beyond historical data available in connection with the relevant determinants (the stronger definitions go farther, of course).[13] To use an analogy from trade, expectations regarding the future price of Brazilian coffee must deviate from the long-

[13] This phrasing should, perhaps, be regarded as the common element in various definitions of rational expectations.

term trend if a natural disaster were to destroy half of the harvest and that fact became known to market participants (cf. Pete 2001). However, it must be stressed that confronting *adaptivity* and *rationality* is not necessarily justified, in other words, there are situations in which following the adaptive scheme is a rational response. As for the analogy referred to above: if neither a natural disaster occurs nor any effect capable of changing the trend of a long-term coffee price is seen emerging, market agents can rely on rational thinking by using all relevant information and thus anticipate the continuation of a former trend. In this case, application of the adaptive scheme will meet all the requirements in Muth's strong definition.

Expectation operator. The relationship between the expected value of an economic variable (target variable) and the information set used in forming this expectation is expressed by the expectation operator. In a broad sense, an expectation operator is a mapping

$$\varphi : \Omega \rightarrow E$$

which creates a link between elements of information set $I_i \in \Omega$ applied in expectation formation and expected values $e_j \in E$ of an economic variable. We can talk about a *mapping* as it is not possible that multiple expectations (several elements of the expectation set) are linked to a certain element of the information set. Elements I_i of information set Ω are sets themselves, that is,

$$I_i \subset \Omega,$$

since market agents *can*, while forming their expectations, exploit several pieces of information (past numerical values of the target variable, past expectation errors, macroeconomic variables, announcement effects, etc.). So, it is known about the elements of the information set that $I_i := \{a_i; b_i; c_i; \ldots\}$. The cardinality of subsets I_i of information set Ω depends on the number of information sources (data sources) applied by agents in the course of forming expectations. According to the hard definition of the REH, this cardinality can even be infinitely high, since expectations are based on all information available. Alternative assumptions of expectations suppose a definitely narrower scope of exploited information stock (Fig. 2.1).

Mapping $\varphi : \Omega \rightarrow E$ is *surjective*, since every element in information set Ω has a corresponding element in set E, that is, an expected value (but only one element, since we are talking about a mapping), and there is not any expectation that is not based on some kind of information. In other words, all elements in set E is related to at least one information subset I_i. However, this mapping is not *injective*, since not every element (subset) $I_i \in \Omega$ is related to different element $e_j \in E$.[14]

---

[14] The latter is not particularly hard to realize. If expectations sets consist of, say, rates of inflation expected for next periods, we can easily find a situation in which the same rate of expected inflation appertains to different information sets. For instance, an x percent budget deficit and a per litre fuel price of y make set $I_1 \in \Omega$; and a z percent budget deficit (where $z < x$) and a per litre fuel price of w make set $I_2 \in \Omega$ (where $w > y$). However, agents may link the same rate of expected inflation to both sets of information.

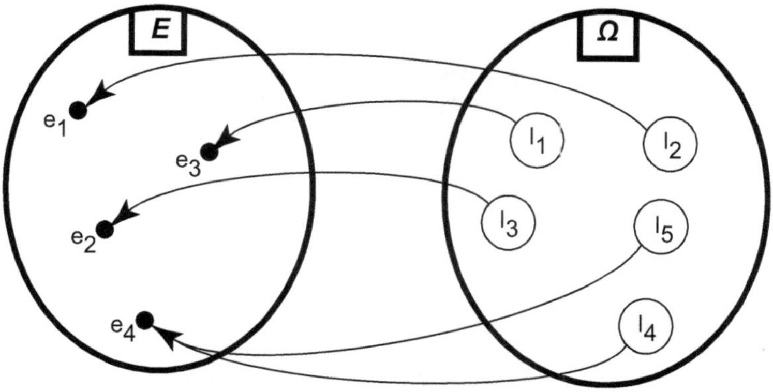

Fig. 2.1 Mapping between information set and expectations

The expectation operator as a mapping is a simple function (that can be written in mathematical expressions with one or more, but finite number of *numerical* arguments) sometimes. Since we have defined the connection between these two sets as a mapping, what follows from this is that in each case the expectation operator is interpreted as a mere function, *in an algebraic sense*.[15] Although it is true that the quantity of applied information makes a distinct difference among the possible methods of expectation formation, but because information and expectation are connected by a mapping, the expectation formation method can always be regarded as a function (*always*, in other words, independently of the cardinality of the exploited information set). It is obvious that the greater the cardinality of individual subsets $I_i \subset \Omega$ is, the more arguments stand in the function. However, it must be stressed that the forming of expectations is not necessarily a functional and stable cognitive action (requiring a mathematical expression) but, on certain occasions, more of a *belief*. In simple cases, there is an *ex ante* function (a mathematical formula) that relates the expected value of the target variable to a finite number of arguments by a stable (simply put: formulized) operation. Therefore to form expectations is to perform operations on the information set, i.e. on the arguments where these operations (intake of inputs) are established by this *ex ante* function. In other cases, the forming of expectations is not governed by an *ex ante* function. A market agent may even process an unlimited volume of information and relate a given expected numeric value of the target variable to a given information set (mapping does, therefore, exist), however, the forming process does not follow a rule (formula, if you like) known by and obvious to the agent. In this case, the forming of expectations can be regarded as a belief.[16] As in this case relation is also driven by a mapping, the *ex post* expectation function can be identified (at least theoretically, i.e. if we know the information set on which the agent formed his expectations). According to the hard definition of the REH, it is not possible to formulize such an *ex ante* stabile relation (function) and primarily not for the infinitely high volume (cardinality) of information exploited in the forming process but because the characteristics of information continuously affect the forming method. In case of a certain information set, a simple extension of

[15] Mathematicians often label functions simply as mappings or operators (see e.g. Rozgonyi and Toledo 2008). It has to be added that functions do not have to be mappings between sets consisting only of numerical elements. In terms of the identification of mappings (functions), the one-to-one character and not the nature of elements has a crucial importance.

[16] However, no consideration precludes the case in which an ill-informed agent forms his expectations as beliefs rather than on the basis of an *ex ante* formula.

the previous trend may be appropriate, but if information changes, another method may be needed.

In a narrower sense, the expectation operator is a function with only a few arguments among the inputs of which only the lagged values of the target variable (x_t) and, perhaps, past expectation errors are present. The simplest extrapolative model can be written in the form of

$$x_t^e = x_{t-1} + \alpha(x_{t-1} - x_{t-2}),$$

where x_t^e denotes the expected value of x_t formed in period $t-1$. It has to be noted that if $\alpha > 0$ then x_t^e is the extrapolation of x_{t-1} reflecting the most recent change in the variable and following the direction of this change. If $\alpha = 0$, it is the plainest naive scheme, that is, agents anticipate a simple continuation of present. In the third case, if $\alpha < 0$, the regressive scheme starts predicting the former value of the given variable.[17] It is the substrate of the adaptive model that the expectation error (i.e. the difference between the expected and the actual values) calculated on the basis of a former (computed usually for period $t-1$) expectation is added to the formation of period t expectations as a correction factor. In this case, our operator can be defined as

$$x_t^e = x_{t-1}^e + \beta\left(x_{t-1} - x_{t-1}^e\right)$$

in algebraic terms, where $\beta > 0$.

Expectation operators also play an important role in econometric modelling as well. By hypothesizing adaptive expectations, you can construct an equation on current and lagged data of income and consumption expenditures which leads you to a numerical estimation of marginal propensity to consume out of adaptively expected income and, hence, to the possibility of testing your hypothesis (Ramanathan 2002). Maddala (2001) offers examples for the treating of naive expectations in econometric terms. However, it is common in all these cases that there is no observed data available for expected values (at best, one can identify the expected values from de facto observations just as it is done by Maddala). If, however, you have access to explicit expectation data,[18] and you are familiar with the

[17] Visco (1984) offers an excellent summary about various expectation operators and the possible applications.

[18] In Hungary, following the methodology established in North America and Western Europe after World War II, the Economic Research Company (GKI) prepares surveys on a monthly basis through which market participants' inflation expectations can be revealed. GKI surveys five groups of market participants (industry, trade, building and construction, services and households) so that expectations regarding relevant future market outcomes and opinions about recent developments can be reviewed. GKI has been providing consumer surveys since March 1993, accompanied by business surveys from January 1996 onwards. The business sector is represented in the survey by being divided into various branches: agents from industry and trade are surveyed on a monthly, while market participants from building and construction plus services were initially asked on a quarterly basis. From January 2002, all the business sector has been examined every month. In the course of the consumer survey, 1,500 individuals are visited by surveyors (the number of those interviewed was 1,000 before May 2001). All in all, the GKI, the Institute of Marketing and Media of the Corvinus University in Budapest also make consumer surveys on a quarterly basis, following the methodology of the University of Michigan (Index of Consumer Sentiment—ICS) the main output of which is "Michigan Consumer Sentiment Index". Sampling is

information set on the basis of which these expectations were formed, you can specify the *ex post* expectation function that governs the formation process.[19]

But what about the weak definition of REH? In its briefest form, it states only that rational agents form their expectations on a relevant (economic) variable through efficiently exploiting all public information on factors which they think are capable of affecting the future path of a specific variable (Horváth 2000). Apparently, this definition reveals next to nothing in terms of the expectation itself,

done in two stages (first: settlements and, second: individuals are picked randomly); the sample should be representative in terms of gender, age, qualification and types of settlement (Vadas 2001). The ratio of successful interviews is, generally, over 90 %. According to econometric investigations, consumer confidence index predicts with accuracy the dynamics in consumption expenditures (consumer confidence index holds additional information about the overall condition of households and the uncertainties of future income stream), therefore the Central Bank of Hungary relies heavily on it for information processing and forecast purposes (Jakab and Vadas 2001).

In the course of business surveys, 5–6,000 enterprises (having legal entity and with 20 employees at a minimum) are asked on a monthly basis via postal questionnaires. Restricting the sample to companies over 20 employees has proved to be a technically and methodologically competent decision. Thanks to this, a greater stability of samples can be guaranteed, while data consistency and reliability continue to be a high priority rather than being something of secondary importance, given that experience has shown that the smaller a firm, the more unreliable (systematically biased) the data provided by it. Small enterprises (with a staff fewer than 20 employees) are units usually relying on family and acquaintance relationships, concentrating on local markets and geared to provide maintenance for the family—therefore they can hardly be separated from the household which provides the infrastructural background and capital sources. The attitude and the goals of these enterprises may radically differ from those of larger businesses (Tóth and Vincze 1999; Tóth 2002). The sample comprises 1,500 processing industry firms. The sampling method is quite unique: the sample available from the previous year is updated every July so that 300 new (that is, previously not involved) firms enter while 300 firms are removed from the sample. This routine biases towards enterprises being already in the sample, which restricts the possibility of deletion—while the sample functions as a quasi-panel. The advantages of panel studies over longitudinal surveys are quite obvious as in this way researchers can gain a better understanding of the changes that affect individual units or groups of those units (cf. Babbie 1989). The number of responses (completed questionnaires) from the entire sample (i.e. not only from processing industry firms) is around 1,000–1,200; propensity to respond is slightly under the EU average, while it is not particularly low compared with former socialist countries. Responses are representative, both territorially and sectorally, i.e. responding firms and the entire sample do not differ substantially. By contrast, there are significant differences between responding firms and the sample in terms of business size as experience suggests that propensity to respond is significantly lower for small enterprises; according to GKI reviews, this is not particularly problematic, since in Hungary medium-size and large companies do, in many ways, play a key role. Active, responding businesses reveal homogeneity to a large extent (both in terms of sectors and individual companies), i.e. the circle of enterprises that fill out questionnaires in each month can be regarded as rather unchanging, thereby helping data processing (Tóth 2002).

[19] The distinction between *ex ante* and *ex post* expectation functions suggested here bears strong resemblance to the potential roles of monetary policy Taylor-rules. A monetary policy setting the key policy rate for the next period may explicitly follow a specific Taylor-rule (ex ante role). Even if the central bank does not follow such a rule, a Taylor-function as a regression equation still describes well the interest rate decisions of the monetary authority subsequently (ex post role). This will be a recurrent theme in Chap. 6.

except a little bit about the formation process—and, according to this definition, expectations regarded as rational will be different from those included in the hard concept, where reference to probability distribution is tantamount to a significant, additional knowledge. As for the coffee price analogy mentioned earlier on, a market player who optimizes his behaviour through comparing costs and benefits, gathers and processes information up to a point when its costs exceed the benefits hoped for (or avoidable losses). This weak definition claims that an investor can, by thinking *rationally*, expect the coffee price to keep in line with the former trend even after the occurrence of a natural disaster affecting the plantations, if the expenses of collecting and processing information required for the changing of expectations are higher than his (potential) loss resulting from an incorrect expectation. However, it is clear that this critical point is different for every investor, as certain investors in the same situation risk losing low amounts due to incorrect estimates, while others might be exposed to a considerably higher loss, depending on the actual amount of investment. And, if this is the case, the convergence of individual expectations will be out of the question, given the discrepancy in the expectations due to the information process having discontinued in various phases (moreover, these expectations may be systematically biased). Therefore, according to the weak definition, a simple trend extension by a small investor may be regarded as being rational. In this case, the only stipulation regarding rationality is that the agent should keep processing information up to the optimal point and expectations formed accordingly will be labelled as rational.

Experimental economics. Experimental economics has discovered the laboratory for economic theories as well—the laboratory which was already known to and used by sociology and social psychology (cf. Babbie 1989; Bungard 1997) or marketing science (e.g. Malhorta and Simon 2009). Following that, economics was no longer in want of experiments. The first attempts were made by Edward Chamberlain (Caginalp et al. 2003), and experimental economics was eventually awarded the Nobel price thanks to Vernon L. Smith, Chamberlain's follower.

At the outset, experimental economics represented a definite methodological programme that can be interpreted as one surpassing the traditional schemes of economics (KVA 2002). According to experimental economics, mainstream economics abounds in theses that are no more than assertions never confirmed. It is, perhaps, nothing more than a belief that macroeconomic systems can be fine-tuned in accordance with the Keynesian suggestions, or that the problems caused by monopolies can be solved by regulatory measures, or, if exposed to certain externalities, the market is doomed to failure. One objective of experimental economics is, therefore, to test the fundamental theses of economics (Smith 1982). Further efforts were directed towards delivering data upon which new theories can be based. Through experiments, certain behavioural tendencies can be identified upon which to build a theory at some later point, capable of interpreting these tendencies in clear and explicit terms. (Smith 1994).[20]

---

[20] This is as trivial in meaning and importance as it is objectionable. At one point, Smith (1994) writes: "Well-formulated theories in most sciences tend to be preceded by much observation, which in turn stimulates curiosity as to what accounts for the documented regularities." This is the very essence of an induction-based theorizing routine. Common theoretical-scientific thinking finds nothing exceptionable in induction—on the contrary, it is regarded as the fundamental

However, to do nothing other than detect and describe the laboratory behaviour of human beings was not the ultimate goal, but to identify the building blocks of which actual market mechanisms and macroeconomic systems consist. Neither astronomy, nor meteorology, nor theoretical economics is able to conduct experiments with the objects (planets, stars, weather fronts or national economies) that they observe. However, the results of some laboratory experiments can be exploited directly in studying astronomical and weather-related data and in interpreting observations, since the same natural laws are valid for the whole physical universe. According to experimental economics, the same is true for the situation when socio-economic systems are investigated as the fundamental elements (self-interest, motivations, risk-aversion) of behavioural effects that can be identified in real-life situations (i.e. reality lived through via direct experience) emerge and also work in laboratory experiments (Smith 1976; 1989).[21] [→ **phenomenology**]

Most important among the findings of experimental economics was, perhaps, the observation and description of unconscious optimization strategies. While theoretical economics assumes humans to be rational, psychology highlights their irrational character (Smith 2003). By contrast, experimental economics, rather than rejecting the concept of rational human behaviour accepted by mainstream economics, is set to fine tune it and also to provide more of an in-depth view of it with the aim of strengthening the consistency between mainstream theory and observations. Mainstream economics postulates the individual's rationality as the result of a conscious, cognitive process. If optimization were an unconscious process, how would it take shape? In the wake of a miracle or casual circumstances? This is exactly the view which experimental economics seems to abandon, while stressing the idea that the concept of market equilibrium cannot be separated from the learning processes that are essential for its creation. (Smith 1991). Despite the discrepancy between experimental economics and neoclassical orthodoxy seems to be serious, there is still no reason to talk about a conflict. Experimental results have often confirmed that a REH-type (or a near to REH-type) equilibrium may be established not only in situations depicted by abstract models but also in laboratory conditions. We might also envisage a set of rational expectations, although this rationality is not the outcome of innate skills, for its emergence takes time and experience, i.e. the observation of market processes and the behaviour of other agents[22] (cf. Smith et al. 1988; Porter and Smith 2003).

At the same time, these experimental results have efficiently supported the mainstream theory. It was realized during these experiments (and, particularly, through post-experiment interviews) that agents achieve the utility and profit maximizing equilibrium of mainstream economics without being aware of this. Although their perception of these experiments, designed as models of real-life economic settings, identified these situations as chaotic messes, an optimal (or near-optimal) state of equilibrium characterizing perfectly free competitive (i.e. theoretical) markets was created without the players being aware of this.

method applied in the process of theorizing. For instance, Popper, on grounds of Hume's philosophy, refuted induction, since, accordingly, there is no pure observation at all which might be independent of any theoretical grounds (as opposed to Vernon Smith's interpretation), therefore it cannot serve as a starting point for theorizing. For him, therefore, observations are selective and are governed by theories (cf. Thornton 2009).

[21] It is interesting to realize that, based on this, Friedman's assertion (Friedman 1953) that social and natural sciences can be hardly distinguished on methodological grounds holds after all. The possibility of performing controlled experiments in economics does not imply that the border-line between social and natural sciences will disappear automatically.

[22] Accordingly, bubbles should be always expected to emerge under life-world conditions since, although practised players (experienced market participants) are fully aware of the fundamental laws of market, there are always agents who, driven by the lack of experience, actively contribute to the emergence of bubbles (Caginalp et al. 2003).

(Smith 1991). Market rationality does not require perfect and complete individual rationality. The only plausible interpretation of these findings is the pervasive presence of the invisible hand. Market agents can realize socially efficient outcomes with no intention to do so (Smith 1994). The pursuit of a state of equilibrium is an unconscious, cognitive process, and the abstraction-based, pure economic theory performed quite well in outcome predictions. [→ **paradigm**]

It is hard to interpret the optimal quantity of information obtained which, as we have seen, may also necessitate the availability of estimates regarding costs incurred through collecting excess information and the benefits derived from obtaining that information. According to a different approach to reach optimum levels, information available should be regarded as optimal as long as it is capable of fundamentally affecting a decision. The only problem is that this optimum can be evaluated only *ex post*. In this case, information—which did not change an earlier decision that continued to be exposed to (repeated) modifications as new pieces of information were being processed—is considered superfluous. However, in order for this to become obvious, it is necessary to obtain information which might subsequently become redundant (in other words, we do not know *ex ante* that a particular unit of information *will* eventually be useless). This definition is unacceptable, since presupposing the $(n+1)th$ piece of information to keep our decision intact might be erroneous, if our expectation is based purely on the fact that the nth item did not force us to modify our expectation. That presupposition implies the view that relevant items are always arranged in an order of importance and if a processed item turns out to be futile (in the sense that it has failed to trigger the modification of our decisions), we can draw the inference that no more relevant information will be obtained. Moreover, optimum can be described as a subjective optimum, in which case an agent, while forming his expectations, *will be led to believe* that he has already collected and processed all necessary information. In order to avoid all these dilemmas, the sociological theory of rational choice uses a simple temporal concept of optimization, i.e. it regards the gathering of information a course of action to pursue only as long as this process does not endanger decision-making (cf. Elster 1989). However, the rule here is confined to stating that it is pointless to search for new information beyond the point at which a decision might have been taken..

It has to be noted that the weak definition of REH refers to the relation between a belief and the grounds on which it is held (we will revisit this problem shortly for further details) and, therefore, this concept is in close relationship with definitions used in the literature of sociological theory. One should not confuse the relationship between a *belief* and *its grounds* with truth, which can be interpreted in the context of the relationship between a belief and what the belief is about (cf. Elster 1989). If we follow this line of reasoning, Othello is seen entertaining a rational belief when he believes that Desdemona betrayed him, since this inference seems to be logical to him on the basis of the information he possesses. Indeed, the weak definition of REH does not venture beyond this. In other words, rationality does not require beliefs to be true (or punctual). Put differently, according to the weak definition, rational expectations are not required to coincide with subsequent de facto outcomes or with the expectations of their previously known probability distribution.

However, the hard definition of the REH goes considerably further, also linking the relationship between belief and its object to the concept of rationality; moreover, it seems to focus exclusively on it, that is, *correctness* is involved in the criteria while *truth* is still not a requirement, since belief is *not* aimed at a current, verifiable fact. If our aim here is to apply a concept in the context of the facts that are verifiable at the present time, one which is as rigorous as the hard definition of REH, we should require rational beliefs to be true: Othello's suspicion cannot be regarded as rational as Desdemona did in fact remain faithful. At the same time, it should not be forgotten that the REH deals with current estimations of future outcomes, and if truth were a prerequisite in this temporal relation, it would be equal to insisting on the PFH—however, a perfect foresight is not possible under stochastic conditions. The present knowledge of the probability distribution of future outcomes is a manifestation of most of the existing knowledge regarding the future—therefore the hard definition of the REH requires an ability to look into the future with utmost precision which, of course, cannot go further than requiring expectations to follow the objective probability distribution of future outcomes.[23] It is the stochastic nature of socio-economic processes alone that excludes truth from our list of criteria.

We can say that the strong definition of the REH offers a criterion on the basis of which beliefs can be classified as *rational* or *non-rational* without a subjective value judgement. Such a labelling would be considerably more problematic in the context of the weak definition, since an external judgement regarding the rationality of a belief implies *knowledge about knowledge*, i.e. the information set, based on which a particular agent made his decision, needs to be known (and the one relying on which he might have made that decision). Mention should also be made of the confusion caused by the augmenting part in Muth's definition. The clause on the shared information set is a visible shift towards the weak definition, since if a cluster of agents has access to less information than the inputs of the relevant model, their expectations formed under these circumstances may be still regarded as rational according to Muth, although they will be biased. We can, therefore, say that both the hard and the weak definition can be traced back to Muth's definition, while highlighting different aspects of it. Applying Muth's (complete) definition would get us into a situation too comfortable to handle (but confusing), since even if the requirements of the hard definition are not met, the weak definition may still be

[23] It seems that the problem caused by temporality is not obvious for sociologists, i.e. they may question the rationality of a decision or belief on future outcomes if it turns out to be false *posteriorly* (!) (cf. Farkas 2006). There is a fable about a person looking for a new job, who accepts an opportunity, thinking that this job is ideal for him in terms of his preferences. He is assumed to exploit all relevant information available before making his decision. However, he becomes disappointed later, since he has less leisure time or earns less than he expected. Some authors think that the rationality of his decision (his expectation) can be doubted—without realizing that what they do is require foresight (i.e. pre-knowing) into the future while even they do not have such a knowledge (since this doubting occurs only posteriorly).

put in place (with all its inconsistencies). In this case, the most serious problem would be caused by the fact that the weak definition does not put forth a measure based on which one can regard an expectation as rational. We have seen a significant difference between requirements implied by the weak and the hard definition, merging these two concept should, therefore, not be advocated. It is not surprising that the new classicals subsequently focused exclusively on the hard definition (accordingly, the REH will from now on be identified with its hard definition).

It is worth considering what the criterion of rationality embedded in the weak definition might be. One can say that an expectation is not rational if a particular agent stops gathering and processing information before achieving an optimum level—in this case, it is pointless to investigate the relationship between a belief and its grounds, since as the information available is insufficient. An agent, confident of success, who is known to have invested huge sums of money in the Brazilian coffee business, can hardly be labelled as rational if the information gathered on future coffee prices was obtained from, say, the evening news or from next-door neighbours (except if his neighbour runs a coffee plantation in Brazil), even if his expectations are in synch with the information set upon which those expectations rely. In this case, the question mark behind rationality does not stem from the assumption of *irrationality* (although the everyday meaning of the term "rationality" undeniably suggests this) but from the fact that the agent in question had no access to the optimal amount of information required. If the weak definition were limited to the scrutiny of the relation between a belief and its grounds, the agent who expects the extension of the previous trend on the basis of an effortless performance aimed to gather information, and is unaware of the natural disaster and the damages caused by it (consider the case of the small investor referred to above) would be thought of as someone behaving in a rational manner. However, according to the weak definition, there is a different option whereby the notion of rationality might be denied. Even if the market agent under scrutiny did obtain an optimal amount of information (whatever optimal means), he could still form an expectation based on that very information, which would contradict his previous knowledge on the matter (in other words, his belief is not in accordance with the information processed). Cast your mind back to the classic example: the attitude of the agent who expects coffee prices to be unaffected after the harvest has been destroyed is not rational, either. Thus the weak definition specifies two criteria without clarifying the connection between them. Moreover, it should also be noted that the assertion (or assumption) formulated in line with the weak definition that agents are rational, is not much use from an analytical perspective. Even if the assertion itself is correct, expectations may well be infinitely heterogeneous, and the analytical apparatus along with the economic policy conclusions of new classical macroeconomics could hardly have been based on this concept.[24] Yet the most

[24] However, there were attempts in the literature to link the implicit acceptance of the weak definition to the conclusions of the hard definition. However, these are not logically consistent efforts (cf. e.g. Shaw 1984).

serious problem is caused by the far too cautious wording of the weak definition and, as a result, its observations do not venture beyond what has been well-known for generations of economists, their view being grounded in common sense (i.e. agents gather information in order to make decisions and form their expectations on the basis of that information).

Although the hard definition avoids all these problems, it creates additional difficulties. The only prerequisite of rationality here is for agents to form their expectations in line with the probability distribution. With a definition like this, to establish a view poses no difficulty: an agent whose expectations coincide with the anticipated outcomes of future processes will form rational expectations. The problem of fudging the issue by the ambiguous interpretations of optimum does not need to be addressed, nor is 'knowledge about knowledge' a must anymore—although, as it turns out, the smokescreen created by the non-clarified background regarding expectations is one of the most serious problems associated with the hard definition.

When dealing with the weak definition, the lack of hard definition-type, unambiguous standards raises the problem of subjective judgement. It is obvious that rationality is either questioned or recognized by an external observer (for instance, a specialist examining the issue from the vantage point of science). Even if it is assumed that this individual is familiar with the information set possessed by the agent under observation, and if it is also assumed that this set is optimal in quantity, a disagreement on the inference drawn on this basis is still possible between the spectator and the agent observed. *The connection between a belief and its grounds is not based on rock-solid stability*. Once again, the coffee plantation analogy comes in handy: even if the investor is informed on frost damages and even if he expects an upturn—to what extent should price fluctuations be considered rational by that agent? If the investor expects a 25 % rise, while the observer predicts a 50 % increase (or *vice versa*), this will inevitably cast a doubt on the rationality of expectations. Accordingly, the weak definition is vulnerable to queries (apart from its weaknesses mentioned already) for leaving the relationship between a belief and its grounds unclarified (while passing judgement on it). At this point, the hard definition makes reference to the relevant economic theory that delivers a well-defined output—therefore, standards in this case are unambiguous. The weak definition, on the other hand, stemming from an ill-defined set of information (since every agent may have obtained dissimilar sets of information) and in the absence of an unambiguous method for the formation of expectations (i.e. reference to the relevant economic theory and macroeconomic model), does in fact make the claim that behavioural decisions can be reached (to use a simile by Berlin 1958) with the help of a slide rule. The attitude of a spectator basking in the scholar's role and assuming the superiority of his observations over the views taken by the agent under scrutiny, may be criticized from various aspects. Moreover, the links between this debate and the one on paternalism are hardly less than direct (cf. Mill 1859).

2.4 Rational Expectations and Unbiasedness

After this brief digression, let us return to the hard definition given by Muth (1961). Simple phrasing can raise doubts. If the distribution functions (and the probability density functions derived from them) of the predictions provided by the relevant model and by the economic participants are *identical*, it follows that the predictions offered by market agents are unbiased (since the expected values are identical), and even the variance of the estimations (i.e. the average of the squared differences from the expected value) will be identical. Therefore, the hard definition of the REH would imply that market participants anticipate the same outcome as the one predicted by the relevant theory. However, this mechanism cannot solidify into a perfect foresight as errors in expectation cannot be excluded. In other words, the claim that expectations of agents are all identical will eventually prove *false* (and we have to admit that a perfect foresight, which rules out the possibility of errors *per definitionem*, would imply that expectations are identical). Our suspicion that Muth believed the predictions of the theory and agents to have identical distributions is supported by the fact that Muth precluded the possibility of better estimates based on the theory. Therefore, there is a fundamental difference between the output of the adaptive model (and other models referred to above), which relies exclusively on past information, and the rational scheme, since a systematic bias from actual outcomes (and from the predictions of the relevant model) *may* naturally occur in the former models—while in the case of rationality this bias would, in theory, be ruled out. However, as we shall see, it is precisely that unbiasedness which is the most problematic element of the REH, the assumption of which can only be confirmed amid grave concerns. The lenient approach manifested in the reasoning that the underlying intention here is to efficiently exploit information and that utility maximization is the only explanatory variable in the process of gathering and handling information (cf. Kantor 1979). This can explain the motives at best. Although there is no denying that the REH highlights the importance of exploiting information, the methodology of the process itself remains unrevealed.

Some additional remarks need to be made with respect to the details outlined in the previous paragraph. Although Muth seemingly precluded the possibility that the theory can lead to *better* (sic!) predictions than the expectations of firms if rationality is assumed, he drops a hint at some point that the predictions of individual firms (agents) may be subject to *greater* errors than the theory itself.[25] Undeniably, this is a cautiously worded assertion, just like the one where Muth stops short of making the unambiguous claim that subjective and objective probability

[25] It implies that the predictions of the relevant theory are not true either, i.e. neither the relevant model nor the aggregated expectations meet the requirements of the PFH. Some definitions of the REH are more daring, making the claim that aggregated expectations are true (Begg 1982).

distributions are identical. All he says is that these two distributions *tend to be* identical.[26] This apparent contradiction will be resolved, since Muth notes that, when expectations are rational, agents' *aggregated* (i.e. not individual) expectations will coincide with the output of the theory (and, therefore, *probably* with the actual outcomes). And although it is true that there are random errors which come between the individual predictions of agents and data subsequently obtained, and that the expected value of errors is zero (to ensure that aggregated individual expectations coincide with the predictions of the theory), complete with orthogonality, a criterion to be tackled later, there must be (according to Muth) a divergence in the absolute values of errors, governed by unrevealed variables. Therefore, one should not remain indifferent (not even on the face of it) to the problem of who those expectations are formed by. *Aggregated unbiasedness* is an existing assumption, anyway; systematic errors are thus precluded.[27] Individual predictions are unbiased at the aggregate level, while the relevant model is capable of delivering unbiased estimations. However, it is quite obvious that hardly anyone is familiar with the relevant model (not in its entirety, anyway) given the fact that the view put forward by Muth on this issue is that there *are* considerable cross-sectional differences in opinion (first conditional is absent from the wording) from which one may even infer the biasedness of individual expectations, because, as it will be seen, there is a strong likelihood that predictions based on incomplete information will lead to biasedness.[28] Muth mentions that the averages of expectations (i.e. at an aggregate level) *are as accurate* (sic!) *as* the outputs of a complex equation system, implying that the model is *not* used in the estimation process. It follows from this that these expectations are generated through a different method. Moreover, Muth also makes the point that his theory falls short of declaring any resemblance between agents' estimation methodology and complex econometric models, although it does say that the way in which expectations are formed depends on the structure of the relevant model describing an economy. *These two kinds of predictions* (i.e. those of market agents and of the relevant theory) *are, therefore, not identical but entirely similar*. This has serious consequences, since apart from having to reveal the structure of the model (and how knowledge relating to that model can be acquired), an explanation is also required to clarify how market

---

[26] "[...] expectations of firms (or, more generally, the subjective probability distribution of outcomes) *tend to be distributed*, for the same information set, about the prediction of the theory (or the »objective« probability distribution of outcomes)" (Muth 1961) (italics added). The question is still, first: what happens if exploited information sets largely differ and, second: whether unbiasedness can still be maintained under such circumstances.

[27] Under stochastic conditions, unbiasedness means that the expected values of parameter estimates are the estimated theoretical parameters and random errors prevent theoretical and empirical parameters from coinciding.

[28] OLS-estimations always guarantee the expected value of errors to be zero—however, one should not infer the unbiasedness of the model or its predictions. Moreover, it is also true that the less information is exploited during the specification of the model (that is, the more relevant variable is precluded), the higher the sum of squared residuals will be. One can hardly achieve unbiased estimations with relevant variables being omitted.

agents can acquire knowledge that in actual fact is not identical with familiarity with the relevant model, only entirely similar to it (at least in terms of its outputs).[29]

In fact, the situation would be less complicated if it *could* be assumed that economic agents are experienced in econometrics and economic theory. In this case, if it is true that all *these* agents and the relevant economic theory possess information sets (including the methodological knowledge that is necessary to specify the model) where each set is equally valued (with random differences in their contents), individual estimations are all unbiased estimations of actual processes and the relevant model will be in the centre of the probability distributions of parameter estimations. However, theoretical parameters are not known—only the applied functional forms (agents would, in this way, both know and apply the relevant model, even if the job of filling it with inputs would be a task left to individual agents). Thus Muth's requirement according to which expectation averages are true (unbiased) would be met. However, this explanation is not consistent with Muth's concepts, since, in his view, the methodology applied by the agents is not based on econometric modelling. This view could explain why agents make major forecast errors, but on the level of individual estimations it would result in bias due to the primitive nature of expectation formation methods—however, it would become clear why errors show cross-sectional differences (since it follows, among other things, from the details above that better-informed agents can make better estimations). However, in this case, no reasonable explanation can address the issue why the average of *biased* individual estimations should be an *unbiased* estimation of actual processes.

These two possible explanations (*agents are well-informed professional econometricians* vs. *agents have naive-primitive expectations*) are differently consistent with the definition of Muth. The first option is consistent in terms of its *consequences*: individual estimations are unbiased. The second option is consistent in terms of its *assumptions*: the method associated with agents is not based on complex econometric methods. It seems that these two points in Muth's definition can be hardly held simultaneously. Moreover, the third option, in which Muth might have suggested the adoption of the outputs from the relevant model by agents, can be ruled out, as in this case there would be no cross-sectional difference in expectations. Thus the *as-if*-argument itself pales into insignificance, since if it were true that agent expectations are being formed similarly to the outputs of the relevant

---

[29] Muth also makes a riddle by claiming that "Averages of expectations in an industry are more accurate than naive models and as accurate as elaborate equation systems" [...]. Now, we are faced with the fact that predictions of market participants are not made on mathematical grounds (at least not by using complex methods). It is exactly that reasoning bluff which was objected by *Benjamin Friedman* and *John Weeks* to new classicals (and to which the obscure hint on the approximating distribution functions mentioned above bears resemblance): expectations are being formed *as if* agents know the relevant model (B. Friedman 1979; Weeks 1989). Remember the F-twist and the parable of rational and mathematically optimizing leaves. Muth simply does not recognize that his concept of market agents could be useful as an ideal type even without this bluff.

model, we would be left in the dark as to where the cross-sectional differences of opinion mentioned by Muth are derived from.

It becomes clear that new classical macroeconomics, by using the REH, comes up with arguments based on the notion of *assuming the conclusion* (that is, *petitio principii*). Unbiasedness is an important criterion because new classicals deny the efficiency of a systematic countercyclical monetary and fiscal policy on these grounds—if there is no bias, there is no room for stimuli via monetary policy, either. And new classicals hold the view that the converse is also true: if the control of output is not feasible with the help of monetary policy, from this we may infer the unbiasedness of estimations (or, to be more exact, we *should* infer it, since the REH postulates *per definitionem* unbiasedness).[30] Our most important question is, therefore, whether unbiased expectations are an over-simplifying assumption or a plausible presupposition by the REH, since if it were understood and accepted that to adhere to this assumption is not a *realistic* option, this finding would considerably impair the relevancy of the REH and, accordingly, the reliability of economic policy conclusions put forth by new classicals.

We cannot turn a blind eye to the fact that there are two kinds of expectations and estimations to deal with during our scrutiny of the REH as we need to make a distinction between the

1. Predictions of the relevant model and
2. Expectations of individual agents.

At the same time, we shouldn't overlook the fact that *the relevant model is in fact unwanted, unless if it is used for the purpose of spearheading the as-if-expectations of agents*. The intention to deceive by a monetary policy is directed towards the actors: the point here is that *their* expectations are unbiased—while the unbiasedness of the relevant model (if there is such a model) is, in actual fact, an insignificant circumstance. Therefore, the problem of what are the prerequisites of unbiased estimates provided by the agents and those provided by the relevant model itself, needs to be viewed in a comprehensive manner. As a starting point, market agents will be assumed to be professional entities producing econometric models, otherwise unbiased individual estimations could hardly be assumed (moreover, to assume that agents have this knowledge is acceptable in a pure theory—the usefulness of these assumptions in a theory aimed at producing a realistic description of the life-world is a different story). If the conclusion drawn from this is that

[30] Here, some attention should be paid to Weeks' reasoning, according to which the REH is also in error by believing that reality is governed by deterministic laws. Accepting it is really problematic. If there exists full knowledge on (an) economy (as the REH states), future outcomes occurring deterministically could be forecast *truly* (and not only unbiasedly). In this case, the REH and the PFH would be equivalent even at the level of agents—but this is not proclaimed by the REH. Therefore, on this account, it is not determinism as far as explicit assumptions considered at least. It seems that the accusation of determinism is trumped-up for rather a prosy reason: it is revealed from Muth's phrasing that not even theory is able to provide true but only unbiased predictions. However, it is a different problem that the PFH is very much in need, say, for operations of the labour market—labour market as depicted by new classicals (for further details on this problem see Chap. 3).

agents are capable of producing unbiased estimates under certain conditions, our criticism should be directed at other areas, i.e. the conditions themselves, by making the point that, say, an individual who runs the local food store can hardly be expected to have a particularly high-level expertise in mathematics. *The existence of the relevant model, the possibility of its successful specification and the unbiasedness of the estimations of agents do not presuppose each other, and in terms of the economic policy consequences, unbiasedness is the only relevant factor*. And if the necessary conditions in the context of knowledge and information are not met, which means that actors cannot be expected to provide unbiased estimates, unbiasedness can only be ensured through the special *as-if*-assumption of the REH (which in turn would obviously need to be justified).

2.5 The Orthogonality Assumption of the REH

So far, we have considered the conceptual difficulties caused by the REH presupposition that expectations are unbiased. It is quite obvious that we are forced to circumvent the issue if we want to follow the reasoning of the theory: by using non-complex methods, individual agents are not likely to end up having unbiased estimates. Agents would need professional modelling skills to achieve this, but Muth precluded this option in advance—his rules regarding modelling techniques will exclude everything except biased estimations. *Unbiasedness is nothing but a presupposition in the narrowest sense of the word.*

However, the REH also assumes orthogonality along with unbiasedness. Unbiasedness means that the expected values of residuals defined as the differences between data (as probability variables) and predictions are zero. By contrast, orthogonality means that expectation errors do not correlate with the information set available. Therefore, according to orthogonality, the *ex post* expectation errors (i.e. those calculated on factual data) cannot be forecast *ex ante* on the information set available and exploited when expectations are formed (Gerrard 1994). On second thoughts, this turns out to be a necessity, since if it were possible to forecast expectation errors in advance, expected errors themselves would become parts of the information set available (ex ante), which would thus be capable of modifying expectations that are being formed—and only random errors might occur (which would fail to take us to the fairly powerful assumption manifested in the PFH). Therefore, according to the REH, the rationality of expectations is confined to the precluding of *systematic* prediction-expectation errors while, as has already been mentioned, random (i.e. unpredictable) differences (*random*, in the interpretation of the REH[31]) may still occur. Therefore, the requirement of orthogonality can be described in the following manner:

[31] This clause stresses that the precluding of systematic errors (that is, biasedness) by the REH is a *statement* rather than a correctly defendable and reasoned *finding*. Thus there are errors—the only question is whether they are random ones.

$$\mathrm{cov}(e_t; \Omega_{t-1}) = 0, \tag{2.1}$$

where e_t is the expectation error of period t (as defined above), and Ω_{t-1} denotes the information set that was exploited when expectations for period t were being formed in period $t - 1$. However, orthogonality leads to a very important conclusion. Assuming that (for lack of alternatives, this is the only line of reasoning to adopt)

$$e_{t-i} \subset \Omega_{t-1}, \text{ where } i = 1, 2, 3, \ldots, \tag{2.2}$$

by doing so we claim that the expectation-forecast errors of $t - 1$ and the preceding periods (e_{t-i}) are all included in the information set (Ω_{t-1}) available in the period in which expectations are being formed. In other words, agents had an in-depth knowledge of errors for the previous periods when expectations were formed for period t. Based on (2.2.) it also follows from requirement (2.1.) that

$$\mathrm{cov}(e_t; e_{t-i}) = 0, \; where \, i = 1, 2, 3, \ldots, \tag{2.3}$$

as all previous errors in expectations were already included in the information set available—in other words, Eq. (2.3) rules out residual autocorrelation. *The learning process is, therefore, completely precluded from the model,* since agents will be incapable of correcting their errors on the basis of their previous errors and, based on the current model, they will be unable to forecast errors for the next period (i.e. the next errors of the current model). After all, the absence of opportunity for learning is caused by the circumstance that, being in possession of the relevant model, only random (and, in this way, unpredictable) errors may occur—that is, learning from these errors is both unnecessary and impossible.[32]

2.6 Unbiasedness: Some Further Considerations

All the details of the previous sections are based on the assumption that knowledge in the context of the economy which can produce unbiased estimations of factual outcomes for the future, exists *a priori*. Simply put, a relevant model is available. In

[32] Interestingly, *Thomas Sargent* and *Neil Wallace* (1975) deny the *possibility* of a systematic economic policy on similar grounds. This is slightly more than refuting its effectiveness, since even the effort itself seems to be impossible in their reasoning. Aggregate supply can be boosted only by unexpected increases in the price level generated by the monetary authority. For a surprise in period t, the non-systematic part of the money supply needs to be unpredictable (the systematic part is, naturally, always predictable) on the basis of the information set that is available in period $t - 1$ and that supports forming of expectations for period t. While market agents and the central bank have access to the same information set, the central bank is unable to set *systematically* the random component (that is responsible for the surprise), since the non-correlativeness of the information available and the unpredictable, random part of the money supply is also true for it.

this case agents can acquire this knowledge by having the relevant information conveyed to them.[33] No other correct, implicit assumption is available, given the fact that individual expectations of poorly informed agents will probably be biased.[34] This is the first time that we can see something reminiscent of inflation targeting, as in this regime, agents can utilize the predictions of the quasi-relevant model (those of the best approximation of the relevant model known to us) with considerable passivity, of course, since rather than using the model effectively to compute predictions, agents possess its outputs—and their expectations will thus coincide with the predictions of the model. It, however, would be a misinterpretation of the situation to assume that this simplifies the case as bearing in mind the inherent limitations of the trial-and-error specification when identifying theoretical parameters, we cannot go any further than confirming the existence of a complete knowledge which can neither be accessed nor acquired. Therefore, knowledge of the relevant model is a comfortable and necessary presumption, not a realistic consequence.

To claim that the relevant model does not exist a priori (it only *may* exist) is a starting point that puts the case into a different perspective, i.e. market agents (and the monetary authority) have to identify it themselves. We cannot circumvent this problem, since even if the relevant model does exist, it needs to be identified. It was John Weeks whose critical remarks on the REH were put forth on these grounds with sharp witticism. This remarkable line of reasoning suggests that the relevant model cannot be identified by the trial-and-error technique, not even if, theoretically speaking, a complex and perfect knowledge of the macroeconomic system is possible. Stochastic data generating process of the economic variable y can be formulated in the following manner:

$$y_t = x_t'\beta + u_t, \tag{2.4}$$

where x_t is the vector of all the explanatory variables (excluding random effects, of course) governing the dynamics of y (containing all relevant information), y_t denotes the numeric value of y in period t, while β is the theoretical parameter vector and u_t is the random term. If there *exists* a perfect knowledge on a given economic system, it is obvious that the relevant model implies familiarity with the theoretical parameters, and our prediction by the model can, therefore, be described in the following manner:

[33] In fact, not only the central bank but any other institution in a central position may be capable of this. Remember the companies analysing market trends or the ministry of finance. Inflation targeting did not *create* expectations, it only *recognized their importance* and makes attempts to affect them.

[34] It should be borne in mind that market agents here are assumed to be professional econometricians.

$$E_{t-1}(y_t) = x_t'\beta, \tag{2.5}$$

where $E_{t-1}(.)$ is the prediction of a given variable. Based on this, the prediction error of the relevant model can be defined by subtracting one equation from the other, as follows:

$$y_t - E_{t-1}(y_t) = u_t. \tag{2.6}$$

The predictions are deviated from actual outcomes by random errors, causing even the relevant model to be incapable of true predictions. Hence, the formula for unbiased individual predictions is this:

$$E_{t-1}^i(y_t) = x_t'b_{t-1}, \tag{2.7}$$

where b_{t-1} is the empirical parameter vector applied for the period $t-1$ prediction. Unbiasedness here is a requirement which translates into $E(b_{t-1}) = \beta$. If that requirement is met, i.e. agents are capable of unbiased predictions of actual outcomes, the expectation error is

$$y_t - E_{t-1}^i(y_t) = \hat{u}_t^i, \tag{2.8}$$

since $E\left(x_t'[\beta - b_{t-1}]\right) = 0$. It has to be noted that errors in the relevant model and those of individual predictions *are not identical*. Although random errors may shift individual estimations from the theoretical parameter vector, these differences will balance each other out aggregately, causing the theoretical parameter to be the expected value of the distribution of parameter estimation. However, an estimation error is derived not only from a specification error, it *may* equally stem from contingencies, but its source cannot be identified in specific cases,[35] which is *precisely* what prevents us from being led to the relevant model empirically. The model in its intermediate (i.e. incomplete and imperfect) state might still offer an accurate prediction due to random effects. In that situation, the modeller may, naturally, assume that his job is done, having acquired some sort of ultimate knowledge. A considerably more interesting situation is one when the output of the model is in error in respect of subsequent factual data (it needs to be borne in mind that these are predictions computed in a stochastic environment). Broadly

[35] It should be stressed here that unbiasedness and the circumstance that the expected value of prediction errors is zero are not interchangeable concepts. As it has been mentioned above, OLS-estimations always guarantee the mean of errors to be zero. It is still possible in case of a biased parameter vector that expected value of the prediction error is zero, if the difference vector of the theoretical and the empirical parameter vector is orthogonal to the vector of the explanatory variables. However, it is a different problem that, in this case, the prediction error is not independent of the explanatory variables—a circumstance that can be precluded only by the unbiasedness of the empirical parameter vector.

speaking, this can be traced back to *two* possible reasons. In case of a model in an intermediate state (which Weeks has underlined is the best to be had):

1. The modeller can never identify which part (if any) of the prediction error should be attributed to the wrong specification (in other words, which is the systematic part) and
2. Which part reflects random effects.

Hence, he will be unable to improve his model. Moreover, if the perfect model exists (although based on what has been mentioned above, no account of its successful specification can be given), errors may occur due to the stochastic environment (since the Muthian concept of the REH does not require the predictions to be true), which, once the modeller realized this, he would obviously start re-specifying his model—which would be a setback compared with the previous situation. Weeks (1989), while side-lining the apparatus of econometric theory aside, makes the relevant point that a gradual improvement of the model would only be possible if we were already familiar with the model which we intend to approach in a step-by-step manner. The predestined impossibility of the learning process, coupled with the lack of familiarity by agents of the relevant model, makes it hard to *realistically* assume the unbiasedness of estimations. Familiarity with the relevant model is, literally, a higher level of knowledge, *deus ex machina*, as it were, and the impression is that the Walrasian omniscient auctioneer is revived in the notion of the relevant model.[36] Naturally, the fact that the relevant model is required to "know" the theoretical parameters, might be exposed to criticism. Seemingly, it necessarily follows from Muth's phrasing, according to which averages of expectations are as accurate as complex techniques[37] (although at this point in the reasoning, as it has been noted earlier on, the fact that according to Muth

[36] Several parallels can be identified between new classical macroeconomics and the preceding classical theory from which Filippo Cesarano highlights some, while concentrating on the effects of anticipated and unanticipated increases of the money supply (Cesarano 1983). Characteristically, classical economists have also paid more attention to unanticipated increases while virtually regarding anticipated changes as ineffective (although it is also true that, in the absence of statistics and information channels almost every change could be labelled as unanticipated—excluding pre-announced actions). The classicals assumed that, as a fundamental tendency, an increase in the money supply triggers only gradual adjustments in the price level, but up to that moment real output and employment also respond (cf. Hume 1752). Imperfect information (and not sticky prices and wages) played a key role in this story, bearing strong resemblance to the new classical view. However, what incomplete information is related to it is not an indifferent circumstance. In this context, the classicals concentrated on changes in the money stock, while new classicals focussed on the problem of distinguishing relative and absolute changes in the price level. Cesarano goes as far as referring to the finding that classical economists have implicitly assumed agents to possess the relevant model, since in his view they could estimate the effects of the pre-announced actions only by being aware of them.

[37] „*Averages of expectations* in an industry are [...] as accurate as elaborate equation systems [...]" (italics added). The fact that "accurateness" is referred to can cause a problem, but it can be dealt with on the basis of our previous findings (*here*, it should be regarded as the synonym of unbiasedness).

agents do not engage in econometric modelling is not taken into account). Since, if agents are supposed to be professional modellers, theoretical parameters will be in the centre of the probability distribution of unbiased individual estimations, the numerical values of which are not known to anybody.

If we do not believe in the existence of a complete and perfect knowledge of the economy, we can only say that, in the absence of this perfect knowledge, every econometric model can function only as a more or less appropriate approximation of reality (i.e. of the actual data generating process) and it is evident that market agents will not be in the same position in any struggle for this knowledge—that is, individual agents will start their investigation with the likelihood of success being different for them. Agents with a stronger likelihood for success are offered the opportunity to approximate the hypothetical relevant model (that does not exist and cannot be specified at all) more efficiently than others. Here, opportunity means access to larger and higher quality data sets. At this point, it should be noted that in the course of this analysis individual agents were assumed to have econometrical skills characterised by professional standards—in the absence of which the lack of methodological and technical experience would further reduce the likelihood of specifying an appropriate model. Mention should also be made of the fact that we require familiarity with the relevant model itself, and with data sets functioning as inputs for model estimations by presuming unbiased estimations at the level of both the relevant model and of individual predictions. Should any of these prerequisites fail to be met, predictions will appear to be systematically biased.[38] The problem of knowing and identifying theoretical parameters has been discussed earlier on, although a few additional remarks still need to be made and the issue of having access to information will be examined at some point later.

Based on mathematical computations, Benjamin Friedman (1979) argues that not even an infinite timeframe can guarantee that agents will eventually find the correct model, therefore their expectations will not be rational in the Muthian sense, i.e. in terms of the hard definition. And now let us re-examine the definition of the individual expectation error under stochastic conditions:

$$y_t - E_{t-1}(y_t) = x_t'[\beta - E(b_{t-1})] + \widehat{u}_t. \tag{2.9}$$

The prediction error on the left is independent of the explanatory variables only if $E(b_{t-1}) = \beta$ is true, and in this way errors reflect only random effects. In other words, expectation errors are independent of the explanatory variables only if parameter estimations are unbiased. So, if $\beta - E(b_{t-1}) \neq 0$, the expected value of prediction errors is not zero, which is possible only if $y_t - E_{t-1}(y_t) = \widehat{u}_t$ (since the

[38] Here, it is worth stressing, yet again, the distinction between the (hypothetical) relevant theory and individual predictions. We have seen that, in a stochastic environment, identifying the theoretical parameters is not possible, not even in principle. Therefore, according to the above interpretation of the REH, there exists a relevant model *a priori*, which is capable of unbiased predictions, and the predictions of individual agents (i.e. *expectations*) are all unbiased estimations of the data generating process.

error term is a white noise) which, however, is equivalent to $\beta = E(b_{t-1})$, hence $x_t'[\beta - E(b_{t-1})] = 0$. A biased estimation of model parameters may follow from various effects: one such factor is the use of a wrong functional form. If the modeller opts, say, for a linear form, any computed error will be dependent on previous errors, i.e. serially correlated—and, in this case, it is not true that there is no prior information for errors, and, by exploiting the knowledge available at the time of expectations being formed, even the residuals (i.e. prediction errors) will become predictable albeit with some limitations. Misspecification of model parameters occurs also when the estimated parameters of certain significant explanatory variables are incorrectly presumed a priori to be zero. This case, in its consequences, is equivalent to the forced omission of relevant explanatory variables, possibly caused by incomplete information. Under these circumstances, prediction errors reflect the effects of the omitted variables, i.e. errors are predictable on the basis of these omitted variables and, because of the resulting autocorrelation, the future dynamics of prediction errors can be approximated by exploiting the information in previous residuals (i.e. expectation errors). Meanwhile, the parameters of the included variables will probably be biased—yet again, the requirement related to orthogonality fails to be met. The possibility of specifying the relevant model through the trial-and-error technique might be seen as being further reduced by these problems—in fact it was known to be impossible right at the outset. Undoubtedly, the unbiasedness of individual estimations, if not assumed on an "as if" basis, implies stringent technical, methodological and information-related requirements.

As has already been pointed out, the availability of information is an equally important issue. The systematic biasedness of predictions resulting from imperfect information may still be the case, even if the relevant model exists and its parameters are known *a priori*. As noted previously, agents fall short of receiving equal opportunities in this race for gathering information: certain data is easily accessible for everyone, while special information sets are available only for narrow circles of agents—both data sets may have an impact on the dynamics of a particular economic variable. Moreover, cost-benefit considerations[39] will add another question mark to the need to have a full range of information. Inflation forecasts by a central bank, for example, contain seeds of information that cannot be accessed by possessing mere data sets. It is a characteristic feature of the Hungarian central bank routine that central bank modellers modify model outputs on the basis of information collected in the course of expert consultations. Several factors that price dynamics are influenced by cannot be identified by analysing previous time series data: if, say, a newbie is expected to join the competition of food distributors within a few months, which present participants are currently aware of (and have estimates of price policy and competition strategy based on foreign examples, upon which expectations of the new player will be based), these developments will also have an impact on current decision as well as those to be taken in the near future. If, for

[39] By the way, it is a serious argument for the effectiveness of inflation targeting, since it may be more economical for actors to adapt and use ready-made model outputs.

example, the new competitor is expected to use lower prices in order to reshape market share conditions, even current price dynamics will change, i.e. competitors will bend over backwards to keep prices at bay in order to prepare for keener competition. What the central bank does by modifying its model predictions on the basis of this information is nothing more than form rational expectations. However, it should also be noted that, generally speaking, central bank modelling gives different priorities to different estimation methods with regards to the relevant time horizon. In the short run, expert predictions are very useful, i.e. model outputs are modified on this basis, while model predictions become increasingly significant on a lengthened time horizon. Equally important is the fact that the predictions of agents in need of additional information will deviate from the central bank forecast, despite the fact that both are rational (at least in weak definition terms) as information unknown (and almost unavailable) cannot be taken into account by the actors.[40] Still, it is unrealistic to assume that market agents forced to do without expert consultations have equal access to knowledge of this kind, for the simple reason that if this knowledge was readily available, central bank economists (as everyday consumers) themselves would have access to it and expert-level consultations would become useless. It is, in fact, unnecessary as Muth's definition itself stopped short of prescribing, that the relevant model (substituted here by a central bank forecast) and individual predictions should be identical. *However, it follows from this reasoning that not even the central bank model is void of bias, let alone individual predictions.*

2.7 Consequences: The Road to Inflation Targeting

Our reasoning so far has been focusing on highlighting the realization that the alternatives of specifying the relevant model are limited, and the full and perfect knowledge of a macroeconomic system can be regarded as a comfortable presumption at best (albeit not a realistic one). Knowledge of the theoretical parameter vector is not feasible, even if high-level econometrical techniques are accessible or an adequate identification of the relevant explanatory variables is performed—all the more so given that the unavailability of certain types of information makes the task of modelling even more difficult to accomplish. Previously, a new concept (or, to be more precise, a characteristic) was inconspicuously introduced. It says that *the biasedness of estimations is not independent of the quantity and quality of information to which the modeller has access.* In other words, estimations are asymptotically unbiased with respect to the exploited information, i.e.:

[40] Barro and Fischer (1976) use the concept of rationality of expectations by providing an interpretation that is very similar as they regard as rational every forecast that can be considered as an optimal estimation of future outcomes on the basis of available information—even if a market agent does not have access to all the information that is exploited in the model.

$$\lim_{i \to i^*} E\left(\widehat{\Theta}\right) = \Theta, \tag{2.10}$$

where Θ is the theoretical parameter, $\widehat{\Theta}$ is its empirical estimation, and i denotes the amount of information processed while estimations were being made. If the latter approaches full and perfect knowledge (i^*), empirical (that is, estimated) parameters will continue to move closer to the theoretical parameters of the data generating process. Consistency can be interpreted similarly:

$$\lim_{i \to i^*} P\left(\Theta - \varepsilon \leq \widehat{\Theta} \leq \Theta + \varepsilon\right) = 1, \text{ for all } \varepsilon > 0. \tag{2.11}$$

On the basis of asymptotical unbiasedness, a moderated[41] version of the strong definition of the REH can be suggested in which familiarity with the theoretical parameters is not a requirement for the relevant model—which, of course, implies that the assumption on the unbiasedness of individual estimations is no longer considered valid. An agent (e.g. the central bank) with access to information along with methodological skills, which are satisfactory in quantitative and qualitative terms alike, will be able to specify a quasi-relevant model (by increasing the quantity of information processed it will be capable of reducing biasedness) describing a specific macroeconomic system, and the more focal its position, the stronger the likelihood to make other agents accept it. All we do is *ease* the assumptions and what (i.e. mere unbiasedness and *not* the knowledge of the theoretical parameter vector) has thus far been required from individual estimations is now a requirement imposed on the quasi-relevant model. If, as a result, individual estimations turn out to be biased, averages of individual estimations will no longer give the relevant model. We need to take only one step to move towards the opportunity of successfully introducing inflation targeting regimes as agents[42] will decide to *adjust* their expectations to the outputs of the (quasi) relevant model. In this way, *individual expectations* become *unbiased* as much as possible, albeit against a background of considerable passivity. If this was not the case, the central bank would be in a position to control the output through deception—and might only refrain from doing this because it is deemed unnecessary, given the manner in which its objection function is specified.[43]

We have seen that *the unbiasedness of individual estimations is likely to follow directly from the circumstances that characterize modelling*, therefore if we want to interpret rational expectations on the basis of asymptotical unbiasedness with respect to information, an appropriate specification of the objection function is

[41] This moderate version needs to be clearly distinguished from the weak definition, since we remain within the scope of the hard definition when using this modified form.

[42] Upon realizing that their efforts to process information have proved useless, since the outcome from those efforts has been limited to systematically biased estimation.

[43] As it will be seen in Chap. 6, in inflation targeting regimes this can be avoided by the central bank's commitment to keeping inflation at a low level.

needed in order for deception-based central bank policies to be precluded. The explanation of inflation targeting can be derived from the modified version of the REH (i.e. one based on the adoption of outputs from the quasi-relevant model), given the central bank's position to control expectations and its commitment to avoid time-inconsistent policies, which prevents it from using potential trade-offs. However, it should be understood that, as for the definition of the REH, inclusive of the interpretation as Muth meant it to be, there would be no room for manoeuvre in this concept for a central bank to exert influence on expectations, since actors forming rational expectations can make unbiased estimations of their own.

It seems, therefore, that the information side of the bias in aggregated individual estimations can equally be understood upon reflection. It goes without saying that estimations are specific functions stemming from information sets adopted. It is not too much of a challenge to recognize the existence of widely and easily accessible information, while some information can be labelled as insider data with a narrower circle of agents. Let us take an example. The assumption is that a national economy is comprised of two distinct groups of agents: by adopting the above example, one group is that of food product consumers, the other group is a distributor of food products (from the perspective of that example, the sector itself is completely indifferent). Both the consumers and the dealer are individual agents. In this case, consumers form their inflationary expectations solely on the basis of past experiences, whereas the dealer takes into consideration all its special knowledge *as well*, which, in the context of inflation dynamics, is also relevant information. Unbiasedness means that future data is couched in individual predictions in such a manner that biases of individual predictions level off—however, a situation like this will only be brought about by accident as a strongly over-represented information set, used in the forming of expectations, can be identified at an aggregate level.[44] By assuming that an agent, having accessed special information (available only to a limited group of agents), is also aware of the knowledge (on past dynamics) acquired by consumers, his prediction will move closer to (later) outcomes (or the expected value of the probability distribution of outcomes), but in order to achieve aggregate unbiasedness, all other predictions (biased *equally* and *in the same direction* as has been described in the example) would need to be biased in a way that these shortcomings will balance each other out. However, this is not possible because biases point consistently in the same direction—moreover, instead of a compensation, aggregated expectations will either underestimate or overestimate factual data. The shortcomings of a prediction, also based on special information,[45] would need to be levelled out by all other estimations in order to form unbiased expectations on the average—while there is no reasoned principle to support the acceptance of this clause as a presumption (unless we are ready to do this in an arbitrary manner). Unbiasedness at the aggregate level (i.e. a condition in which a later outcome or the theoretical parameter vector appears as the average of

[44] Being over-represented means that it is taken into account as a significantly high priority.

[45] This is still not the relevant or quasi-relevant model.

individual estimations) may emerge accidentally at best, and even if such a situation occurs (i.e. estimations are unbiased at the aggregate level), its recurrent occurrence should not be expected in any way. Unbiased aggregated estimations are unlikely to result from biased individual estimations—something not even Muth denied given that his definition prioritized the amount of information available. However, this informational symmetry can only be assumed with some difficulty as biased individual estimations might be the outcome of various factors. An additional problem is that certain assumptions in the REH are not consistent with each other, since unbiasedness as opposed to expectation formation assumed to be based on non-professional techniques, preclude each other (here the *as-if*-reasoning renders useless the theory: how can cross-sectional differences in opinion be explained when expectations are formed in a manner similar to that of the relevant theory) and the estimation of the relevant model, based on familiarity with the theoretical parameters, is an almost insurmountable task if undertaken empirically.

All this eventually takes us to inflation targeting from Muth's hard definition of the REH. To get to this point, all we needed to do was to realize that agents cannot be assumed to make unbiased estimations of future outcomes, unless an *as-if*-reasoning is adopted. However, assumptions of the REH can be investigated not only as a pure theory but on grounds of the intention to describe reality in a realistic way. As we have seen, for reasons due to methodological difficulties, knowledge on the theoretical parameter vector cannot be acquired via the traditional trial-and-error technique. Even if it is accepted that agents can have outstanding macro-modelling skills, individual estimations are still biased due to limited accessibility to information, thereby creating an opportunity for a systematic monetary policy based on deception. According to the moderate version of the REH introduced here, a sufficient degree of unbiasedness of expectations is facilitated by circumstances in which market agents adopt the outputs of the model that performs the best (i.e. of the quasi-relevant model), and in this way predictions of the model and those of the agents will coincide. In the absence of this, agents can only make biased estimations, making a systematic countercyclical central bank policy possible. If, following the moderate version, the hard definition of the REH is interpreted in terms of inflation targeting, the need for prohibition on deception will need to be taken into account: a central bank capable of and ready for commitment can anchor expectations, since it does not support short-run output maximization by controlling expectations. Since the unbiasedness of individual estimations (or, to use a cautiously worded language: the absence of biasedness reaching the greatest possible extent) is not the consequence of the performance of individual modellers, the lack of deception by the central bank can be interpreted only as a voluntary commitment. Without this the functionality of inflation targeting would be jeopardized.

There is yet another circumstance to highlight. For new classicals, the relevant model is the framework investigated in Chaps. 4 and 5.[46] The question is, and this book itself has been devoted to the way in which this problem should be addressed,

[46] This important aspect was brought to my attention by Prof. László Vígh in a conversation.

whether an economist who is not a follower of the new classical school of thought, should accept the relevancy of this model. The line of reasoning that we follow is that this is feasible only with serious caveats. It seems, at the same time, that the mechanism of inflation targeting is described well enough by the hard definition of the REH if (as is clear from the above details) the rigour of assumptions is loosened and we rest satisfied with a setting in which the relevant model is transmitted to the agents by the central bank. Moreover, one does not need to be a monetarist to believe in the functionality of this mechanism: a relevant model is not necessarily motivated by a monetarist approach. Sure enough, another problem needs to be dealt with: in an orthodox Keynesian model, for example, price dynamics is explicitly defined, leaving no room for expectations. *Therefore, not every macro-model will be consistent with the mechanism based on communication by the central bank*, not every model can be the quasi-relevant model in the focus of central bank communication—except those (kinds of) models that derive price changes from expectations (or those that leave room for expectations in price dynamics). What a central bank thinks about its own transmission mechanism does carry significance. These problems will only present themselves if the new classical theory is not thought of as a monolithic mass but rather as a puzzle. This is the only case where the relevant model may be something else than the new classical construct.

There is no easy answer to some questions. Is the realization that even the quasi-relevant model of the central bank is not necessarily unbiased conducive to the finding that public expectations anchored by that model are also biased? Should we infer from this that the central bank anchors expectations erroneously, which would lead to a systematic bias? If the rationality of expectations is disregarded, the answer is that the introduction of the quasi-relevant model of the central bank may cause public expectations to have systematic biasedness. This shortcoming cannot be fully eliminated by using the trial-and-error technique, therefore, the quasi-relevant model will cause a (prima facie) disequilibrium state. Whether real output will fall below or rise above the potential (equilibrium) level for longer periods or if these cycles will be only random pendulum swings—it will hinge on the nature of expectation errors. However, this very quality of errors is what should be compared to the (alleged) rationality of expectations. Should any kind of systematic error occur, it will be recorded by the central bank modellers and the public alike and, accordingly, model predictions will be modified—this was exactly what we pointed out earlier on in the context of expert predictions. Errors of the quasi-relevant model can, therefore, only be labelled random, resulting, among other things, in the volatility of real output, which is higher than what would be experienced on the basis of the relevant model.

It seems that new classical macroeconomics is capable of giving us considerable assistance in understanding the mechanisms of inflation targeting. Economists, reading the pertinent literature, may be inclined to think that there is considerable experience and technical knowledge behind inflation targeting regimes. This is true, although that knowledge is mainly focused on direct, factual and observed mechanisms (i.e. interactions). Hardly any mention is made of principles that underscore and, as it were, facilitate the functionality of inflation targeting. Literature on inflation targeting has a distinct concept of humans, while reflecting clear assumptions on the structure of macro-systems and the motivations of agents—while turning a blind eye to the fact that these elements are not axioms. The particulars of this system of presumptions can be modulated, its existence might be argued for or against and, seemingly, it can be traced back to underlying premises. It is not enough to assert that central banks are capable of anchoring public expectations—it also needs to be explained *why* they are capable of this. That is where new classical macroeconomics and the REH prove helpful. It is precisely what has been proved in Chap. 2: market agents are incapable of doing the job of central banks. Based on this alone, though, rationality cannot be argued against, since the availability of the most reliable prediction is of crucial importance from the point of view of economic activity. Rather than questioning the rationality of agents due to their shortage of skills, an institution needs to be introduced—one capable of adopting functions such as modelling and forecasting. If market agents are committed to forming rational expectations, we have serious arguments to support the idea that expectation formation is suboptimal and the adoption of easily accessible data capable of a satisfactory performance will be preferred to individual procedures. In this way, the result can be identical with what is required by new classicals: the probability distribution of market expectations will coincide with the outputs of the (quasi) relevant model. Eventually, inflation targeting results in a partial manifestation of the essential new classical doctrines.

References

Anderson RG, Kavajecz KA (1993) A historical perspective on the Federal Reserve's monetary aggregates: definition, construction and targeting. Federal Reserve Bank of St Louis Review 77 (3–4):1–31

Arrow KJ (1973) Information and economic behaviour. Federation of Swedish Industries, Stockholm

Babbie E (1989) The practice of social research. Wadsworth, Belmont, Hungarian edition: Babbie E (2000) A társadalomtudományi kutatás gyakorlata (trans: Kende G, Szaitz M). Balassi, Budapest

Barro RJ, Fischer S (1976) Recent developments in monetary theory. J Monet Econ 2(2):133–167

Begg DKH (1982) The rational expectations revolution in macroeconomics–theories and evidence. Philip Allan, Oxford

Berlin I (1958) Two concepts of liberty. In: Berlin I (ed) (1969) Four essays on liberty. Oxford University Press, Oxford, pp 118–172

Bronk R (2009) The romantic economist–imagination in economics. Cambridge University Press, Cambridge

Bryant J (1983) A simple rational expectations Keynes-type model. Q J Econ 98(3):525–528

Bungard W (1997) A proszociális magatartás empirikus kutatásának módszertani problémái. In: Lengyel Z (ed) Szociálpszichológia. Osiris, Budapest, pp 23–36

Caginalp G, McCabe K, Porter D (2003) The foundations of experimental economics and applications to behavioral finance–the contributions of Nobel laureate Vernon Smith. J Behav Finance 4(1):3–6

Cesarano F (1983) The rational expectations hypothesis in retrospect. Am Econ Rev 73 (1):198–203
Dickinson DG, Driscoll MJ, Ford JL (1982) Rational expectations, random parameters and the non-neutrality of money. Economica 49(195):241–248
Elster J (1989) Nuts and bolts for the social sciences. Cambridge University Press, Cambridge, Hungarian edition: Elster J (2001): A társadalom fogaskerekei (trans: Helmich Z, Szántó Z). Osiris, Budapest
Erdős T (1998) Infláció. Akadémiai, Budapest
Farkas Z (2006) A racionális cselekvés, a társadalmi cselekvés és kölcsönhatás. Miskolci Egyetem, Miskolc
Fischer S (1977) Long-term contracts, rational expectations, and the optimal money supply rule. J Polit Econ 85(1):191–205
Friedman B (1979) Optimal expectations and the extreme information assumptions of 'rational expectations' macromodels. J Monet Econ 5(1):23–41
Friedman M (1953) The methodology of positive economics. In: Friedman M (ed) Essays in positive economics. University of Chicago Press, Chicago, pp 3–43
Friedman M (1968) The role of monetary policy. Am Econ Rev 58(1):1–17
Friedman M (1970) The optimum quantity of money and other essays. Aldine Publishing Company, Chicago
Gerrard B (1994) Beyond rational expectations: a constructive interpretation of Keynes's analysis of behaviour under uncertainty. Econ J 104(3):327–337
Hahn FH (1980) Monetarism and economic theory. Economica 47(185):1–17
Harasztosi P (2004) Racionális várakozások és a Monetáris Konszenzus. Fordulat (Spring 2004):64–75
Hoover KD (1990) The new classical macroeconomics: a sceptical inquiry. Blackwell, Oxford
Horváth L (2000) Monetarizmus. In: Bekker Z (ed) Alapművek, alapirányzatok, vol II. Aula, Budapest, pp 433–455
Hume D (1752) Of money. In: Hume D (ed) (1994) Political essays. Cambridge University Press, Cambridge, pp 115–125
Jakab MZ, Vadas G (2001) A háztartások fogyasztásának előrejelzése ökonometriai módszerekkel. Central Bank of Hungary, Budapest
Kantor B (1979) Rational expectations and economic thought. J Econ Lit 17(4):1422–1441
Keuzenkamp HA (1991) A precursor to Muth: Tinbergen's 1932 model of rational expectations. Econ J 101(9):1245–1253
Király J (2000) Robinson és a közgazdászok. Café Bábel 10(2):107–113
KVA (2002) Foundations of behavioral and experimental economics–Daniel Kahneman and Vernon Smith (Laudation). The Royal Swedish Academy of Sciences, Stockholm
Maddala GS (2001) Introduction to econometrics, 3rd edn. Wiley, New York, Hungarian edition: Maddala GS (2004) Bevezetés az ökonometriába (trans: Békés G, Cseres-Gergely Z, Csorba G, Reiff Á, Szentpéteri Á, Szilágyi K). Nemzeti Tankönyvkiadó, Budapest
Malhorta NK, Simon J (2009) Marketingkutatás. Akadémiai, Budapest
Mátyás A (1984) A polgári közgazdaságtan napjainkban. Kossuth, Budapest
Mill JS (1859) On liberty, 2nd edn. John W. Parker & Son, London
Mises L (1981) The theory of money and credit. Liberty Classics, Indianapolis
Muth JF (1961) Rational expectations and the theory of price movements. Econometrica 29 (3):315–335
Pete P (2001) Bevezetés a monetáris makroökonómiába. Osiris, Budapest
Porter DP, Smith VL (2003) Stock market bubbles in the laboratory. J Behav Finance 4(1):7–20
Ramanathan R (2002) Introductory econometrics with application, 5th edn. South-Western, New Delhi, Hungarian edition: Ramanathan R (2003) Bevezetés az ökonometriába alkalmazásokkal (trans: Marcinkovics R, Nagy G, Petrimán Z). PANEM, Budapest
Rozgonyi T, Toledo R (2008) Halmazok és függvények. Nyíregyházi Főiskola, Nyíregyháza

Sargent TJ, Wallace N (1975) "Rational" expectations, the optimal monetary instrument, and the optimal money supply rule. J Polit Econ 83(2):241–254
Shaw GK (1984) Rational expectations–an elementary exposition. Harvester Press, Brighton
Smith VL (1976) Experimental economics–induced value theory. Am Econ Rev 66(2):274–279
Smith VL (1982) Markets as economizers of information–experimental examination of the "Hayek Hypothesis". Econ Inq 20(2):165–179
Smith VL (1989) Theory, experiment and economics. J Econ Perspect 3(1):151–169
Smith VL (1991) Rational choice–the contrast between economics and psychology. J Polit Econ 99(4):877–897
Smith VL (1994) Economics in the laboratory. J Econ Perspect 8(1):113–131
Smith VL (2003) Nobel lecture: Constructivist and ecological rationality in economics. Am Econ Rev 93(3):465–508
Smith VL, Suchanek GL, Williams AW (1988) Bubbles, crashes, and endogenous expectations in experimental spot asset markets. Econometrica 56(5):1119–1151
Thornton S (2009) Karl Popper. In: Zalta EN (ed) The Stanford encyclopedia of philosophy, Summer 2009 edition. http://plato.stanford.edu/entries/popper. Accessed 11 Nov 2010
Tóth IJ (2002) Vállalati és lakossági konjunktúra felmérések Magyarországon. Central Bank of Hungary, Budapest
Tóth IJ, Vincze J (1999) Hihetünk-e a vállalati felvételeken alapuló kutatások eredményeinek? Stat Szle 77(10–11):844–855
Vadas G (2001) Túl a makró-változókon: a lakossági bizalmi index és a magyar háztartások fogyasztási kiadása. Central Bank of Hungary, Budapest
Varian HR (1999) Intermediate microeconomics. W.W. Norton, New York/London, Hungarian edition: Varian HR (2001) Mikroökonómia középfokon–egy modern megközelítés (trans: Bara Z, Temesi J). KJK–KERSZÖV, Budapest
Várpalotai V (2003) Disaggregated cost pass-through based econometric inflation-forecasting model for Hungary. Central Bank of Hungary, Budapest
Vigvári A (2004) Pénzügy(rendszer)tan. KJK–KERSZÖV, Budapest
Visco I (1984) Price expectations in rising inflation. North-Holland, Amsterdam/New York/ Oxford
Weeks J (1989) A critique of neoclassical macroeconomics. Macmillan, London, Hungarian edition: Weeks J (1998) A neoklasszikus közgazdaságtan kritikája (trans: Szalai Z). Aula, Budapest

Chapter 3
The World View of New Classical Macroeconomics

now this is such an age it's now the fightplace
no flippant paltering no antic side-face

(Mari Falcsik: Now then straightest-out)

Two things awe me most, the starry sky above me and the moral law within me.

(Immanuel Kant)

3.1 Introduction

We might have dealt with too many issues in the previous chapter. In Chap. 2, an element (i.e. the most influential and central theme) of REH and new classical macroeconomics was selected to demonstrate its absurd consequences. The effects of these thoughts on monetary (and fiscal) policy were also highlighted. It is time to centre our discussion on investigating what the followers of REH and new classical macroeconomics teach about systematic (countercyclical) economic policy—and, first of all, the aspectual-logical grounds on which they formulated their theories. The statements made in Chaps. 1 and 2 on pure theories and on mainstream economics lay the foundation for the discussion here. Our next task is to reach the monetary and fiscal policy recommendations of new classical macroeconomics starting from our insights into the attitude of mainstream economics and into the theses (statements) of REH. Before analysing these thoughts in Chaps. 4 and 5, here we are going to examine how the principles already known emerge in neoclassical/new classical macromodels. So, the logical framework is studied here in which new classical macroeconomics originated and, simultaneously (because of their common roots), we can also gain new knowledge of the way mainstream theory (as a broader concept also incorporating new classical macroeconomics) thinks. Our primary task is to review the logical consistency, sometimes even to check the plausibility of new classical thoughts.

Before doing so, in the following section, the ideal of freedom in economics is subjected to scrutiny. The role of freedom in enhancing welfare and efficiency, as we will see, is one of the most fundamental doctrines of mainstream economics that precedes and establishes all other thoughts. The meaning and importance of this thesis in economics and the role it plays within the theory will be investigated later

P. Galbács, *The Theory of New Classical Macroeconomics*, Contributions to Economics, DOI 10.1007/978-3-319-17578-2_3

on—here, its roots will be traced and, as a result, further threads will be found that link economic theory to the platonic traditions.

3.2 Freedom as a Tradition of Economics

History of economic thought has formed a narrative by today that can be easily grasped and outlined. This view describes the history of economics as a story of successive schools that replace each other in terms of validity. From time to time, these histories describe the chronography of economic thoughts—fulfilling a trivial undertaking. Accordingly, economists' fundamental question focuses on the way authors of different centuries thought about economic issues. In this history, the main emphasis is perhaps placed on the issues that were regarded as problems worthy of considering and solving by economic thinkers. Economics, as it is known today, is a science of a socio-economic setting that was mainly unknown before the eighteenth century, so it is evident that a science investigating *this* scenery could hardly evolve before this period.[1] However, when Adam Smith experienced the mature capitalism of the eighteenth century, he could justly designate the ideal of freedom as the principle of economics. Starting from this point, for economics, freedom *functionally* is an axiom: though the economic benefits of freedom can be (even mathematically) supported and reasoned, therefore it should not be regarded as an axiom in the literal sense, economists, given their education, hold and proclaim the superiority of freedom as a premise over every other socio-economic setting. By virtue of our education, we do not try to prove the economic gains of freedom but accept it as a fact supported by the authority of our professors. The only problem is that Smith appeared as a follower of an intellectual tradition several centuries long when he introduced freedom as a fundamental ideal of economics. It seems that the ideal of freedom is perhaps the most fundamental intellectual foundation having an existence independent of economics itself—the conventional narratives, however, do not reveal the evolution of the ideal of freedom.

So, freedom has become one of the cardinal ideas of economics by now. Stressing freedom as a value permeates economic thinking from entry level courses. On the basis of even the simplest economic models, one can formulate the statement that the most significant consequence of economic freedom (absolutely free competition) is the most efficient allocation of resources. It means that the welfare of societies can be maximised through the allocation made under the conditions of market freedom: as we near the state of free competition and market freedom through the stages of the distorted market structures (monopoly, oligopolies—which are *not* atypical formations according to Galbraith 1967, 1973), the extent

[1] Some experts are of the opinion that mainstream economics is not relevant from the point of view of our present economic problems, because, in its approach, it adheres to 18–19th century conditions on which economics was built in the beginning (cf. Galbraith 1973).

of market grows (the quantity of goods available for consumers becomes greater and greater) and, simultaneously, prices decrease and consumers' surplus rises as a response to each step—introductory courses predestine economic evaluation of freedom through these provisions. Market competition that emerges as a manifestation of freedom is the source of welfare for mainstream economics. Of course, freedom is a fundamental value not only for mainstream (being a pure theory) but for institutional economics as well. Market freedom is the foundation even for the development of an efficient and viable public sector, since the conflict between these two, often rivalling sectors (public sector and market) usually results in a situation in which one overcomes the other and the one forced to adapt will necessarily bear the signs of the determining attributes of the dominant sphere. A public sector that is efficient and capable of tuning the attitudes of market agents in a positive direction can only develop if the *correction* of the failures inevitably emerging even in a strong and healthy market system (there is a wide range of government actions can be listed here from countercyclical demand management to proper targeting of allowances) and *not* an unnecessary and centralizing *replacement* of market is the purpose and the object of reflex-like sense of duty of state interventions (cf. Muraközy 2009).

It is not an overstatement to say that the whole history of nineteenth—twentieth century economic thought can be narrated as a contest between views on freedom. The complementing relation between neoclassical-monetarist and Keynesian theory (it was already mentioned in Chap. 1 and we will return to it in Chap. 7) emerges with relation to the freedom topic in a special way. While neoclassical (pure) theory could confidently stress the efficiency of free competition on the basis of its own abstract environment (Walras 1874), Keynes (1936) called the attention to the fact that presumptions of the pure theory are not often supported outside its artificial settings, so market needs to be supported and corrected.[2] It is not suggested in the orthodox Keynesian theory that the public sector should replace a well-functioning market mechanism—and this is exactly the recognition that constitutes the fundamental question of public economics: where do the borders of the efficient market lie beyond which *market failures* necessarily emerge and, hence, efficiency-enhancing state interventions can be entirely justifiable (Stiglitz 1988; Vigvári 2008b; Barr 2004) in order to eliminate the inherent crisis tendencies of capitalist socio-economic systems (see Habermas 1973).

[2] Some theories tried to pedantically define the exact line between state interventions and free market mechanisms. For example, Kalecki (1935) argued for a marketization of state interventions suggesting the option of paying transfers instead of direct government expenditures. In such a framework, allocation of excess demand would be a market responsibility. It is easy to realize that Kalecki's purpose was to maximize the scope of market mechanisms by virtue of an underlying belief in market efficiency. However, we should not infer that the problem of a strict and well-defined border-line between market and state intervention is definitely and irrevocably solved now—actually, there are excessive revelations on both sides that are in contradiction to even the widely known and evident facts (cf. Vigvári 2008a).

As it might have become obvious by now, freedom *as an ideal* fundamentally permeates economic thought. Here, ideal means that importance of freedom, depending on interpretation, points beyond freedom itself. The worth of freedom, whether it is *political* or *economic* freedom (that is especially important for us now) assuming and supporting each other, does not primarily lie in the fact that we can merely declare ourselves as free (manifesting through it the simple opposite of an unfree state fraught with constraints), which is, as we will see, only a necessary but not sufficient condition—but in achieving a position (of a *higher level order*, as it will be labelled below) that emerges as a result of a state being free of constraints. So, in this section, an attempt is made to outline the meaning of the concept of *freedom* also used by economics, without the slightest chance to offer ultimate answers. The impressions and conjectures published here have only the hope for such an answer. It cannot be told where the border-line is between formal reasoning on the one hand and mere belief and conviction on the other. If any attempt was made to find this line, it is to be feared that the Reader would relocate it somewhere else.

Since Isaiah Berlin (1958), the interpretation of freedom has been often made within the framework of the contrast between *negative* and *positive liberty*. In this dichotomy, negative freedom means the simple absence of constraints. The deficiencies that follow from describing freedom as the absence of constraints have been well-known for long. The most common example is the parable of an individual who does not have any desires or intentions that can be made impossible by external restraints. If, being a philosopher, the only goal of my life is to contemplate sitting in a barrel, and I need only some food from the external world to do so, then it is indifferent for me whether my barrel is placed in a totalitarian dictatorship or in an open society, provided that my daily ration is always received on time, of course. An introverted individual completely turning inward becomes an island to such an extent that the degree of restraint that would deprive him of his freedom in the negative sense cannot even be imagined.

Another example may be an individual who is *incapable* of utilizing a state free of confinements for the benefit of himself or society. Recently there has been extensive world press coverage of the story of a 23 year old North-Korean young man, who, together with his family, was an inmate of a labour camp near Pyong yang. He spent his life suffering from tortures and starvation irrational for us—however, for him, it was reality itself and he could not even imagine that there could exist another kind of human life (this story almost directly reminds us of the cave allegory of Plato; Republic, VII 514a2–517a; see Slings 2003). Evidently, he did not even know that the Earth is not flat. To alleviate his agonizing and gnawing hunger, he testified against his family hoping to receive more food. His family was executed on the basis of his testimony—while he did not show any sign of repentance for his deed, as it became apparent from his statements after escape. A man being socialised in this way was hardly able to feel anything more than fear, hunger, thirst or pain—it is not an overstatement that he could not be considered as a human being in the social sense of the concept. Well, it seems to be obvious that an individual spoiled in such a way cannot be judged to be free even after the removal of constraints, since, even possessing his (negative) freedom, he remains incapable of utilizing his new conditions in order to create a higher order state that can be regarded as the result of the disappearance of restraints—at least as a partial result, since it is argued here that something more is needed than a mere fall of walls to establish and achieve a state of this kind. Here, for obvious reasons, the case is not elaborated that this individual, in a state of freedom interpreted as the absence of constraints, probably becomes able to be guided by his animal instincts and desires, breaking with everlasting starvation. On this account, he should not be regarded as free, since appealing to our basic biological instincts is a desire, effort and motivation we share with the animal kingdom. If its possibility was a state of affairs labelled as freedom, it would cease freedom to be a specific human status. We should not suppose a concept to be an adequate framework for describing human and social actions on the basis of which these free states can be considered as elements of a common set shared with instinctual and basic activities of the animal kingdom.

Even economics offers several examples to underpin our reasoning that the concept of negative freedom alone is not suitable for describing and judging freedom. In the literature, it has become a generally accepted view by now that a fundamental source of global financial crisis of 2007–2008 was the deregulation of overseas monetary system (e.g. Király 2009; Gál 2010), i.e. an exaggerated degree of liberalization in which moral hazard was encoded in advance through perverse incentives. If freedom was the mere absence of constraints, the attempt at ethical judgement would be embarrassing, since freedom itself, i.e. the absence of constraints, triggered countless destructive effects in this case. How can it be explicated that the same freedom (the absence of constraints) can be regarded both as a favourable and as an unfavourable state? *Freedom* (i.e. whether the absence of constraints is actually freedom or not) *can only be judged on the basis of the order that evolves from and created by it*. The absence of constraints is not freedom for the individual and society (group, etc.) that is unable to fulfil the inherent potential for improvement after the fall of the walls—this potential may be described either through moral (for example, as it will be seen, *dignity*) or efficiency (an efficient resource allocation) categories, but there is a common element: the removal of (e.g. physical or legal) barriers is only a necessary but not a sufficient condition for their evolvement. The framework of formal public law is filled with substantive and eventual content by the members of societies: and this actual content determines whether the absence of constraints should be regarded as freedom or not. Moreover, the absence of constraints is not freedom either when the individuals and societies filling the empty frames evolve a lower order state. This is the well-known and dread case of anarchy—that only *seems* to be freedom. Deregulation of the monetary system *has not turned out to be* freedom, since this situation cannot even be judged by the mere absence of constraints. This deregulation evolved to be anarchy (that is, it *became* anarchy, and it *was not* that in advance, as the absence of constraints, i.e. in itself), since not a higher order state was developed from it by market agents. If we cannot understand it and do not interpret the emergence of a state of anarchy as degeneration of freedom or as an unavoidable consequence of too much freedom, we inevitably fall into the trap of eliminating freedom itself (or its possibility) in order to cease an unfavourable and harmful setting. This is what happened when re-regulation of the monetary system started as a consequence of and a response to the crisis that was nothing else but a (re)building of some barriers. It seems that such a case is rather a moral than a legal-regulative issue—and if it is not regarded as such, it is easy to fall into the trap of legal enforcement of morality that proved to be a serious attack against individual freedom.[3] By eliminating a badly (e.g. harmfully) utilized freedom from restraints, even the sole possibility of freedom fades away, replaced by the state of coercion believed to be a medicine, from which a higher order setting cannot emerge, since its prerequisites are missing (the barriers have been already built). The root of the problem is not the absence of constraints but the goals and purposes that a given socio-economic system wants to achieve through it. Regulation should be directed at sanctioning the unwanted consequences (and at creating the incentives to prevent unfavourable outcomes) and not at ceasing the initial circumstances (i.e. the absence of constraints) offering the possibility for both favourable and unfavourable results. So, possibilities for innovation (again, the absence of constraints) are to be complemented with appropriate incentives (Júlia Király 2009 probably suggested this setting). From this point of view, *it is significant whether the prerequisites of freedom or its destructive exploitation are sanctioned by a society*. (Of course, regulation is never definitive but a continuously evolving-accommodating and extending system. For example, the prudential regulation of banks explicitly serves the purpose of making financial innovation and the willingness to assume risks part of a balanced institutional behaviour; cf. Erdős and Mérő

[3] There is a problem still under debate. Quoting Mill (1859), Herbert Hart (1989) argues that individuals *must* have a moral sphere into which law does not break, and any kind of interference should be justified by the harm principle. On the contrary, Patrick Devlin, who basically echoed the views of Durkheim (1895, 1897), states that morals being the integrating norms of societies can be enforced by law, since this coercion is only the manifestation of the self-defending efforts of a society (for the sake of good order: even those theories that tailor individual freedom as narrow as possible still maintain a minimal area in which self-determination is free–the only questions are what the extent of this scope is, who decides on its size and what it is bordered by). This debate, though it seems to be far from our investigation in its theme, bears relations to the freedom ideal of mainstream economics to be clarified. It seems that there is an implicit assumption (which is often lost) in the background of our freedom ideal that is fundamentally optimistic about the abilities and instincts of man: accordingly, man and society are able to take advantage of their freedom (emerging as the absence of constraints). Bretter (2004) gives a detailed analysis of Mill's pedagogical principles from which one of the key principles of liberal philosophy, i.e. the demand for the absence of constraints follows. However, for Mill, these constraints are built in individuals due to the internalizing effects of education, and, therefore, societies can justly count on that individuals use their legal playground in a right way. That is the reason why we can say that regulation can be more permissive and liberalization-deregulation can be more extensive in case of an optimistic man concept–this is true for both individual morality and, for example, regulation of the monetary system, as we have seen it above. It seems as if there was a non-highlighted assumption behind our freedom ideal that supposes individuals and societies to be capable of evolving higher order states in the lack of freedom-eliminating constraints. This problem is directly related to economics through the case of paternalism.

2010. However, no one can know whether regulation in its current form can be regarded as ideal and final.)

It seems that the concept of negative freedom on its own is not capable of judging complex phenomena such as the nature of market freedom (or its mere existence) or of an adequate distinction between anarchy and a liberalized setting. It is exactly the variety of outcomes why we should not be satisfied with identifying, for example, market freedom with the simple absence of constraints (e.g. the constraints in entering a market). Since freedom from these restraints in itself has the potential for evolving outcomes filled with either a negative or a positive moral content (*anarchy* vs. *free market order*): the mere absence of constraints must be morally neutral, while what is labelled by economists as freedom and what (as we have seen) results in optimal welfare outcomes is morally definitely positive. However, positive freedom, thanks to its special semantic content, cannot be applied to describe spontaneous effects (cf. Hayek 1988) such as the cases in point. In the simplest case, positive freedom can be described as freedom of will and self-determination (for a great summary see Carter 2012), and it requires an active doer and pre-action considerations, i.e. a rational action—which should be obviously precluded in the case of spontaneous social-economic processes.[4] Discussing the evolution of market order in terms of the positive freedom would lead us to inherent contradictions.

Distinguishing *anarchy* on the one hand and *market freedom* or liberalization (or deregulation) on the other hand poses several problems. Firstly, for a liberal thinker such as Hayek[5] (1988), anarchy represents a definite value—even if not in itself but as a starting point of a shift towards a well-ordered state or of an improvement. But anyway, it seems to be an interesting macrosocial experiment to let everything go its own way in an anarchic-chaotic situation expecting that it might even be a starting point of a higher-order state (naturally, that cannot be judged either by us or society—just there and just then). Seemingly, this is a conflict between static and dynamic approaches of the problem. From a dynamic perspective, the elements of a shift towards a well-ordered state can be recognized in principle. To avoid the positive connotations related to the concept of anarchy, in the context of capitalism also examined by Hayek (1988), it is better to refer to this state, following Weber (1934), as *ruthless acquisition*. Perhaps this is the point where the significance of regulation and night-watchman state can be grasped: this is a regulation that precludes the emergence of certain outcomes in advance, defining, he standard playground. Market did not appear as an anarchic-chaotic and unregulated state even for Smith (cf. e.g. Bayer 1998), and, in this aspect,

[4] The case of *fallacy* is not a relevant aspect here. According to the concept of positive freedom, an individual is free if he is able to act in his own interest. Weber (1964) reminds us of that an agent does not necessarily realize his interests and may act erroneously as a consequence—the process (action itself) is conscious, however, it is not an easy question whether this action is not free in terms of the positive concept or positive freedom itself is conceptually not relevant in such cases.

[5] Barr (2004) uses a different, political-historian categorization. For him, Hayek and Friedman count as libertarians with the caveat that Friedman labelled himself as liberal.

neoclassical-monetarist theory followed him (Friedman 1962). The economy-wide auctioneer and its necessary function in regulation and supervision of market processes cannot imply the presumption of unregulated state; cf. Vígh (1997). *A good market is hardly an unregulated market.* Of course, even if we study, from dynamic perspective, the difference between an anarchic-chaotic state and an order evolving from it (or, in case of their comparison), distinguishing anarchy/chaos and order is not quite an easy (though a distinct) task, even theoretically, since the so-called sorites paradox (in other words, the paradox of the heap) arises. If one carries grains of wheat into a mound, when does the non-heap turn into a heap? Moving from an anarchic state towards an order, at which point of the continuum does the former end and does the latter start? In a general equilibrium model, is it the final and actually achieved equilibrium that only counts as an order—or is the tâtonnement-searching process already moving towards equilibrium an order? And if equilibrium-seeking process is also an equilibrium (i.e. a kind of equilibrium), which is the first time when a chaotic-anarchic initial state turns into an order—into a state that can be somehow regarded as an order? These questions can hardly be easily answered. Moreover, we also have to take that case into consideration in which anarchy and chaos plus order simultaneously exist—and order is not evident for and not experienced by agents but perceived as chaos by them (Smith 1991, 1994). This case is not quite far from that of the invisible hand of Adam Smith. It is an inevitable question: *who regards order as order?*

The idea that freedom is much more than the mere absence of constraints (and this lack is a necessary but not sufficient condition for freedom at best[6]) is suggested by Simmel (1900) perhaps the most clearly. So, it seems to be better to regard freedom as the higher order (and, for the time being, very mysterious) state which evolves in the lack of obstacles[7] (e.g. after they are ceased). Regulations are wrong if they interfere with the process towards the desired higher order state—but they are still needed to define the playing field and to guarantee the equal opportunity. It is the sense of freedom that seems to be the most deep-rooted in western thought and, illustrating it, that also inspired economics. If freedom is not described and interpreted by the mere absence of constraints but by the order that evolved from this absence, it implies at the same time that the quality and extent of freedom cannot be judged on the basis of the existence or non-existence of these constraints. The extent of freedom can hardly be measured by the nature of institutional (e.g. of public law) frameworks. As it has been mentioned above, these institutional frameworks are filled with actual content by the agents of the socio-economic system—since we are talking about *frameworks* which must be filled, per definitionem. The extent and nature of freedom directly depend on the content whose emergence accurately reflects the way individuals and societies, in order to

[6] The caveat of "necessary but not sufficient" seems to be an acceptable approximation with regard to the present formulation of hypotheses, but it is not probable that one cannot find better terms.

[7] As if Walras (1874) also suggested it, saying that: "Freedom procures, *within certain limits*, the maximum of utility [. . .]." (italics added)

establish a state of order, apply these frameworks as empty and formal rules of game.

This higher order state emerging in the absence of constraints appeared in various forms during the history of philosophy. Among others, it was the dignity of man in the platonic heritage (cf. e.g. Barcsi 2011). This thought is not said to be one of the early appearances of the freedom ideal, though the problem of freedom was already posed in ancient Platonism (see Bene 2003). According to this philosophy, individual human souls directly relate to the One appearing as the ultimate source and creator God through the phenomenon and process of emanation[8] and human beings utilize this relationship, embodying the ideal of contemplation, to break their bodily ties and to ascend to the sphere of the platonic ideas and, then, to unite directly with the source of every good and wisdom, as a result of the meditative elevation. The essence of this theory can be well grasped by the philosophy of the Plotinusian hypostases (a special hierarchical structure of the world) bearing direct platonic influences (see Stead 1998; Wallis 1972). The first emanation of the One is Reason; from Reason proceeds the World Soul that, eventually, places the individual souls to form the physical bodies of human beings (*Enn.* IV.8,6).[9] Accordingly, Reason is located above the directly experienced reality, which is the sphere of platonic ideas (*Enn.* V.9,5–6): individual human souls can reach here by a contemplative lifestyle from which a special, theoretical-conceptual knowledge, i.e. the knowledge of the experiencable world follows (as we can remember from Chap. 1, according to the basic thought of Platonism, the objects of sensory experience only reflect the Forms and are not really existing things—this is the essence of the theory of the platonic Forms). The utmost and ultimate level, the One can be only reached by an emotional-intellectual identification-union but not by consummating conceptual thinking.

The scheme that can be distilled from this beauteous philosophy is very simple. A human being, or his individual soul, physically constrained by and locked in his own body is not capable of consummating the pre-defined program determined by his divine origin: to ascend to the intellectual sphere. The fundamental idea of the new platonic philosophy is that every being *instinctively* (i.e. involuntarily and not voluntarily) longs to return to its source (and, accordingly, individual human souls sigh after Reason and the One)—and that this return can be completed by an

[8] It should be mentioned here as a technical detail of secondary importance that the present summary is based on the philosophy of the platonic Plotinus (c. 204/5 – 270). His commentaries to Plato (the Enneads—or, the individual essays in an abridged form, simply *Enn.*) became the primary source of western thought to the philosophy of Plato. The influence of Plotinus can hardly be overestimated, he left his mark on the entire history of the philosophies following him. In addition to Plato and Aristotle, he is said to be among the three greatest ancient philosophers. His approach was the foundation for the interpretation of the platonic philosophy even in the nineteenth century (Gerson 2012).

[9] The writings of Plotinus were edited and arranged by his student, Porphyry. He was the commentator who arranged the short essays, written at various dates, into six units—these are the six Enneads which are referred by the Roman numbers from I to VI. The first Arabic numeral denotes the treatise while the second indicates the chapter within, respectively.

introverting and contemplative spiritual activity, by a special self-contemplation, so the returning to the origin and the contemplation upon oneself is the same activity (*Enn.* VI.9,2; cf. Wallis 1972). Happiness of the soul, as a higher order state, depends on this ascent, i.e. on the success of this return (cf. Maróth 2002). In this philosophy, constraints are of simple physical nature,[10] and the higher order state is, firstly, the spiritual-intellectual ascent attained by contemplation and, secondly, the resulting happiness of the soul. It is not of secondary importance either that the definition of the program, i.e. that of the higher order state to be accomplished in the absence of constraints does not originate from individuals (or, as it should be referred to as, individual human souls) but it is originally given to them as a calling, and consummating this plan is not necessarily a conscious action but comes from the instinctive longing for the ultimate origin of the human soul. It is not the individual either who creates and declares his relationship to the ultimate creator or who intends to achieve the intellectual ascent as a result of a deliberate decision.

By the time Adam Smith formulated the freedom ideal that later directly determined economic thought, *this* notion of freedom had had a tradition of more than 1,500 years. For Smith, freedom was the same as it had been in platonic philosophy: a higher order state that evolves in the absence of constraints (i.e. in absolutely free market competition) that is briefly referred by today's economists, as *efficient allocation of resources.* For Hayek and, following his teaching, for Vernon Smith freedom means the same: the evolution of capitalist socio-economic system.[11] The meaning of this pre-defined program was most thoroughly elaborated

[10] The body is a prison and a cave (sic!) for the soul from which it has to reach the realm of the ideas, i.e. the intellectual sphere, since the things, that one can see in his cave, are only imprints of the Forms (ideas) as mere shadows (accordingly, Plotinus himself regards human body as a cave as well: e.g. *Enn.* IV.8,3–4). This beautiful philosophy is not as abstract and distant as it seems at first sight, since it says nothing more than human beings have lower level basic needs and focussing on satisfying them may deprive them of devoting themselves to intellectual pleasures and delights. Although, seemingly, mainstream economics does not pay too much attention to this thought, the ultimate goal of the abstraction creating homo oeconomicus is to deprive human beings of their higher level needs reducing the horizon of economic decisions to mere consumption and satisfaction of bodily needs. Moreover, this philosophy is also capable of explicating, e.g. the backwards bending shape of labour supply curve. According to this theory, above a certain income level, individuals respond to any increase in income by lowering their labour supply; meaning that we turn to higher level demands (which are mainly of non-physical, but mostly intellectual nature) after satisfying our basic (physical) needs. It is obvious that the concept of the hierarchical structure of soul is in the background of Maslow's well-known hierarchy of needs that is widely accepted not only in the various specialized branches of economics (e.g. Kotler and Keller 2006) but also in social sciences (e.g. Perrow 1986). The common saying, *when poverty comes in at the door, love flies out of the window*, is probably also rooted here. As far as the latter is considered, it may be enough to mention the parallel (or just the slight association) that both Plotinus and the Renaissance Platonist Marsilio Ficino, following Plato, regard the craving for the One as *love*.

[11] When Max Weber distinguishes the objective-technical prerequisites of capitalism (Weber 1923) and the capitalist ethos (Weber 1934) which he relates directly to protestant practical ethic, he similarly stresses the spontaneous evolution of capitalist socio-economic system. Protestants supported this evolution without the intention to do it and without being aware of it. Although Hayek (1988) did not quote Weber's theory, his notion bears close resemblance to it. Protestants

by Friedrich Hayek for economics. For him, capitalism designated by the name of extended order[12] was a spontaneously evolved and not a consciously intended state that emerged because societies started following certain behavioural rules.[13] Eventually, these rules rival each other and more successful rules are selected in the selection process—together with societies that maintain them. The scheme is the same as it was in the initial platonic heritage: after the removal of impediments, a human being, responding to an unrealized calling and being unaware of the ultimate meaning of his acts, makes efforts to achieve a state the urge for inducement of which is not the result of voluntary considerations.

The platonic tradition talks about the *recognition* of natural longing and urge (and the necessity to recognize) rather than an arbitrary-voluntary purpose. In this context, we can remember the way Hayek (1988) articulated the spontaneous nature of the emergence of capitalism, according to which the process will never be rationally plannable (and, otherwise, this rationality would be definitely destructive according to Hayek). It would be an obvious parallel that juxtaposes the Plotinusian concept of efforts made to achieve the state of freedom with the theory of the cultural evolution. However, this aspect (viz. the absence of rationality) of these two kinds of philosophy should not be overstated—because, e.g. the explicitly meditative New Platonism and its Renaissance revival (Ficino[14] 1576) deliberately urged to lead a contemplative lifestyle to accomplish the spiritual-intellectual ascent. Here, the main point for us is that freedom in both theories *is a higher level order that spontaneously evolves after the removal or elimination of constraints*. For Plotinus, freedom does not mean a simple fall of barriers but that, *after*

following religious orders were not aware of the fact that they live in accordance with practical ethic that fills the objective-technical frameworks of capitalist mode of production eventually with a special ethical-moral content–this is the spirit of capitalism.

[12] As it was highlighted above, freedom is regarded here as a *higher level order* emerging in the absence of constraints, obviously following Hayek's terminology (capitalism as *extended order*). It can be said that Hayek's theory of capitalism is a manifestation of a (more) comprehensive theory of freedom (or its application in economics); it explains the terminological resemblances.

[13] In these terms, new classical macroeconomics seems to be an exception, though, as a radical side-shoot of monetarism, it also had to share the freedom concept universally characterizing mainstream economics. The roots of the confusion is that (referring to the findings of Sect. 3.2) the new classical interpretation of labour market considers equilibrium as the result of conscious and voluntary efforts of labour market agents. New classicals evade this problem by applying the as-if-reasoning of Friedman (1953) (cf. Weeks 1989). According to this argument, we do not need to assume labour market agents to be economic designers having professional econometric skills and seeking general equilibrium—it is well enough to presume their spontaneous behaviour to result in an outcome *as if* they were so. This theory asserts nothing more than that was incorporated in the abstract-idealizing Walrasian theory (Walras 1874) with its equilibrium-seeking auctioneer: these two concepts are on the same ground, which means the presumption of equilibrium. Anyway, market order is still the result of the spontaneous behaviour of agents governed and arranged by the invisible hand.

[14] Four essays of Ficino should be highlighted here: *Argumentum in Platonicam Theologiam*; *Compendium Platonicae Theologiae*; *Quaestiones quinque de mente*; *Quid est felicitas, quod habet gradus, quod est eternal.*

the removal of constraints, a higher level order develops the structure and nature of which are independent of the individual—that is, of his will, desires, intellect and personality (cf. Bene 2003). At this point of the discussion, it is worth briefly referring to the Hart-Devlin debate again. In that context, freedom belongs to the individual who was prepared in advance by education: i.e. freedom is not the absence of constraints in this case either, but a state that emerges from this absence as some creation or achievement of the individual who can creatively utilize this absence. The comparison between platonic texts and the freedom ideal of economics seems to be quite direct.

Of course, evolving the order requires will, but for ancient philosophy, as it will be seen below, every kind of order involved a deliberate ordering process. In any case, it is still undoubted that the ascent to the One, i.e. the way beyond Reason and the theoretical understanding of the world is not achieved by intelligence—of course, it is also intellectual, since suppressing instincts and bodily needs plays the main role here though it cannot be planned, understood, urged or not even described by human reason.[15] The rationally incomprehensible and unachievable ascent to the One (only the intention can be rational) suggests a comparison with the also non-rational cultural evolution the result of which (in our reasoning: as a manifestation of freedom), for Hayek, became the capitalist socio-economic order. Of course, these parallels need a more detailed analysis. Even if reason has a more important role for Plato than it has in the Hayekian concept, then probably, we have to face what was also mentioned by Hayek (1988): the society that was the bed of Greek philosophy was fundamentally different from the social order of early modern times of modern history. Society and the entire world seemed to be

---

[15] Plotinus also highlighted that the One, transcending the Intellectual can be attained by abandoning intellect (*Enn.* V.5,6), since one who would be able to grasp (intellectually) the One in its concentrated fullness that must be identical with the One (*Enn.* V.5,10). Regarding the case in point, Plotinus is the clearest when says: "The main part of the difficulty is that awareness of this Principle [viz. the One] comes neither by knowing nor by the Intellection that discovers the Intellectual Beings but by a presence overpassing all knowledge. [...] knowing and knowable must all be left aside; every object of thought, even the highest, we must pass by [...]." (*Enn.* VI.9,4) An approach governed by the intellect will seem to be obviously impossible if we remember that the concepts of thinking cannot be harmonized with the One—however, an intellectual approach would be performed among these concepts. The One is not attainable by the intellect: "[...] we must withdraw from all the extern, pointed wholly inwards; no leaning to the outer; the total of things ignored, first in their relation to us and later in the very idea [...]" (*Enn.* VI.9,7). Probably Wallis (1972) would agree with the impossibility of an intellectual understanding as well, saying that: "[...] all theoretical discussions of the One are finally inadequate, since its true nature is revealed only in the mystical union [...]." In this context, László Bene (2003) remarks that *will*, to Plotinus, meant something different than to us: to him, will is part or an aspect of intellect, *an intention based on rational considerations*–that is why the ascent to the One can be regarded as the work of intellect, even if it is not performed by the usual conceptual thinking. As far as the case in point is considered, all these are just technical details of secondary importance.

comprehended even by one mind. Interests and purposes of the individual seemed to be fully plannable.[16]

As we can see, one of the most characteristic features of the extended order (capitalist socio-economic setting or free market competition) is the *spontaneous nature of its emergence*: members of societies follow some rules without knowing and understanding the rules themselves, their importance or the way they evolved. One can hazard the remark that the selection of the rules that support societies in surviving and in achieving capitalism as a higher level state takes place in a cognitive environment labelled by phenomenological social sciences as *natural attitude*. There have only been the first attempts to analyse this natural attitude in economics. Moreover, these early efforts are mainly directed at applications covered by phenomenological sociology—so, one can talk about the examination of the possibilities of adopting these sociological studies in economics, rather than establishing brand new phenomenological economics. Sociologists have achieved significant advancement in describing how the natural attitude of everyday life turns to reality in order to experience and to understand it: the implicit and underlying typifying assumptions of life-world actions have to be stressed first of all (cf. Schütz and Luckmann 1975). These types, in a broader sense, are schemes that are unconsciously in the background of our everyday life-world actions—and this is exactly the feature that places the norms that are important for us now next to these cognitive structures. Moreover, this problem is part of a major issue relating to the economic role and evaluation of the subject in which philosophers have shown interest since Kant at least. It is about subject's importance that precedes and establishes everything else. This issue seems to be settled in economics by the subjective theory of value—but, actually, it is exactly the sure and safe sense of this settledness that has prevented us from grasping and addressing the root of the problem. As far as the subjective theory of value is considered, one can talk about a combination of at least two traditions: though referring to the subject as an ultimate ground had emerged, economics built it into its general mathematizing

[16] At this point of the discussion, the ancient tradition of European thought can be well grasped, according to which in the case of any kind of order, presupposes someone who has made this order. In other words, we tend to infer the presence of a person or mind creating order from order itself. In his *Philebus* (28d–e), Plato says: "Shall we say [. . .] that all things and this which is called the universe are governed by an irrational and fortuitous power and mere chance, or, on the contrary, as our forefathers said, are ordered and directed by mind and a marvellous wisdom? [. . .] But the assertion that mind orders all things is worthy of the aspect of the world, of sun, moon, stars, and the whole revolving universe [. . .]." One can meet this thought in other texts of his (e.g. *Timaeus* 30a–b; see Archer-Hind 1888; cf. Maróth 2002). This conception is also popular today: it is enough to mention the most fundamental theory of modern cosmology that links the Creator to the (hypothesized) fact of the Big Bang (Singh 2004). Moreover, one of the special directions in modern fractal theories is directed at searching for natural occurrences of the infinitely self-similar (mathematical) patterns (see Bourke 2012). It is obvious that the mathematically formalizable natural-superficial patterns can be really considered as a piece of work of an organizing mind.

tendencies. The result was an *objectivised subjective theory of value*[17] (sic!) that investigates the allegedly natural laws of value attributing acts and not the processes of this value attribution taking place in the conscious mind. Following this line of thought it is also justifiable by the fact that exactly recognizing and highlighting the importance of the subject became the most significant intellectual foundation for the freedom ideal: this idea, the modern current of individualism started off by Renaissance philosophy, which was directly inspired by the antique heritage (cf. e.g. Trinkaus 1983). Subject as a free entity and as the foundation preceding and establishing objective reality have common philosophical roots.

Now, let us turn our attention to the grounding theses of new classical macroeconomics. As we will see, freedom and the resulting macroeconomic equilibrium have particular significance in this theory and the way the theory tries to interpret equilibrium and market freedom is very special and unique. The analysis starts with the new classical concept of the labour market then focuses on further elements.

3.3 The New Classical Interpretation of the Labour Market

The literature on REH is full of statements intended to summarize the essence of REH and new classical macroeconomics as briefly as possible. In this context, full employment and neutrality of money assumed to hold even in the short run have special importance for us first (cf. Weeks 1989). For new classicals, unemployment means the emergence of the natural rate of unemployment (and, accordingly, of natural-equilibrium output). Actually, this is the direct renewal of the neoclassical concept of full employment.[18] It follows from the assumed flexibility of all prices and from the presumption that the parties in the wage negotiation processes know the relevant model that describes the given macroeconomic system.[19] So,

[17] We said "objectivised" in the sense that László Tengelyi (1995) writes, following Kant, about how mechanical philosophy inspired scientific attitude even in fields where it is not justifiable by the adequacy of the topic under investigation. Economic theory is an excellent example for a mere functionalizing of a factor regarded as subjective, undermining the possibility of understanding the nature and effects of subjectivity. Of course, it does not imply that the theory of ordinal utility was a more successful attempt to conceptualize the value attribution process: in these terms, both ordinal and cardinal theory poorly perform–possibly, it was not the purpose at which they were aimed.

[18] The shifting of stress occurred in economics (from the demand and unemployment considerations of the Keynesian era to the inflation-centred thinking of the 1970s) is not independent of the fact that monetarists started from an assumed labour market equilibrium. For them, inflation was a problem to be solved, so labour market was deleted from among the potential sources of problems. They tried to underpin these efforts by the idea of the increasing stability of capitalist economies. Moreover,, this line of development is stressed by most of the pieces of work written in the history of economic thought (cf. e.g. Mátyás 1979, 2003).

[19] On the grounds of the preceding chapter, we can remember that this assumption implies the existence of this knowledge.

unemployment is a phenomenon for new classicals the emergence of which (even if it is to be reckoned with) does not disturb the evolution and permanency of labour market equilibrium. Of course, there is still room for cycles in real output—their mechanism will be investigated in Chap. 4.

At this point of the discussion, it is worth considering the natural rate of unemployment within the context of the Walrasian economic equilibrium.[20] Let us recall the infinitely simple and, hence, extremely abstract Walrasian macroeconomic system for a moment—setting the robust mathematical apparatus aside. Naturally, prices are flexible and equilibrium is guaranteed by a phantom (who actually aims to ensure the idea of the existence of a centralized economy) referred to as an *auctioneer*, who announces price vectors in the course of his searching process again and again. Given these prices, market agents declare their intention to trade (supplies and demands) and if supplied and demanded quantities do not coincide (i.e. because of excess demands and supplies the equilibrium state cannot emerge), the determination of further price vectors is necessary. The failure of the act of trade in the presence of a non-equilibrium price vector is not the result of deliberate considerations: false trade (i.e. transactions at disequilibrium prices) is precluded by the auctioneer through an explicit prohibition. This process lasts until the economy-wide equilibrium emerges (strictly speaking: until the whole economy arrives at equilibrium through the equilibrium of its markets), i.e. until the auctioneer finds the only price vector that implies this equilibrium.[21] Precluding false trade implies that a given macroeconomic system does not have any participants who want to trade at a given price vector but who cannot find a partner (that is, for any market, excess demand or excess supply cannot emerge in a trade situation). *Either everyone or no one can trade*. Accordingly, *involuntary unemployment cannot occur*: if excess demand or excess supply emerges in the labour market at a given wage, equilibrium (through the adjustment of wages) is going to be established. So, *only those do not work who do not want to*.

This analysis is as elegant as meaningless, since it only holds if any *actual job-seeker* is supposed to regard working as a completely accidental effort: he sells his labour in exchange for an appropriate wage, but, otherwise, he can easily forgo it. Moreover, he can do it without any considerable long-run consequences, since when voluntary unemployment starts becoming troublesome for him, by lowering his wage claims, he can find a new job at any time (and even if this decision of his, i.e. the emergence of a labour market excess supply, decreases the actual wage level, others can still voluntarily decide to work or not to). Talking about *an*

[20] In its simplest form, Walrasian equilibrium means that demands and supplies of markets of an economy are equalized in a state of an economy-wide equilibrium—i.e. supply and demand in *each* market are equal (Kopányi 2003).

[21] This formulation is misleading, so it needs to be specified. The ratios of components of the equilibrium price vector (i.e. price ratios) are important. Therefore, any multiples of the equilibrium price vector is also capable of establishing and implying the equilibrium. However, equilibrium vectors can be computed from each other through scalar multiplication, since multiplying with a scalar does not alter the ratios of the components of the price vector.

equilibrium wage and an equilibrated labour market means declaring that the unemployed do not work due to voluntary decisions. Labour market clearance requires flexible wages, since if John Smith, becoming tired of idleness, determines to find a job, he can only be regarded as voluntarily unemployed if wages flexibly respond to the fact that he has changed his mind. Otherwise, his status should be considered as involuntary unemployment, since if wages are rigid, John Smith may think that, at the given and fixed wages (moreover, perhaps even at a lower level), by changing his previous decision, he will be inclined to work. This would be equivalent to a labour market excess supply. At the given wage level, however, employers have a given demand, and Mr. Smith‘s labour is redundant at this level (and a lower wage level cannot emerge, since wages are rigid). This would be involuntary unemployment (in other words: labour market disequilibrium) and false trade. Avoiding false trade is only possible when prices are flexible—or if someone (e.g. the auctioneer) precludes that. *The Walrasian auctioneer is a non-real abstraction created by analogy to (strictly speaking, replacing) completely flexible prices.*[22] Unemployment is purely voluntary only then if wages flexibly respond to changes in demand and supply (or, in the lack of this flexibility, someone ensures their proper modifications).[23]

Our doubts are further strengthened by the explanations of natural rate of unemployment, according to which, broadly speaking, one chooses to be unemployed because offered wages do not cover costs of employment, either opportunity costs or actual expenses incurred. Under this interpretation, John Smith voluntarily withdraws from the labour market if he can only find unfavourable employment opportunities, because he would have to wake up early on a daily basis or have to move into an other town. Of course, it might be disputed whether declining a job opportunity located thousand miles away is a voluntary decision or the result of exigency. In any case, talking about natural unemployment explicitly implies that unemployment is a real phenomenon—the terrifying picture of which, however, is tried to be presented in a more favourable light by monetarists through declaring this unemployment to be exclusively voluntary. On a second thought, there is nothing else monetarists can do, since how could we talk about macroeconomic equilibrium if involuntary unemployment was part of the big picture? Talking about full employment is only possible if the unemployed are ignored (or, at least, regarded as people with lazy or too material attitude and, hence, self-destructive)

---

[22] As far as *the* pure theory is considered, the axiom of completely flexible prices or the auctioneer as their substitute are not troublesome at all—so, *it is not* a critique of neoclassical economics.

[23] In a stochastic environment, expectation errors may similarly result in false trade, that is, exchanges may occur at disequilibrium prices. However, these errors are atypical cases. This is the thesis that is cited by new classicals when talking about short-time period disequilibrium states of the real system. In Chap. 4, we will examine whether a stable equilibrium can emerge in a stochastic environment at all. It will be argued that, because of the accidental and momentary nature of equilibrium, there is no sense in exclusively talking about voluntary unemployment at all. Perhaps, it would be worth making the picture more complete by detailing the nature (e.g. permanency) of involuntary unemployment.

and only the employed are taken into consideration, i.e. excluding somehow the rest of the world. It is justifiable *to think that the new classical denial of the possibility of systematic countercyclical economic policy is ideologically rooted somewhere here*. If the unemployed themselves cause for their own penury, why would an economic policy to enhance employment be needed? Regulators of the economy would take the responsibility for the salvation and financial welfare of people who do not worry about their own destiny. New classicals do nothing else but almost outlaw the unemployed—and doing this in order that the operative (economic) policy makers can go to bed untroubled every night believing that all their tasks are well performed. Although there are people who are unemployed, economy is in the full employment equilibrium—however, this is mere self-delusion.

The way in which new classical macroeconomics views and describes wage formation is interesting. New classicals set aside the assumption of flexible wages necessary for labour market clearance and for maintaining the strictly voluntary nature of unemployment as a corollary. Instead, it is stated that *wages are set in advance in a manner that they possibly ensure economic equilibrium* (i.e. there should not be employees who would take up employment at a given hypothetical wage level, which level would, however, result in demand deficiency—and *vice versa*, there should not be employers who would demand labour at a given wage level but could not find candidates). In a deterministic environment, in principle, it would succeed, but, under stochastic conditions, efforts can only be made at setting wages that are *expected* to equilibrate labour market—of course, noises from the economic environment may interfere with it. In such cases, labour market would move out of the equilibrium state for a short time. Moreover, trade unions[24] are also included in the story (exactly for bargaining wages) to make the story more credible, since, without them, all employees would have to enter into individual negotiations and their intention to clear the labour market would imply that all the workers know the relevant model (that would be an obvious exaggeration—though it could be still solved through the *as-if*-reasoning). No wonder if we are amused at the details of the original role meant for trade unions by new classicals. Up to that time, organization of labour force was equivalent to keeping wages rigid and, hence, unemployment unnecessarily high (cf. Galbraith 1967; and, for instance, Chikán 2003). Suddenly, for no particular reason, concerns over the united working class start governing trade unions—of course, we could disregard this concern through a minor distortion, assuming all employees to be organized[25] (it also

[24] Under some interpretations, trade unions have been already included in Friedman's concepts of labour market. According to them, temporary effectiveness of countercyclical economic policy is ensured by the assumption that trade unions, partly responsible for wage setting, form adaptive expectations (cf. Shaw 1984). Friedman actually presumed that the natural rate of unemployment is affected by actions of trade unions (Friedman 1968). For one of the most explicit references of new classical macroeconomics and REH to trade union involvement see Shaw (1984).

[25] The assumption of fully organized employees could be acceptable in the pure theory, since, in this case, it would be a significant aspect of analysing actual situations to record the fact of a partial labour force organization and to take its consequences into account. Considering neoclassical

seems to be an excessive assumption judged in reality). Trade unions *and* employers try to set the wages for the next period to equilibrate labour market, and if someone does not work, he does so taking a voluntary decision. [→ **trade union**] This is the revival of the auctioneer as a blue-collar worker wearing a beret. The analogy is made quite clear and obvious by the interpretations of Frank Hahn (1980) who, investigating the Walrasian foundations of monetarism, does not mention the auctioneer (becoming embarrassing), but instead, assumes rational expectations in the Walrasian framework—i.e. the auctioneer becomes redundant if there are rational expectations.[26]

Trade union. Trade unions have evolved in various ways in various countries. Although their fundamental function was to protect the (financial) interest of employees, trade unions of different periods and different parts of the world could only fulfil this task with huge differences.

In Eastern Europe, the views on and the notions of trade unions are connected to socialist state power, not without reason. Trade unions had a share in Eastern European history from almost the birth of the socialist-communist ideas. Although trade unions always enjoyed a special status, this status did not make it possible to really fulfil these basic functions. Accordingly, scientific socialism could not easily find the proper role, even theoretically. As the theory of socialist societies had promised, a conflict of interest would cease to exist, therefore there is no need for safeguarding the separated interests of classes. If power belongs to the working class, what role can trade unions play? Trade unions also became subordinated to Communist Parties (as it was required) according to the Marxist-Leninist ideology: they supported the economic and political decisions made by the Party and engaged in their implementation (Buza and Vas 1980; Hámori 1986). It was simple logic. Since the socialist state serves the workers' interest, trade unions can do the most for employees by supporting the Party and the political elite and, moreover, the state itself interwoven with and subordinated to them as an administrative body. In these times, trade unions mainly served educational-training purposes in order to support replacement for labour supply. Employees could do the most to improve their living conditions by making every effort to build socialist state.

The subordinate role and position of trade unions did not change after World War II either. Their existence and functionality became less significant. Their task, in association with the Party, still was to enhance material production and to encourage workers. After the 1950s, there have been some reform efforts to transform the feudalistic state socialism but they mainly failed.[27] Trade unions remained part of the state machinery and could only

interpretation of labour market, Weeks (1989) passes the remark that keeping the wage above its equilibrium level is consistent with the contentment of the vast majority of employees. It is evident that such a situation would be regarded as favourable by only those who can sell their labour force even at this higher wage level. So, it evidently follows that the intention of setting the wage to its equilibrium level is conceivable assuming fully organized working class.

[26] Sometimes, the literature reproves new classicals for (over)simplifications (e.g. Greenwald and Stiglitz 1987). It is obvious that our analysis does not (and, moreover, cannot) give a favourable description of new classical macroeconomics starting from its labour force theses, but one can never be too careful (as it will be argued throughout the chapters below). It will be shown in the forthcoming chapters that *a more comprehensive analysis and interpretation can place new classical macroeconomics in an unusual activist role in economic policy.*

[27] Of course, these reform attempts emerged with various intensity in Eastern Europe. For instance, in the Soviet Union, only anxiety eased at best, and leadership executed by a close elite replaced the autocracy inherited from Stalin–but the regime, as a whole, remained totalitarian. Everything

assist in meeting central requirements and aspects. Meanwhile, the bottom-up self-organizing approach (that is, relying on the members) of trade unions almost ceased to exist. According to some contemporary commentaries, trade union leaders were regarded as members of the management[28] rather than delegates of employees to safeguard their interests (cf. e.g. Sinkovics 1989). Taking the circumstances into consideration, it is out of the question that the genuine protection of interest was impossible since it was sill supported by the former ideology that the socialist state serves the best interests of the working class.

After political transformation, a completely new situation emerged, naturally. Trade unions received legally guaranteed rights to soften the changes that unfavourably affected crowds of workers. So, employees became part of institutionalized societal dialogues and, hence, social and labour tensions became manageable. During this period, governments, of course, did not lose the control over the participants in social and labour conflicts. So, trade unions survived political transformation and, as the new (independent and free) trade unions also emerged, they were built into the new economic-social-political order together. From this time, trade unions have been actors in the society-wide dialogue. However, it is interesting that labour market and in-company positions of trade unions have weakened. Their role in the new and dynamically developing sectors has not strengthened and the members do not come from the widest possible circles of society. Trade unions responded to these changes by increasing their political[29] and economic power (for further details see Thoma 2002).

The way trade unions developed in Western Europe and overseas was completely different, of course. The relations, firstly, between trade unions and the state and, secondly, between trade unions and the companies also varied. Trade unions have the longest traditions in Great Britain (Diederiks et al. 1994) and in the United States—in other countries, their history could only begin later, which was interrupted several times (e.g. during World War II). The number of unionised workers also varies: 20 % of workmen is unionised in the United States, while in Sweden this figure is 90 %. Their formation dates back to the industrial revolution when productive capital (by applying capital-intensive technologies) gained disproportionately enormous power over the workers (Cameron 1993)[30]; trade unions, through organizing workers, tried to eliminate this disproportionality, but could rarely adopt the real role of a decision-maker or consultant (the only

---

still was under the control and sanction of the Communist Party. Romania was also successful in opposing all kinds of reform attempts–while, on the contrary, Hungary gave room to some market mechanisms at a relatively early stage in order that political power could remain easy to centralize (Davies 1996). However, Hungarian efforts towards marketization failed one by one, efforts to establish a socialist market economy were abortive (Csákó 1992).

[28] Over and above the manager, party and YCL (Young Communist League) secretaries and a labour-union steward (or secretary) were delegated into management.

[29] Building an economic policy and/or political power is not an Eastern European speciality, but was extant in the Scandinavian countries as well (cf. Giddens 1993). Although it would be evidently a forced parallelism to mention Solidarity of Poland, it is obvious that the entire Polish transition was built on trade union activity (for further details, see Szokolay 2001), which itself can be regarded as strong political involvement.

[30] No economic history can reproduce and describe the inhuman conditions of the emerging and developing manufacturing industry in a rough and direct way as the novels by Charles Dickens (especially Dickens 1854). In these terms, Friedrich Engels (1845) can only compete him As long as an economist does not meet these monumental and mournful books, industrial revolution remains a sad and distant episode for him. Prof. Taxner-Tóth (1972) gives a comprehensive review of the radicalism and gloominess of Dickens' voice and of how the outstanding pieces of work fit into the biographical background.

exception is collective agreements which constrain both the employers and the employees). Thanks to this, we still associate trade unions with strikes, i.e. with the most vehement and utmost form of protesting (Giddens 1993). However, Western European trade union model, though these organizations also had to face the decline of their power and the demographic determinants of the membership (Akyeampong 1997), became the basis for theoretical approaches, since these trade unions, as opposed to the Eastern European counterparts, could really struggle for the improving of working conditions or wage level. This role is mainly characterized by the conflict with the entrepreneur-side representing the interests of productive capital. Of course, some cooperation is also needed, since, when the playing field narrows, wage claims can be only satisfied by *joint* improvement of productivity and of work intensity.

The existence of equilibrium is mere tautology. Equilibrium exists because economic agents want it to exist. *There is equilibrium because it exists.* There were debates lasting for decades or even centuries in the literature on whether the assumption that macroeconomies have tendencies towards equilibrium is acceptable. New classicals quickly close this debate. For them, equilibrium is evident since it is wanted by everybody (even if several problems are to be ignored for its presumed achievement).

Naturally, the *PFH* vs. *REH* debate cannot be set aside here either. As we have seen, new classicals cannot be persuasively criticized for reputedly introducing PFH disguised as REH into their models—not persuasive at least in the sense that new classicals never talked about determinism explicitly (of course, it is another question that assuming determinism may be needed for their models, which was left unsaid by new classicals). In this context, it is very illuminating how Begg (1982) writes about wage negotiation: "[. . .] the equilibrium expected real wage at the date the nominal wage bargain is made is assumed to be set in the expectation of clearing the labour market. [. . .] I assume that nominal wages are set each period to produce an expected real wage which is expected to generate measured unemployment at the Natural Rate". Moreover, Begg explains the expectations-augmented Friedmanian Phillips curve by using this mechanism: labour market can equilibrate at the natural rate of unemployment because agents set the agreed nominal wage to form and to maintain the natural rate. This notion furthermore is very close to what Hahn (1980) writes on the meaning of rational expectations. According to this, considering the occurrence of a given situation as a condition, agents forming rational expectations expect prices, that will clear the markets when this situation actually occurs. So, agents presuppose the clearance of markets in advance and form expectations accordingly—the only factor of risk *may* be the events themselves.[31] Hahn takes a further step when identifying continuous market clearing with rational expectations.[32]

[31] It is worth recalling here what was said on stochastic and deterministic models in Chap. 1. Assuming determinism, REH is equivalent to PFH.

[32] It is not needed to investigate here the circumstance of not marginal importance that other conditions are also necessary for continuous market clearing—namely, flexibility of prices, or the auctioneer who sets the prices to their equilibrium value. For us, the most important detail is that,

Begg only talks about the intention of equilibrium—however, it is worth considering under what conditions can an actual equilibrium follow from this intention. Perfect foresight only makes sense under deterministic circumstances—if there are contingencies, predictions are probability distributions, therefore expectations cannot be true (if so, it is due to chance) only unbiased. In the case of the wage negotiation mentioned above, it means that agents forming unbiased (but not true) expectations may err in later outcomes, so they cannot determine market clearing wages in advance (and, under these circumstances, even the accordance between employers and employees is not guaranteed concerning the expectation of future equilibrium wage level). If that is the case, labour market will not necessarily clear, i.e. involuntary unemployment or, on the contrary, labour market excess demand may occur, which is, after all, equivalent to allowing false trade. Disequilibrium, for new classicals, is a short-run state that will be very quickly eliminated by agents forming rational expectations. The situation could be even simpler if there was an accommodating period—i.e. a searching process in the course of which parties would try to set the market clearing wage. In this case, the story would continue after the aggregate wage bargaining in a manner that the negotiating parties, setting the wages for the next period that are expected to clear the markets, withdraw to observe real economic outcomes and (especially) whether their expectations prove to be true or not and whether labour market equilibrates or not. If they err, as it could happen because of the probability distribution of predictions, they modify the agreement and withdraw to observe again. This searching process would last until the desired labour market equilibrium emerges—and, in model terms, labour market excess demand and supply would fluctuate around the equilibrium state with a decreasing amplitude (here, it is necessary to highlight that recurrent noises may disturb this adjustment-searching process again and again, so this convergence is not necessarily steady). Unless the economy equilibrates, we have to talk about a disequilibrium state—*stable equilibrium in a stochastic environment cannot mean anything else but the effort and the ability to tend towards the equilibrium state; but not permanent and long-lasting equilibrium.* It is believed here that it is necessary to introduce (and recognise) these limitations. Precluding false trade in a real macroeconomic environment could hardly be accepted—that would mean that a general stoppage (i.e. a strike) would prevail in the given national economy until equilibrium wages are set. If that was the case, then a searching process itself would become impossible, since further modifications of wages are only possible through observing equilibrium-disequilibrium states. Compared to it, new classicals are much more rigorous. For them, equilibrium *actually* emerges in the period taken as relevant in the course of wage negotiations. It follows from regarding unemployment as strictly voluntary rather than from Begg's view mentioned above. It also has to be noted that this rigor is only needed in order that unemployment should

---

according to the theory, rationality of expectations implies the clearance of markets—and, accordingly, if equilibrium does not emerge, it is due to an inappropriate change in prices.

remain of voluntary nature for new classicals and real output should be at equilibrium level.

Insistence on to these concepts poses several problems. One should imagine a situation in which PFH is set aside and in which it is assumed that both employer- and employee-sides are able to give unbiased predictions of future outcomes. It is not too easy to assume! If there is not PFH, predictions of the parties, though unbiased, may differ. One should assume that, in such a case (i.e. a simple difference of opinion of the negotiating partners[33]), the wage (expected to clear the labour market) for the next period is the simple (or even some weighted) average of the two predictions. If this prediction is *true*, labour market will really clear (due to contingent effects). However, under stochastic conditions, it is not guaranteed that predictions turn out to be true. If we believe, as followers of REH do, that anticipatory wage settings will clear the market, then this equilibrium, thanks to the differences of predictions, never even emerges and the economy starts random and never-ending walking around the equilibrium—and, of course, equilibrium also fails if labour market negotiators (whoever they may be) are only capable of biased estimations. However, this cannot occur because of the exclusive nature of voluntary unemployment. Under stochastic conditions, we have to take even long-lasting involuntary unemployment into account (permanency only reflects contingent effects). Setting equilibrium in advance is only possible under PFH and deterministic conditions—otherwise we would face further logical difficulties. In a stochastic environment, complete and perfect knowledge of the future can be achieved by waiting for the future to actually come. Without PFH, wages that *will* equilibrate labour market *in the future* could be continuously set and with absolute certainty if chronological and model time were allowed to separate: i.e. if negotiating partners (and, since the objective functions of both aim at equilibrating the market, this task can even be completed by one of them) were allowed to *get to know* the future (and this knowledge is *not* anticipatory, i.e. *not* based on expectations) the real time demand of which, however, would equal zero and to set equilibrium wages in the knowledge of future outcomes. Over and above the fact that this assumption is absurd, it means introducing PFH—that is, no progress is made. The assumption about perfect foresight is necessary for precluding involuntary unemployment by anticipatory wage negotiations. If we do not want it, involuntary unemployment is to be reckoned with in the story.

Introducing the auctioneer as a trade union leader does not solve the problem, it only places its root somewhere else. This logical manoeuvre can be presented through an example of natural science. Physics and astronomy have believed for a long time that the universe is endless—however, this is not a satisfactory (rather a troubling) answer for human intellect. Accordingly, some recent research tries to overwrite this piece of our knowledge. Relying on the theory of Big Bang dating back to approx. 13.7–14.5 billion years, physicists infer that we, human beings, can

[33] Caused, for example, by the fact that there are contingent differences in the data sets of an explanatory variable of the relevant model.

sight the borders of the universe if we strain our eyes a bit (and this effort is manifested in building higher- and higher-powered telescopes). We may think that, now, human intellect can free from the burden of facing the concept and realization of infinity, since it may be discovered that the universe is not infinite. However, those who are optimistic do not mention that these expected discoveries result in a further question instead of a problem seemingly answered. *What lies beyond the universe?* Nothing has been solved, we make no progress even after our discoveries said to be revolutionary.[34] Answering the question just places the problem somewhere else. If trade unions endowed with the necessary abilities and intentions exist, we do not need to be concerned with flexible (or inflexible) wages or the possible rigidities following from the labour contracts, since the objective function of the trade union (and of capital) supports setting the equilibrium—and the same outcome, assuming unorganized workers, can only be achieved by the debatable *as-if*-reasoning. By involving trade unions, new classicals tried to approximate their models to directly experienced reality. In this case, there is no need to apply the *as-if*-reasoning: in a world lacking trade unions with appropriate objective functions, workers left to themselves could only attain the setting of the equilibrium real wage in the course of negotiations with cooperating employers in one step if they were assumed to be ready for and capable of it *per definitionem* (this is the essence of the *as-if*-reasoning). If we disregard this, wages could only converge towards their equilibrium level during a disequilibrium period. In the latter case, of course, flexibility of wages (i.e. a negligible requirement if trade unions are present[35]) and allowing false trade are also necessary. All this is not required in case of appropriately defined trade unions—though there are other problems to be reckoned with.

Weeks (1989), for example, remarks that if both employee- and employer-sides have a stake in setting the equilibrium wage, then, eventually, it is nonsense to talk about wage negotiations: this "negotiation" may even be unilateral, i.e. one of the "bargaining" parties establishes the equilibrium wage that will naturally satisfy the

---

[34] Of course, there are researchers who apprehend this conflict and try to by-pass it. For them (e.g. Hawking 1988), universe is self-contained that can be described as the finite universe borders on itself (using an illustrative example, a ray of light stepping out of the universe at one side immediately returns on the opposite side–of course, it is easier to describe than to imagine-visualize it). It is still a question whether this concept can be more easily digested than the concept of the infinite universe.

[35] Since, instead of the flexibility of prices, trade unions guarantee that market clears through setting wages (i.e. prices). It is argued here that the auctioneer (particularly if the auctioneer appears as a trade union leader) and perfect flexibility of prices (wages) are not analogous concepts, since the auctioneer deliberately *sets* prices, so these prices are not allowed to freely *evolve*. As it will be seen below, the auctioneer in the Walrasian setting *substituted for* the market mechanisms, but the outcome of his activity is not the result of the invisible hand but of voluntary actions. Equilibrium in the Walrasian world emerges because the auctioneer has a stake in this equilibrium–and the same was the task of trade union and of employer-side in the new classical system.

needs of the other side. It is not of marginal importance either, as we have seen, that the anticipatory setting of the equilibrium wage is possible under PFH (of course, we would also have to carry its burden in the story of the employees mimicking trade union). Introducing trade unions implies that predictions of workers are replaced by the trade union estimation—to highlight the two opposing extremities, we have only one prediction instead of innumerable estimations. If there are countless unbiased predictions for a future outcome, the true numerical value stands in the centre of the probability distribution of the predictions. However, it does not follow that the trade union (or employer-side) estimation turns out to be true (since it is one of the innumerable predictions, all of them is burdened with random errors). Setting the equilibrium wage in advance and clearing the labour market are only possible under PFH (either with or without trade unions).[36] The difference between REH and PFH may be grasped here the most expressively. *Forming rational expectations means expecting an outcome that will happen with the highest probability—on the contrary, perfect foresight means expecting an outcome that will actually happen.* As far as their purposes are concerned, new classicals tried to accept and propagate the first hypothesis—however, the mechanism of the labour market assumed by them revealed that they left little room for the stochastic developments of the world of REH. Although, in principle, one can err, everyone becomes infallible in the new classical world (since markets are in equilibrium, with the exception of some unimportant episodes), and this is the implicit assumption of PFH. It seems that the trade unions augmented model did not become more realistic—one could only say that this version of the model is closer to directly experienced reality if one were ready to admit that a dragon with two heads is less fictitious than the seven-headed one is.

[36] It is interesting how Weeks (1989) formulates: "Walrasian market clearing requires the assumption of perfect foresight (to avoid False Trading), either in the form of the PFH itself or that same assumption embodied in a mythical auctioneer". Indeed, continuous market clearing requires perfect foresight—however, the auctioneer is not mythical anymore. Auctioneer appearing as a trade union leader is as real as employees or employers are. Auctioneer is really a fictive (abstract) actor in the original (Walrasian) concept of general equilibrium economics and his activity really *substitutes for* market mechanisms. A market itself does not do anything, equilibrium is not the result of markets—only the auctioneer is responsible for it. However, as new classical macroeconomics made the auctioneer be an actual actor, we cannot say it anymore that the activity of the auctioneer substitutes for the perfect foresight of market agents. Trade union of REH does not *substitute for* perfect foresight of agents but *realizes* (and, hence, replaces) that. In brief, we cannot state that perfect foresight of agents is not necessary for labour market clearance if, simultaneously, we refer to the perfect foresight of one of them. It seems that new classical macroeconomics tried to approximate its theory to reality by reducing the issue of perfect foresight to a numerical problem, saying that the perfect foresight of two agents is more plausible than that of millions.

3.4 Neoclassical Macromodels and Equilibrium

In the preceding paragraphs, major new classical theses on the labour market were reviewed. Now, some attention will be paid to the neoclassical traditions revived in new classical macroeconomics in order that we can understand the theoretical foundations (and can check whether they are correct or not) on the basis of which new classicals talk about general equilibrium and continuous market clearing. In this context, the discussion on labour market was far from unnecessary, since the idea of general equilibrium strongly relies on the view formed on labour market.

A critic of new classical and, generally speaking, monetarist economics is in trouble. He faces concepts such as Walrasian equilibrium that are rarely cited explicitly by monetarists[37] and that are rooted in neoclassical theory. It is not an overstatement if one says that the critique of monetarism is equivalent to the critique of neoclassical economics. Labelling new classical macroeconomics as "monetarist" naturally follows from the fact that Lucas and his group accepted the theses of flexible prices, market clearance, neutrality of money or of natural rate (cf. Mátyás 1984). Moreover, monetarism itself is linked by innumerable lines to the pre-Keynesian orthodoxy. It is enough to mention the presumption of equilibrium (cf. Vígh 1997) or the inevitable acceptance of Walras' law, which follows from the application of the technique of simultaneous (i.e. general equilibrium) analysis. So now, we can direct our attention to neoclassical models.

Below, the theoretical-logical difficulties of neoclassical models will be demonstrated through some textbook examples. In this context, it has to be highlighted that *our charges are not against these textbook models—they are merely well-known examples*. So, what are the most characterizing features of the macromodels rooted in neoclassical theory? The *single-commodity character*, for example. Modern econometrics represents this approach, even if economists treat this fact briefly and talk about models specified for real GDP instead of single-commodity economies. A further characteristic feature is the assumption and handling of *two factors of production*. In the methodological literature, there are rare examples that regard energy or natural resources in addition to capital and labour as explicit explanatory variables—the examinations of the effects of these factors are not completely carried out yet. [→ **ecological economics**] The functional form is usually of Cobb-Douglas style, though CES functions (with constant elasticity of substitution) were also introduced in a significant part of the analyses. "Properly sophisticated" models somehow based on the Phillips curve apply this theory that is well explained in basic textbooks. These features characterize both textbook and high theory models. *Therefore a critique presented through textbook models is justifiable*. This critique would only be *wrong* if such logical-content elements were also included in these models simplified because of didactical considerations that are

[37] For one of the few examples, see Friedman (1968). However, Frank Hahn (1980) mentions that the Walrasian framework cannot capture the market failures that economists have been interested in from the middle of the twentieth century.

excluded from high theory models (however, as far as our topic is considered, such elements are not known to us). If these textbook models were not criticized, then, *reductio ad absurdum*, we should also accept the statement that if we construct a *wrong* high theory model simplified later, then we can eventually provide a model as a result that could be vindicated from a certain point of view—that, at least, should not be criticized exactly because of the applied simplifications. It seems that this reasoning both theoretically and didactically would take to the wrong direction. If a high theory construct is wrong, then its simplifications will be wrong as well—however, in the meantime, we should not forget that even these logically wrong models can produce predictions consistent with reality. Logical order is not dependent on the number of equations. To take a representative example: the most simplified, textbook-level versions of IS-LM model could also be criticized, *if* this criticism is made against the replacement of the Keynesian causal relations by the simultaneous technique—since this feature characterizes both textbook and high theory variants of the IS-LM model.

Firstly, let us consider a single-commodity barter model. There is a

$$Q = K^d \times L^{1-d}, \text{ where } 0 < d < 1 \text{ and } K, L > 0 \tag{3.1}$$

Cobb-Douglas aggregate output function underlying the model ensuring both constant returns to scale and diminishing marginal productivity.[38] One of the first caveats in textbook exercises is keeping capital fixed at a given level $\overline{K}$ that serves, in addition to solvability, the didactical purpose that students should unambiguously distinguish growth models allowing capital changes from business cycle models precluding these changes and only concentrating on short-run real economic fluctuations (if such fluctuations are built in a given model). Equation

$$w = Q_L^{'} \tag{3.2}$$

describes the formation of real wage that is, according to the assumption borrowed from microeconomics, the marginal product of labour (which is, formally, the first partial derivative of the aggregate production function with respect to labour). In the present context, real wage is

[38] Our example is a first-order homogenous equation that is formally expressed by the sum of the exponents of the independent variables. In the case of this functional form, multiplying the inputs by k also multiplies output by k. So, the order of homogeneity characterizes the quality of returns to scale at the same time (Zalai 2000). It is exactly the feature that is highlighted by constant returns to scale—moreover, a simple linear production function could also ensure this feature. Diminishing marginal productivity justifies this Cobb-Douglas form, according to which the gradual increasing of one input while keeping the other fixed regressively increases the output. Linear functions are not able to ensure this.

$$w = (1 - d) \times \left(\frac{K}{L}\right)^{d}, \tag{3.3}$$

that is, keeping *K* fixed, a *decreasing* function of employment—and that also gives the demand for labour (since employers demand less and less labour as real wage increases). According to equation

$$l_S = g(w), \text{where } g_w^{'} > 0 \tag{3.4}$$

labour supply is an increasing function of wage. As the phrasing reveals, an equilibrium level of employment and, hence, an equilibrium real wage emerge that characterize labour market—both are determined as the coordinates of the intersection of labour demand and supply functions in the two-dimensional space of wage and employment.[39] Equation

$$r = Q_K^{'} \tag{3.5}$$

describes (real) interest rate which similarly to wage, is the first partial derivative of the aggregate production function with respect to the appropriate variable. Here:

$$r = d\left(\frac{L}{\overline{\overline{K}}}\right)^{1-d}. \tag{3.6}$$

Equality

$$Q = r\overline{K} + wL \tag{3.7}$$

is the "adding-up" equation that, keeping the above in mind, states nothing extraordinary but income is equal to the sum of products of utilized inputs and factor incomes. However, Euler's Theorem automatically guarantees holding of the adding-up equation, since it is assumed, on the one hand, that the (unit) income of labour and capital is their marginal products and, on the other hand, the first-order homogeneity of the production function (cf. Sydsæter and Hammond 1995; Kopányi 2003). Saving and investment are described by equations

[39] Authors do not stress that the neoclassical concept of labour market lies as far from reality as the two-input production function does. Although Patinkin (1989) remarks that, according to the common assumptions, there is a homogenous (composite—i.e. one kind of labour service) good present with a single price in the labour market, however, he keeps quiet about the obvious fact that it follows from the nature of the aggregate production function that only uses one kind of labour in the manufacturing of the only product. The neoclassical concept of the market for labour services is none less abstract than the production function.

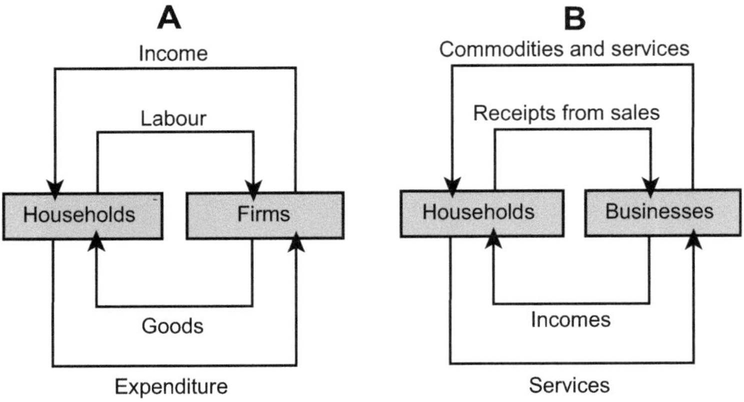

Fig. 3.1 Two concepts of flow of macroeconomic incomes (**a**) Mankiw 1997; (**b**) Branson 1972)

$$s = MPS \times Q + ar \text{ and} \quad (3.8)$$

$$i = AQ - br, \quad (3.9)$$

where $MPS, a, A, b > 0$. So, saving and investment are functions of interest rate and output.

It is obvious that aggregated production function does not take material inputs of production into consideration, so, it seems that there is no need for raw materials in the production process of outputs (and, in terms of the production function, this is not mere appearance)—hence, aggregate production function only describes the value added (generated in production). [→ **ecological economics**] In this context, it seems to be typical how certain textbooks on macroeconomics conceive the flow of incomes. We have two examples in Fig. 3.1.

These concepts are very similar—the differences in the names of items are negligible. Part *A* of Fig. 3.1. is from the introductory section of a modern American textbook, while part *B* appeared in another one some generations earlier. Part *A* was outlined assuming that only one input, i.e. labour, is needed in production, since this is the only factor flow that points towards firms. Processes can be started at any point. Income gained in return for labour services is the only earnings of households, so this is the only resource that can be allocated to cover expenditure (and it must be fully spent, since we cannot talk about savings in the story). Therefore, household takings and expenditure are equal. And it is also necessarily true for firms since we are talking about a circular flow; i.e. what is paid by firms as salaries that appears in households as receipts, to fully pay for household consumption. Eventually, business expenditure and revenues are also equal. So, firms do not make profit—however, it is more important that there is no sense in suggesting a demand failure as a possibility: unsold goods cannot emerge in this circulation. The reason for this is that we are talking about the manufacturing of *products* instead of *commodity*-production—i.e. a commercial process is precluded from the concept (Weeks 1989). Products manufactured are the incomes of employees at the same time that they directly receive as their special income transfer. *The special characteristic feature of circular flow of income described above guarantees that all the products are sold.* Indeed, it is no use keeping the lower and upper loop of the figure separated, since there is no money built in the system, i.e. we cannot talk about consumption expenditure (on the part of the households) and paid salaries (on the part of the firms). Employees, in return for worked hours, receive the total quantity of the single product[40] as income, that is, production and the realization of incomes are not separated.

Ecological economics. In certain cases, the relationship of neoclassical paradigm to natural environment is fiercely criticized by economists sometimes. This relationship is adequately described by the concept of weak environmental sustainability (in other words: the neoclassical concept of sustainability). According to it, societies should endeavour to maintain the total amount of capital available. It is evident even at first sight that this concept of sustainability assumes different kinds of capital (produced, natural and human capital) to be interchangeable. Therefore, destruction of natural capital is not forbidden as long as the depletion of this type is compensated by, for example, produced capital. We are allowed to clean all the forests if, in their place, factories, schools and airports will be built.[41]

Weak environmental sustainability directly follows from applying Cobb-Douglas type production functions. Neoclassical theory identifies sustainable development with a constant output/consumption path[42] (Solow 1974, 1986; Hartwick 1978), and if per capita macroeconomic output (q) is described through a general

$$q = k^{\alpha} l^{\beta} n^{1-(\alpha+\beta)}$$

Cobb–Douglas function, in which k is per capita produced capital, l is per capita labour

[40] John Weeks (1989) gives a comprehensive analysis of the single-commodity character of the model discussed here and of neoclassical macromodels generally. He underpins his view, firstly, by aggregation problems and, secondly, by the fact that, in a multi-commodity case, there is no unique relation at an aggregate level between inputs and outputs. The fact that new classical macroeconomics regards natural rate of unemployment and natural level of output as synonymous concepts is revealed by Shaw (1984), but Begg (1982) makes it clear as well. Mankiw (1997) also uses the well-known two-factor aggregate production function, not mentioning the problems. Moreover, Mankiw gives a slight twist to the formulization of his income flow model with respect to one product (which is correct), but, later, this restrictive characteristic is omitted when he introduces the aggregate production function (or, at least, it remains unveiled, since sometimes he talks about GDP and sometimes a bakery specialized in the production of only one commodity).

[41] This interchangeability holds for the various sustainability indices as well, when scores achieved along different sustainability dimensions compensate each other. It is also characteristic for HDI (Human Development Index) computed on the basis of indices of life expectancy, knowledge and education, and standard of living (Kerekes 2007). It is similarly true for Genuine Saving emerged as an alternative approach that also tries to capture investments in (and depletion of) environmental and human capital through only a single index—so, possibility of compensations is still given (cf. Hamilton 2000). Adjusted Net Savings is interesting for the same reason: this measure is explicitly built on the concept of weak sustainability making different kinds of capital interchangeable (Bolt et al. 2002).

[42] Its ultimate foundation is the circumstance that welfare is estimated by consumption in this framework. The more the agents consume (produce), the higher the welfare is.

input and n is per capita natural capital (and where $0 < \alpha < 1$ and $0 < \beta < 1$ in a way that $1 - (\alpha + \beta) > 0$), then maintaining q at a given constant level does not encounter difficulties as long as the decrease in n (where n may tend to even zero) is offset by, for example, a corresponding increase in k.[43]

This aversion to the neoclassical paradigm resulted in ecological economics which starts from the finite character of natural environment and tries to fit human economic activities into nature taking these constraints into consideration. The emphasis (sometimes deliberately radical) of ecological aspect cannot be placed on the neoclassical concept of sustainability, so, in this new notion, preserving natural capital is the declared purpose. The neoclassical interchangeability between natural and produced capital is not assumed here and more and more considerable emphasis was placed on natural services (not only the absorption capacity but also, for example, the recreational function of nature). Replacing natural capital by produced capital has been criticized on both moral and technological grounds. The economic and social goal of (plain quantitative) growth has been questioned (e.g. Daly 1972, 1974) and "sustainable growth" deteriorated to be a simple oxymoron.

The situation has not significantly changed, if, as in part B, physical (i.e. productive) capital is also needed for production (we still do not need to take raw materials, semi-finished goods and other inventories into consideration, since our production function excluded them) but the consequences are even more interesting. Since revenues and salary payments of firms are necessarily equal, firms still do not make profit—i.e. they do not retain any proportion of revenues. It follows that firms are *not* market participants that posses capital assets (however, this objection can be refused by referring to firms as, eventually, privately owned ones).

The lack of demand failure is not ceased by introducing savings either, since the definition of the system unambiguously determines the handling of investment and savings. A possible (but lacking) disequilibrium situation would be the case when aggregate demand and aggregate supply fail to equilibrate—however, this possibility cannot be realized because of the way the equation system is formulized. So, when aggregate demand and supply are equal, it is also true that $i = s$ (and it cannot be overstated that the single commodity serves the purposes of consumption, saving and investment). Because of the characteristics of the system, total income (i.e. aggregate demand) and total quantity of products (i.e. the aggregate supply) are *equal*—and, which is a stronger statement, they are *identical*.

[43] The relation of neoclassical growth theory to natural environment is well represented by the following rows: "An economic policy that makes higher rates of economic growth achievable is not necessarily successful. This is also highlighted nowadays by worsening environmental damage. [...] The attitude of economic policy towards economic development cannot remain intact. This is taken seriously by the author writing these lines. Now, however, he has to deal with the problem, following from the central topic of the present publication, of the relation between economic policy and the growth rate; how the latter could be higher or lower, supported or hindered by an economic policy." (Erdős 2006) Interchangeability of production factors is discussed in a more general framework, on grounds of various production functions, by Prof. Zalai (2000).

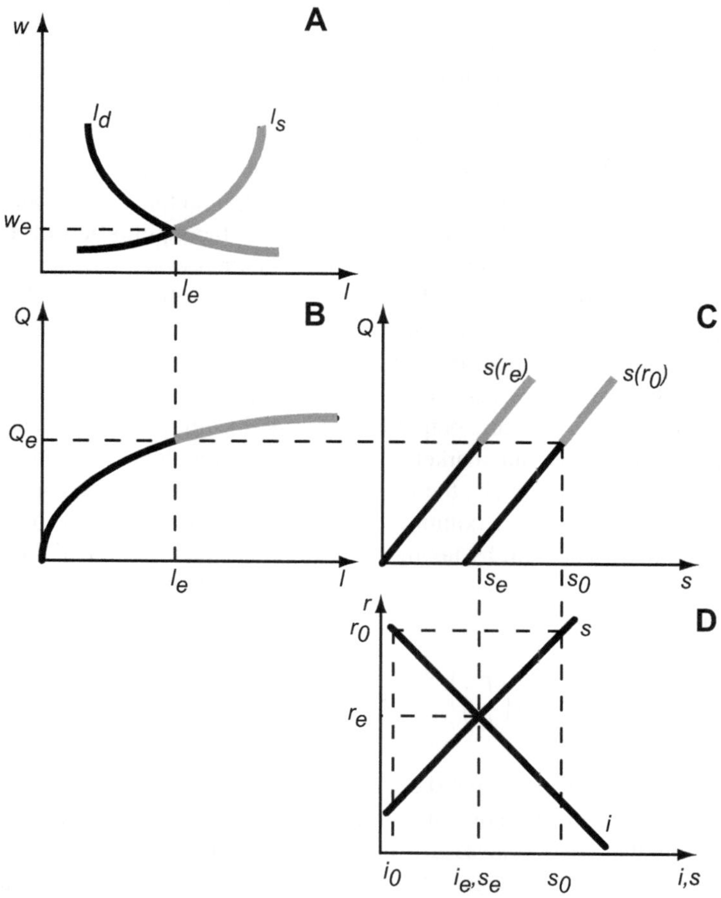

Fig. 3.2 Components and equilibrium of the single commodity barter model (Weeks 1989)

The lack of the demand failure eventually means that the selling of products is always guaranteed—so, it is undoubted that the entire system will move towards the state of full employment, since the higher the incomes are the higher the labour input is. Income level is dependent on the decision of households. Mechanism of the labour market, and that of the entire real system can be summed up by Fig. 3.2.

Figure 3.2 illustrates the mechanism of the labour market (*A*), the development of aggregate output (*B*) and of savings (*C*) and the relationship between savings and investment (*D*). The general equilibrium of macroeconomy is represented by the labour market, since this is the only market that is investigated in this model. Diminishing marginal product of labour (that makes the demand for labour a decreasing function of wage) and the increasing opportunity cost of working designate together the level of employment (l_e) and the level of real wage (w_e, expressed in quantities of the single commodity) that characterize the equilibrium of the market for labour services.[44] In addition, if the equilibrium of the labour market is given, the level of output-income (Q_e) appertaining to the equilibrium is also unambiguously given because of the fixed capital (and the single-commodity character). Moreover, the right of this income level, the output function cannot be interpreted at all (and, assuming flexible wages, the system cannot fix to the left of it permanently). Distinguishing *equilibrium output* and *output appertaining to equilibrium* is not a mere semantic play. The level of output in this model (but it is also true for the dichotomy model) is not established by the equilibrium of aggregate demand and supply—talking about an equilibrium output would only make sense in this case. The level of output is set by the equilibrium of labour market through the partial production function—so, the special and important numerical value of output appertains to the labour market equilibrium.[45]

The key element is the market for labour services: if wage was either higher or lower than its equilibrium rate, output (and, hence, income) would fall short of the level appertaining to labour market equilibrium. So, employees are interested in setting a wage that equilibrates the labour market, since it (through Euler's Theorem) ensures to achieve the maximum of income. In addition, it is easy to realize the validity of Euler's Theorem by taking the relation between "national" income and the level of employment into consideration. The adding-up equation can be reformulized as follows:

$$Q = d\left(\frac{L}{\overline{\overline{K}}}\right)^{1-d} \times \overline{K} + (1-d)\left(\frac{\overline{K}}{L}\right)^{d} \times L. \tag{3.10}$$

It is evident that national income (keeping *K* fixed and assuming the given production function) is the function of employment *only*. We also have to take the fact into consideration that, under these conditions, interest rate is also determined by employment; cf. Eq. (3.6). It should also be recognised that income is an increasing

[44] A consequence, being far from unimportant, also follows from it: employees are capable of enhancing employment through lowering the wage claims they required for additional working. This lowering is nothing else but the reconsideration of the claims for the compensations for increased opportunity costs of increased working: if employees are ready to accept lower wages for additional work, the level of employment becomes higher. This act of reconsideration, in formal terms, is symbolized by the moving of the labour supply function to downwards-right. Therefore unemployment is strictly voluntary. It is even truer since, eventually, the utilized quantities of both capital and labour are possessed by households (i.e. the level of employment is a decision variable controlled by them). Weeks (1989) is brilliant when stresses that there is no need to draw the labour demand and supply function the right from the equilibrium point (i.e. the intersection), since those sections are only graphical representations of quantities demanded and supplied indicated at given wage levels; but they have no role in determining the level of employment. Employment is set by labour demand at a wage that is higher than its equilibrium level and by supply at a wage lower than the equilibrium level.

[45] When, analysing the barter and the dichotomy model in this chapter, we talk about "equilibrium output/income" or "equilibrium level of output/income" for simplicity, we always refer to "the output/income appertaining to labour market equilibrium".

function of employment—but it is easy to see, since, after possible simplifications (exactly because Euler's Theorem holds), Eq. (3.10) is congruent to the aggregate output function. Additionally, it is true that, keeping K fixed, the partial (i.e. univariated) Cobb-Douglas output function is strictly monotonic increasing and its slope is given by exactly the marginal product function of labour. Although diminishing marginal product holds for the latter, that is,

$$\lim_{L\to\infty}(1-d)\left(\frac{K}{L}\right)^{d}=0, \tag{3.11}$$

an intuitive interpretation reveals that national income, in decreasing-eliminating steps of course, can be increased to infinity. This argument should not be forgotten if we want to understand how the new classical labour market works. *It is worth lowering wage claims because, doing so, employees can enhance (their) income for sure*, that is, higher-level employment, through flexibly responding labour supply, results in a higher income. The signs of this line of reasoning can be found in all of the following neoclassical models. Maximization of national income gained in the macroeconomic system can be implemented by setting employment to its equilibrium rate—for which, of course, a complete exclusion of demand failure is also needed.

What holds at an aggregate level that does not necessarily hold for individual market participants. Although we can underpin our reasoning by formulizing and solving elegant maximizing problems, our arguments seem to be powerless if we want to try to justify the behaviour of individual employees following *this* line of reasoning. We have seen that *it is needed to lower real wage in order to enhance employment and, through this, national income can be increased.* Let us consider the case of an imaginary employee who can maximize his utility by, for example, working 60 h a week. Enhancing national income contradicts the demands and intentions of individual workers, since *maintaining the previous level of income, given a lower real wage, is only possible by an increase in the amount of working hours*. Therefore, employees (to be precise: the actually employed) will do their best to fix market real wage above the equilibrium wage rate (since the employed could preserve their living standard, given a lower wage, only by excess working, and taking higher opportunity cost—moreover, keeping the wage rate above its equilibrium level is also in their elementary interest, since they can also enhance their utility through it). On the contrary, though the unemployed would be ready to take jobs even at the prevailing wage rate and though they would be ready to accept even lower wages, trade unions are good at preventing this external pressure from making disturbances in working conditions of the employed (seemingly, rather *this* is the real function of trade unions—though we have to admit the function one attributes to trade unions is always dependent on belief and word view). Now, it is quite clear why the notion of a fully unionised working class is important for new classicals. However, there is another option that makes the case of full employment underpinned by maximization acceptable as an assumption—this is, if the employees (i.e. economically active population) are regarded as monolith totality, *as a single entity*. In addition to the single commodity production, being deprived of material inputs, we face a further idea: economically active population is concentrated in a single body, in a single economic actor (and, hence, we do not need the comprehensive umbrella-attitude of trade unions anymore). [→ **trade union]**

Parts *C* and *D* of Fig. 3.2 is merely of marginal importance, since maximal output is guaranteed even without (the knowledge of) the investment function. It should be noted that the aggregate production function and equilibrium employment unambiguously establish the *interest rate appertaining to equilibrium*[46] (r_e) as well (which can be established directly on the basis of the level of employment or, depending on formulation, indirectly on equilibrium wage rate). Of course, it unambiguously determines the location of the savings function in part *C*.

Interest rate has no role in setting the aggregate output and national income, since equilibrium employment and maximal output are ensured through labour market. If interest rate was still higher than its equilibrium level (r_0), it would, on the one hand, move savings function to the right in part *C* (i.e. the saved portion of income would be higher at a given level of income) and, on the other hand, would lead to unsustainable (and *inexplicable*) disequilibrium between savings (s_0) and investment (i_0). In this case, interest rate should lower to its equilibrium rate as the only option that would set the amount of savings to its equilibrium level (i.e. savings s_e appertaining to output Q_e would emerge again) and, simultaneously, would restore the (necessary) equilibrium between savings and investment. *Flexible (i.e. flexibly developing) interest rate does not establish anything, its numerical level is unambiguously set by other elements of the system, hence, it is not able to equilibrate investment and savings, it is only able to indicate their equilibrium.* If parameters of equations were changed in the system, labour market equilibrium would still last (or emerge again), and a change in the rate of interest would only modify the distribution of income between consumption and savings (i.e. their ratio); or, the distribution of aggregate demand between consumption and investment.

Although the market for labour services unambiguously determines the interest rate appertaining to equilibrium, we should not talk about a simultaneous solution: finding the equilibrium can be still carried out by starting from the labour market. One can notice that part *D* of Fig. 3.2 depicts savings and investment by separate graphs, though these two acts (i.e. savings and investment) are *identical* in the income-expenditure model. The analysis solves this problem by assuming (*declaring*, literally) one part of the households to be borrowers and the other part to be lenders and the rate of interest to equilibrate their quantities (Weeks 1989). The role of interest rate is to divide income between consumption and savings, on the one

---

[46] It is argued here that distinguishing between *equilibrium rate of interest* and the *interest rate appertaining to equilibrium* is not a mere semantic play either. Interest rate is set *outside* the market for interest rate dependent variables (i.e. investment and savings), that is, it is a mere corollary of the equilibrium of another market.

hand, and to divide aggregate demand between consumption and investment, on the other hand: *savings* and *investment* are two different names of the same activity. What households do not consume that will be saved by them—but since they do it in physical quantities of the single product, they also perform the act of investment at the same time. In addition, the rate of interest is a source of considerable problems in the real model. If, for example, savings and investment are *not* equal at equilibrium level employment and output, the flexible adjustment of interest rate is not able to restore the (general) equilibrium since the level of the rate of interest is unambiguously set by employment according to Eq. (3.6). A given level of employment establishes a given level of the rate of interest; accordingly, savings and investment are either equal or not—but if not, the system cannot allow the equilibrium state to recover. Weeks (1989) argues that if savings exceed investment at the level of interest rate appertaining to the equilibrium employment, then a fall in the interest rate creates equilibrium. For him, its only consequence is that an increase in investment (which is the consequence of the fall in the interest rate) automatically triggers a decrease in consumption (and an increase in savings) since aggregate income cannot change as it is determined by the level of employment. If the starting point is the labour market, the interest rate cannot flexibly adjust, depending on the relation between savings and investments. Additionally, we cannot carry the adjusting process of the interest rate over the labour market (that is, we cannot allow a decreasing interest rate to result in a decreasing employment), since we have derived the state of the labour market from the wage, so we could not explain the emergence of a level of employment lower than the previous one. If we tried to introduce another interest rate that is different from the interest rate determined by employment in the system, the adding-up equation would also fail.

The difficulty with the interest rate follows from the fact that equations of the barter model refer to the rate of interest at two different places. As we may remember, the rate of interest is presumed to be the marginal product of capital. Well, the equilibrium level of interest rate is established through it—and it *literally* means that equilibrium of the savings and investment functions in part *D* of Fig. 3.2 is not capable of determining a level of the interest rate differing from it. However, it means that only such investment functions can be consistent with the savings-investment market modelled in part *D* that give an interest rate (at the intersection of the savings function) that is predetermined through the production function. If the equilibrium of savings and investment functions does not give the interest rate determined *by* and *as* the marginal product of capital, the entire system proves to be inconsistent. So, it should be noticed that this system tries to derive the *same* interest rate from two different sections—which is a source of possible inconsistencies. Firstly, already *existing* and utilized capital pays its owner according to *the* interest rate that corresponds with the marginal product of capital. Secondly, an interest rate also emerges through the equilibrium of investment *to produce new physical capital* and savings. These two interest rates should be equal, but this requirement is not necessarily met. It should be also noticed that, since the savings function in part *C* already divided the output into consumption and savings on the basis of the interest rate set by and as the marginal product of capital, and since

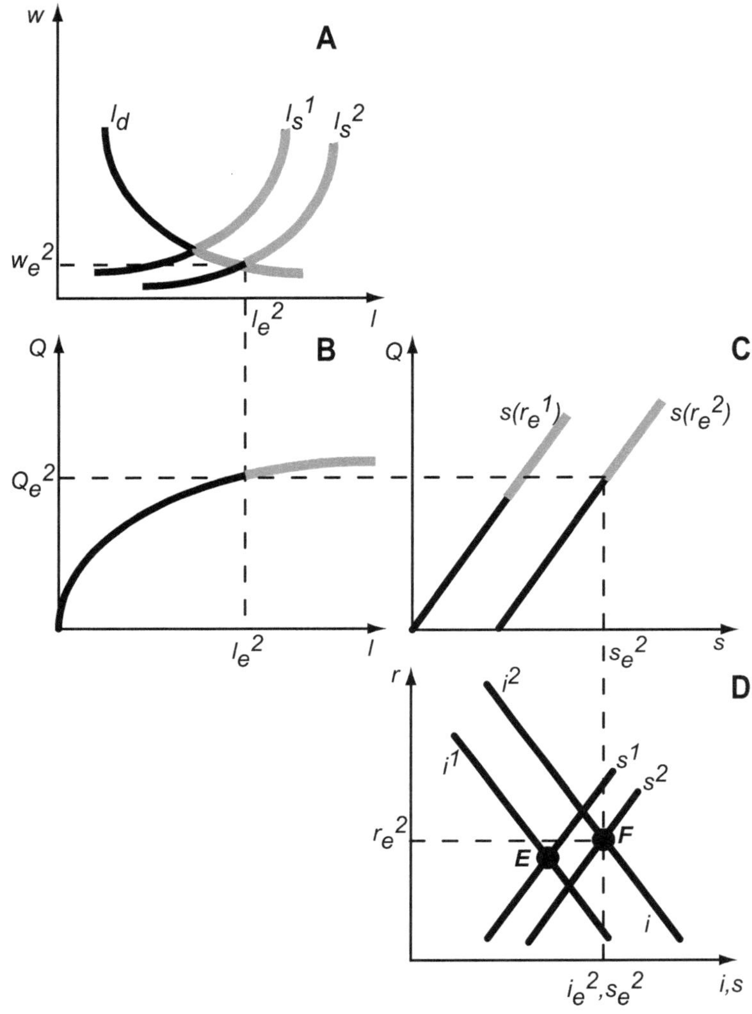

Fig. 3.3 Transition to a new equilibrium in the single commodity barter economy

savings and investment are necessarily equal, the equilibrium that can be investigated by part *D* is actually redundant and only introduces a possible inconsistency in the system. This reasoning will also be true for the pseudo-monetary dichotomy model: in that context, the inconsistency emerges as the contradiction between the interest rate set on the basis of the IS curve and the interest rate interpreted (or to be interpreted) as the marginal product of capital.

The secondary role of the rate of interest can be well demonstrated by a situation in which the emergence of a new labour market equilibrium is required. Such a situation occurs when, for example, employees reconsider the opportunity cost raised by additional working and lower the real wage at which they are still ready to work. In formal terms, this act is symbolized by the moving of their labour supply curve downward (from l_S^1 to l_S^2) that necessitates a new equilibrium. Effects of this rearranging process can be well illustrated in Fig. 3.3.

In the new labour market equilibrium (*A*), the level of employment increased (l_e^2) at a real wage lower than the previous one. As a result, taking the aggregate production function into consideration, national income (value added) also increased (Q_e^2 on part *B*). Let us recall the direct relationship between the level of employment and the interest rate: it explains that savings function in part *C* moved outwards ($s(r_e^2)$) establishing the new level of savings (s_e^2), since we drew the previous savings curve on the basis of the interest rate appertaining to the prior equilibrium. The new equilibrium rate of interest following from the employment and savings (and investment) appertaining to the new equilibrium income set together the new equilibrium point (*F*) in part *D*. However, the system in part *D* plays a passive role again in the evolution of equilibrium levels, since all of them were derived from the new labour market equilibrium. The only development in part *D* is that savings-investment system moved to the right upwards (curve i^1 to i^2 and curve s^1 to s^2). Equilibrium relocated from *E* to *F*, that was caused by assuming both investment and savings curves also to be the functions of national income (however, without it, the system cannot be consistent with the new equilibrium, since the equilibrium in part *D* would remain unchanged, though both employment and output increase). The equilibrium of savings and investment does not have any significance, since both employment and national income are independent of the interest rate, so there are no repercussions.

A characteristic feature should be highlighted, since it will be there when we try to transform this barter system into a money economy (by spreading a *veil* over it—a veil, that is *money*). Now, we have a straight relation between interest rate and aggregate output that defies common sense (we expect a reverse relationship). However, it is a consistent consequence of the model. Notwithstanding, we should bear in mind that the direction of the effect (i.e. the causal relation) does not point from the rate of interest towards output (this sort of relationship is precluded even as a possibility, since, as we have seen, a level of the interest rate that does *not* appertain to the equilibrium state only triggers its own adjustment), but exactly on the contrary: a higher interest rate is the *consequence* of the increase in income. There is, however, an underlying common factor behind the development of both (i.e. that of output and interest rate): this is employment. Therefore, the relation between the rate of interest and output is indirect. The necessary equality of savings and investment and the ineffectiveness of the changes in the rate of interest in the other parts of the system (excluding, of course, the ratio of consumption to savings/investment) make it very difficult to introduce the concept of an interest rate dependent aggregate output, that is, the IS curve.

Before continuing, it is worth summarizing our previous findings. Of course, we have not dealt with many issues yet, since only the macromodel rooted in neoclassical theory has been examined so far. The most important circumstance to highlight is the neoclassical idea of strictly voluntary unemployment. Setting the level of employment (and, through it, that of national income) only depends on how workers determine their labour supply, that is, what wage they require for additional working. Increasing employment definitely enhances national income until the relocated equilibrium point is reached—and this increase in employment is not constrained by a demand failure, since the acts of production, distribution of income and exchange are not separated. *The whole quantity of actually manufactured products are sold.* It is not necessary to say that the features of the model (single-commodity character, the lack of money, setting material inputs of production aside, disregarding transactions between firms) make it quite difficult to generalize the conclusions drawn from the mechanisms of this model. In any case, we have to bear in mind in the course of the following analyses that *full employment automatically emerged* and is permanently fixed in the frameworks as the model specifies them.[47] This model well illustrates the absurdities in (neo)classical way of thinking that were also highlighted by Keynes (1936). On the first pages of his *General theory*, Keynes states that (neo)classical thinking does not separate the theorem under which the income derived by all the elements in the community concerned in a productive activity has a value exactly equal to the (monetary) *value* of the output from that the one under which the costs of output are always covered by the sale-proceeds resulting from demand, well known as Say's Law. Simply saying, Say's Law states nothing else but demand and supply change together (i.e. simultaneously and correspondingly), since the ultimate purpose of all sales is purchasing and since each seller is a buyer at the same time (Mátyás 2002). Adding-up equation shows that the first statement naturally also holds in our single-commodity barter economy—characteristics of the model, however, ensure that Say's Law also holds, since the identity of production, distribution and exchange (and the lack of demand failure following from it) does not allow demand and supply to move differently. Additionally, the change in the rate of interest can only induce a change of the ratio of consumption and investment within demand—that was also regarded by Keynes (1936) as a speciality of the way of thinking based on Say's Law.

If money is introduced in the system above, the nature of equilibrium and the way it is set do not change, i.e. *we still have to talk about real equilibrium*. As a first step, we should believe that, here and now, we face nominal variables instead of real quantities. Labour demand function is still derived from the marginal product of labour (i.e. based on the aggregate production function), the monetary value of which, that is, the value of marginal product of labour fulfils its role. So, the equation of nominal wage (W) that underlies our new labour demand function is:

$$W = P(1 - d)\left(\frac{\overline{K}}{L}\right)^{d}, \tag{3.12}$$

which is, eventually, the real wage function of the previous model transformed into monetary value (multiplied by the price level P). Moreover, the way the

[47] If auctioneer was also introduced in these frameworks (that would be equivalent to precluding false trading), setting of the equilibrium would be totally *inevitable*.

equilibrium nominal wage was created can also be understood in formal terms if we transform the previous aggregate production function into monetary value:

$$Y = P\overline{K}^{d} L^{1-d}. \tag{3.13}$$

Weeks (1989) presents a model in which we have to assume labour supply to be exogenous, i.e. it is invariable with respect to wage ($\overline{L}$). It is unimportant in the analysis whether we start from a rigid or flexible labour supply. Consequences are only of technical nature, since this feature only determines the order of the polynomial that comes from the equation representing the equilibrium of labour market (and that is to be solved). Notwithstanding, in his textbook, Mankiw's (as another example) also assumes a rigid labour supply (Mankiw 1997). Since the equations describing labour market behaviour are functions of the real wage (only the labour supply may be exogenous, i.e. real wage independent), *introduction of money does not (cannot) affect the mechanisms of the real system, equilibrium emerges independently of (the quantity of) money.*[48] As we will see, its explanation is simple: since (the quantity of) money that has to establish prices is let in an economy that is defined on the basis of real variables and that equilibrates (i.e. that is equilibrated) on their grounds, a change in the quantity of money induces a proportional change (i.e. a change that is corresponding to the change in the quantity of money) in prices. *All* the prices *have* to change in a way that real quantities (and, since equilibrium depends on labour market developments, it is the real wage that should be thought of first of all) necessary for the equilibrium cannot assume non-equilibrium numerical values—since, if it was not the case, labour market developments would not depend on real wage ultimately. Equilibrium of labour market interpreted in terms of the real wage emerges, then a price level following from the quantity of money is *assigned* (sic!) to this equilibrium, so nominal wage can be also computed, *afterwards. Real equilibrium is primary compared to price level.* Since prices (and, hence, nominal wages as well) are flexible (or the auctioneer is handy in substituting for this feature at least), this change does not come up against a technical difficulty. Since the price variable in labour market is the nominal wage (W), that, itself, also is a price, a change in prices induced by the quantity of money generates a change of exactly the same proportion

[48] This, i.e. a definite separation of nominal and real variables of a macroeconomic model, is exactly the essence of the classical dichotomy. This intention culminates in the theorem of the neutrality of money, according to which the real equilibrium, after a change in the money supply, re-emerges keeping real variables unchanged. If it holds, the roughest form of the quantity theory of money also holds that is devoted to demonstrate that a change in the money supply (quantity of money) only affects the price level (if velocity is presumed to be constant which is not quite an extreme assumption with respect to the short-run), since real economy is fixed in the state of full employment and the (equilibrium) output appertaining to it. However, if money does not matter, the modelled system is still a barter economy, in which absolute levels of (nominal, monetary, etc.) prices are unimportant, only their ratios matter. That is exactly why it is not right if this economy (that already incorporates money) to be analysed below is referred to as a *money economy*.

in nominal wages (since all prices equally alter and immediately). So a change in the price level does not affect the size of real wage, since nominal wages alter together with and to the same extent as other prices. If labour market is equilibrated by the emergence of the proper real wage, this real wage will remain unaffected by the changes in the price level, since all the prices are only equally allowed to change (however, it should be noticed that we can only talk about two prices: the nominal wage in the labour market and the price of the single commodity).

Dichotomy model investigated by Weeks (1989) tries to avoid difficulties by a clever trick. We can remember that, in the barter model, we specified the way both production factors are rewarded, i.e. both capital and labour are assumed to be paid in accordance with their own marginal products. It is not the case in the dichotomy model. As far as labour is concerned, this stipulation still holds (the only modification is that it is the value of marginal product of labour we are talking about here—though it is not too hard to realize that the real wage defined as the quotient of nominal wage and the price level is still equal to the marginal product of labour), but the rate of interest has to be derived from somewhere else. The rule available that can be supported by arguments from microeconomics was disregarded in order to introduce a new function, i.e. IS curve. The purpose of neoclassical economists was to prove that national income flexibly responds to changes in the rate of interest—but this intention could not be supported by the barter model, since, as we have seen, the flexible interest rate had no role either in the labour market equilibrium or in setting aggregate output. Here, the only problem is that the markets (functional relations) establishing the equilibrium of the dichotomy model still lack explicit reference to the rate of interest. The system incorporates the following three equations in order to prove the interest rate dependency of the national income:

$$c = c^* + MPC \times Q, \text{where } c^*, MPC > 0; \tag{3.14}$$

$$i = i^* - g \times r, \text{where } g > 0; \tag{3.15}$$

$$s = Q - \left(c^* + MPC \times Q\right). \tag{3.16}$$

Equations (3.14–3.16) are the functions of consumption, investment and savings (in real terms), respectively. It is worth stressing that savings are not interest rate dependent, which is a very serious modification for the barter model (c^* and i^* are the autonomous consumption and investment, respectively, and g is the parameter that describes the interest rate dependency of investment). After all this, deriving the IS curve is simple, since such points are to be found in which the commodity market is in equilibrium (i.e. $c + i = c + s$ holds true), i.e. savings and investment are precisely equal. So, IS curve is:

$$Q = \frac{\left(i^* + c^*\right) - gr}{1 - MPC}. \tag{3.17}$$

Now, we have the inverse functional relation between interest rate and national income. Nothing could indicate the unimportance of this function better than the fact that the setting of equilibrium still starts from the labour market, from which the equilibrium employment (determining aggregate real output) also enters the IS equation (of course, a rearrangement is needed) and provides a certain level of the real rate of interest as a result.

The arbitrary character of the equation of IS curve (meaning the stipulation according to which, firstly, consumption/savings are independent of the rate of interest and, secondly, investment is independent of national income) cannot be stressed too much. As we have seen, for example, a straight relation between the interest rate and income would follow from the equality of the savings and investment equations of our single commodity barter model,

$$Q = \frac{a + b}{A - MPS} \times r, \tag{3.18}$$

(if $i = A\bar{Q} - br$ and $s = MPS \times Q + ar$, where all the previous conditions for the parameters still hold), that determines the shift $\overline{EF}$ of Fig. 3.3 (an increased employment, an increased income, a higher rate of interest and increasing savings). However, it should be remembered that *the direction of the causal relation in this latter model is the same as it was in the barter model*: it is not the interest rate that determines the equilibrium level of income, but a certain numerical value of the real interest rate is *assigned* to the equilibrium level of income, based on the parameters of the system, that can be regarded as "equilibrium" (real rate of interest) with some distortion.[49]

However, the way IS curve is derived is a source of further problems. The interest rate established on the basis of the IS curve is not necessarily equal to the marginal product of capital (on the basis of which the rate of interest was interpreted in the barter model). Since the reward for capital inputs is not presumed to be the marginal product of capital (or its nominal value), the adding-up equation fails, which means intuitively that the monetary value of aggregate output (total value added) and total nominal income of agents (and real values of these variables) are not necessarily equal, not even in an equilibrium state. This inconsistency can only be resolved in one way: interest rate cannot be the reward for unit exploitation of capital anymore. So, the rate of interest is *something else*—but we do not learn exactly what it is. Of course, this characteristic feature is not emphasized by neoclassical theorists and they tacitly assume total real income realized in the economy to be equal to output.

The simplest section of the analysis is the transformation (or rather pretended transformation) of the system into a money economy through the quantity equation.

[49] So, what was said on the distinction between *equilibrium rate of interest* and the *rate of interest appertaining to equilibrium* when analysing our previous model, also holds here. However, since aggregate output also depends on the equilibrium of labour market (its level is to be derived from equilibrium employment), similarly, it is worth talking about the output *appertaining to* (labour market) *equilibrium* (or, in other words: *maximal* output).

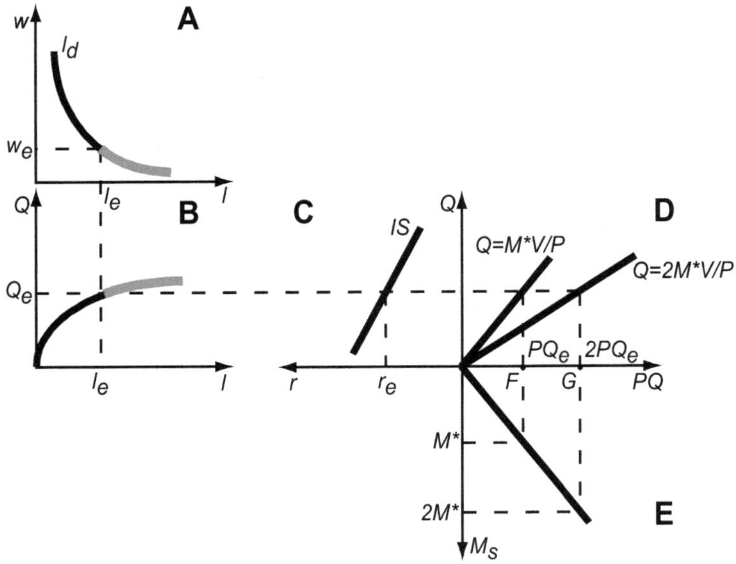

Fig. 3.4 Equilibrium in the dichotomy model (Weeks 1989)

If, following the spirit of this equation (where M^* is the exogenous money supply and V is the velocity, assumed to be constant), we state that

$$M^* \times V = P \times Q, \tag{3.19}$$

then the theorem of setting the price level through the exogenous money supply evidently follows from the stable (maximal) output appertaining to equilibrium employment and the velocity of money presumed to be constant. The ultimate goal of the quantity equation is to build financial market into the system, since it is written as

$$M^* = kPQ, \tag{3.20}$$

where $k = 1/V$ (in other terms: k indicates the portion of nominal income the public chooses to hold in cash balances), then it becomes clear that the right side of the equation describes the money demand for transaction purposes. Nominal values of real variables (wage, rate of interest, aggregate output) are given by a correction with the price level (simple multiplication). The line of reasoning of the analysis can easily be comprehended by Fig. 3.4, disregarding the assumption of a flexible or inflexible labour supply.

Parts *A* and *B* are already known, equilibrium of real economy only depends on labour market adjustment (the "realness" of the equilibrium is so true that part *A* gives the level of employment in the function of the real wage). Part *C* depicts the objected IS curve. What was said previously becomes quite obvious here: the interest rate adjusting flexibly has no role in the setting of equilibrium employment and equilibrium output.

Applying the real wage in the model also has technical reasons. It is true that, given a certain quantity of money, only one numerical value of the nominal wage level is compatible with full employment, but, in the state of labour market equilibrium, an increase in the quantity of money has a proportional effect on the nominal wage, keeping the real wage unchanged. If part *A* was specified in terms of the nominal wage, the system should have a feedback from the monetary sphere depicted in parts *D* and *E*, taking those details into consideration that were said on labour market equilibrium, or real and nominal wage. Equilibrium can be found in the dichotomy model exactly in the same way as in barter economy: at a real wage above its equilibrium level, the reduction in this wage to the equilibrium level through moderating wage claims depends on employees. If nominal wage reduces, it enhances employment and, hence, output that results in a decrease in price level (of course, it is very hard to explain outside this model why an *increase* in output is followed by a *drop* in price level). However, real wage must decrease because of the diminishing marginal product of labour. Weeks (1989), in connection with the irregularities of exactly this mechanism, stresses that claiming a nominal wage higher than its equilibrium value and the resulting decrease in employment trigger an increase in real wage—a development that is hard to interpret considering life world mechanisms, since we expect that wages above their equilibrium value generate movements in the price level that negatively affect real wages.

Parts *D* and *E* of Fig. 3.4 incorporate the monetary side of the model. According to part *E*, the higher the monetary volume of the output is, the higher the quantity of money allowed into the economy is (the higher the exogenous money supply M^* is). That is, a given output (appertaining to full employment) has a higher numerical value of *PQ*. Of course, it is also true that the monetary value of output must be unchanged at a given money supply and velocity: if output decreased from its equilibrium level, then, because of the quantity equation, price level should increase—the strangeness of which is not to be highlighted here. Part *D* contains the same relationship, with reference to the connection between real output and its monetary value. Neutrality of money and classical dichotomy can be easily presented: increasing the money supply from M^* to $2M^*$ only induces changes in segments of the model economy depicted in parts *D* and *E* (price level doubles—see movement $\overline{FG}$), while *there is no modification in any real* in the meantime. As far as the labour market is concerned, neither employment nor real wage change (the only adjustment is that, evidently, nominal wage, as a price, doubles). The whole aversion of the neoclassical paradigm to government interventions and fiscal policy boosts rooted in the assumption of a permanent equilibrium of the labour market. If the economy is already in full employment equilibrium, how can a fiscal expansion improve this state? Equilibrium and full employment have already been achieved—this optimal situation cannot be further amended by government interventions; and, hence, it is not necessary either. Monetary policy is treated with similar scepticism.

Increasing the quantity of money does not affect the equilibrium of the real system, there is only inflation—therefore, any intervention is also superfluous and detrimental in this sphere.

Concerning this model, the literature primarily highlights the logical inconsistency that occurs between Walras' Law and the quantity equation. Don Patinkin has shown that *quantity equation cannot be applied to a real economic system in which Walras' Law holds without a necessary and unavoidable burden of a logical inconsistency*. Walras' Law implies that the sum of the excess demands are zero in the system from which it follows that excess supply of money and the excess demand for the commodities (products) circulating in the economy are necessarily equal. Using simple mathematical apparatus, Patinkin has proved that the excess demand for commodities and the excess supply of money change to the same extent due to the influence of the change in the price level, while this mechanism does not work if excess demands and excess supplies are derived from the quantity equation. Accordingly, Weeks (1989) designates this model as a "false dichotomy model", referring to the untenable inconsistency between Walras' Law and the quantity equation.

However, if we ignore the "minor" flaw in this logical inconsistency, money proves to be strictly neutral in this second pseudo-monetary system, that seems to be a money economy: employment and output are at the equilibrium and maximal level, respectively. Of course, all this requires that wages flexibly adapt to demand and supply conditions, clearing the labour market and setting maximal output.[50] *We face a fundamentally different situation if flexibility of wages cannot be assumed*, and, accordingly, money wage is stuck above the equilibrium level. Within a simple algebraic framework, Weeks (1989) reveals that if there is a disequilibrium money wage (that is, if $W_0 > W_e$), then the demand for labour is determined by money supply together with the velocity of money and the production function. In other words, if equilibrium wage is equal to the value of marginal product of labour and we can use that

$$\frac{Q}{L} = \overline{K}^d L^{-d}, \text{from which} L^{-d} = \frac{Q}{L \times \overline{K}^d}, \tag{3.21}$$

then Eq. (3.12) can be rewritten as

[50] We have empirical experience, though to limited extent, of how firms set prices and wages. Only a small portion of firms respond flexibly to market developments: they make pricing decisions following a pre-fixed pattern, similarly to their Western European counterparts. Quick adaptations are only triggered by sudden events greater in extent and more turbulent in effect. Most Hungarian companies changes their prices on a quarterly basis (Tóth and Vincze 1998). Pula (2005), for example, reported the increasing rigidity of nominal wages, analysing Hungarian business environment.

$$W = P(1 - d)\frac{Q}{L}, \tag{3.22}$$

from which the demand for labour (solving for L) is

$$l_D = P(1 - d)\frac{Q}{W}. \tag{3.23}$$

Applying the expression

$$Q = \frac{M^* \times V}{P} \tag{3.24}$$

from the quantity equation, we get

$$l_D = \frac{(1 - d)M^* \times V}{W_0} \tag{3.25}$$

as the demand for labour services. *Therefore, neutrality of money can only be assumed in the case of full employment with flexible wages*, since, *in the case of sticky wages, an increase in the money supply enhances the demand for labour services* which is, however, a real variable—so, money is not neutral under these conditions. However, Eq. (3.25) also shows that, in the state of labour market equilibrium (if money wages are flexible), a change in the money supply proportionally modifies money wages—so, a change in the money supply does not modify the demand for labour (since labour market functions will not alter in the real balance effect model either and in the interest-elastic money market model, money will not be neutral in those models either, *assuming sticky wages*). Additionally, it is also a valuable lesson that, *with wages above the equilibrium money wage, by lowering their wage claims, workers enhance the demand for labour and, hence, employment.*

We can briefly summarize the conclusions drawn from this section: the dichotomy between real and nominal factors of economies is false, since we have two different and contradicting functions of the excess demand for money. Moreover, neutrality of money *only* holds to a limited degree, with flexible wages. According to a witty commentary from philosophy of science, we can hardly expect a logically confusing and inconsistent theoretical construct to fail merely on these grounds—its incompatibility with life-world processes should also become obvious (Horváth and Szilágyi 2004).[51] So, a fail requires a twofold defeat. But anyway, the conclusion relevant for new classicals is given again: *if* wages are *not* rigid, economy settles in the state of full employment.

[51] Kuhn (1962), in his book on scientific revolutions, stresses that a paradigm shift occurs very late, when the prevailing theoretical system is so inoperable as that cannot be concealed even for the followers. The first hints of inconsistency do not lead to the replacement of the paradigm in any way.

We will return to the issue of false dichotomy in connection with the island models of Lucas. We will see that, Lucas finally solved the problem posed by the veil thrown as money onto the economy. *In his system, money is not neutral anymore, but it does not imply that monetary policy could successfully control the dynamics of real economy by the quantity of money*: it does not follow from the non-neutrality that there is a Phillips curve applicable for monetary policy. Lucas has never invalidated the quantity equitation, he only highlighted that it should be used to describe some long-run processes (cf. Lucas 1980).

The inconsistency between the two excess demand equations that come from Walras' Law and the quantity equation are first eliminated by Don Patinkin through his neoclassical model that also includes real balance effect. The mechanism can be very briefly summarized: if money is the only form in which the public can treasure up wealth, then the numerical value of the real balance M^*/P they posses (being a real wealth after all) is decreased by an increase in the *price level* (and not simply in inflation), and *vice versa*, and since both real income and real balance are assumed to affect consumption demand, the price level can also influence consumption through real balance. In this model calculating with real balance effect, changes in the price level help find and restore the equilibrium. Since the consumption function $C = c^* + bQ + g(M^*/P)$, where $c^*, b, g > 0$, explicitly refers to the real balance (over and above autonomous and income dependent component, the third term on the right is a straight function of the real balance), an increase in the price level reduces demand through reducing the real balance, resulting in an excess supply of the single commodity that finally triggers a corresponding decrease in the price level, restoring equilibrium. However, equilibrium is still of real character: full employment and, through it, maximal output determine the further variables characteristic of the economic system—the only problem is that the price level is included in the equilibrium, that is, only one numerical value of P can be consistent with the equilibrium state[52] (i.e. unlike the previous models, not only the elements of the real system but also the price level receive their numerical value in a simultaneous setting—and not as a veil, once the real equilibrium is already found). Figure 3.5 outlines the mechanisms of the model.

At any rate, Patinkin successfully prevented the inconsistency of investment and savings from hindering full employment.[53] However, adjustment is not incumbent on the rate of interest but on the price level and, through it, on savings (and consumption as its complementary): according to parts *C* and *D* of Fig. 3.5, demand

[52] It is highlighted by Weeks (1989), and, as it seems, this feature makes the Patinkin model troubling for him. However, it seems to be an ignorable feature, since we can also draw the same consequence from the false dichotomy model: with maximal output determined by labour market equilibrium and with *exogenous* (i.e. pre-given) money supply, the price level can only have one numerical value also in this model. The point is the way in which the price level emerges (even if the result remains the same): money was a veil hitherto, *but now, money is included in the structural equations through real balance*.

[53] In the dichotomy model, it could happen that the full employment equilibrium of savings and investment requires a negative rate of interest. Since it is not possible, full employment cannot emerge under these conditions.

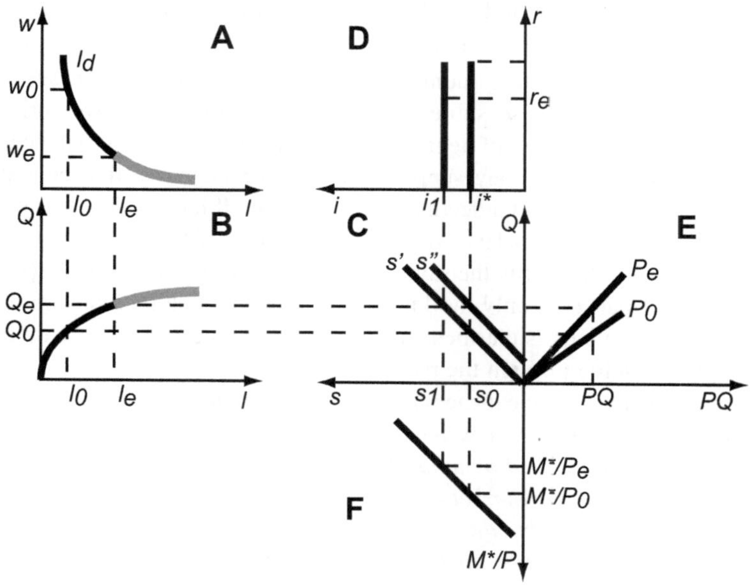

Fig. 3.5 Mechanisms of the real balance effect model (Weeks 1989)

deficiency in the commodity market is resolved by temporary adaptation of production and permanent adaptation of the price level[54] (and, hence, by the transmission of real balance). In the dichotomy model, a decrease in output caused by a commodity market disequilibrium make prices rise, since the exogenous money supply determines the monetary value of output. This freak still holds in the real balance effect model (cf. Weeks 1989). In this model, the demand for money is divided into a transactions component and a real balance part and, accordingly, the equilibrium of demand for and supply of money can be written as $M^* = kPQ + fM^*$, where $0 < f < 1$ and $k > 0$. The first component on the right is the transactions demand, while the second one is the demand for real balance—the latter simply means that the public holds a proportion f of the rigid and exogenous money supply as an idle balance. However, after realigning, we get $(1 - f)M^* = kPQ$, i.e. the monetary value of output is continuously unchanged, assuming a constant velocity (and, hence, k), as it was in the dichotomy model. It should also be noticed that the setting of the price level is still based on the quantity equation (though, on its real balance augmented version). If an economic system is in equilibrium with investment i_1 and savings s_1 (that is, s' is the prevailing savings

[54] It should be kept in mind that changes in the price level did not have any role in determining the equilibrium so far–since, in the previous models, we did not have an equation (except for the quantity equation which, however, is irrelevant as far as equilibrium is considered) that referred to the price level when setting a numerical value for a real variable (even if indirectly, through the real balance).

function) and if investment independent of interest rate autonomously drops to i^*, this change dislodges labour market from equilibrium; investment i^*, assuming savings function s', generates output Q_0 and employment l_0 and it seems as if excessive wage claims cause sub-equilibrium employment. A lower price level appertaining to a lower level of aggregate demand triggers the real balance effect: assuming an exogenous money supply, real balance increases that generates an increase in consumption and a decrease in savings at all income levels (i.e. agents consume more at *any* given level of income, because of the higher level of real balances they hold) that shifts the savings function of part C to the right (since the savings curve of part C is only the function of income—the relationship between real balance and savings is clarified in part F). So, the equilibrium of investment and savings is ensured through the real balance effect.

This model conveys a clear message about the labour market: *workers lay the foundations for the drop in prices and an increase in the real balance and, through it, a rise in consumption and a cutback in savings by lowering their wage claims. The equilibrium of labour market can be set only in and from the labour market itself—that is nothing else but the presupposition of the conclusion. The equilibrium of labour market is created by the equilibrium of labour market*, which, moreover, requires the flexibility of wages and prices.

However, this solution raises some peculiar issues. Although the decrease in investment resulting in disequilibrium triggered decreases in aggregated demand and, hence, production, and although a *drop* in the price level is needed to start the real balance effect (generating an increase in the real balance) that is necessary to restore equilibrium—or, at least *should be needed* (demonstrated by the increase in real balance in part F in Fig. 3.5 from M^*/P_e to M^*/P_0), since the characteristic feature that the monetary value of output is constant does not allow these two events to simultaneously occur (since the price level of output, decreased by the dropped aggregate demand, increases from P_e to P_0 in part E). Price level cannot increase and decrease at the same time. A decrease in investment results in a drop in aggregate demand indeed, however, output cannot actually fall, since it would lead to a rise in the price level (i.e. the price of the single commodity). If the price level increases, then consumption demand cannot grow because of the decreasing real balance, that is, consumption would not be capable of compensating the drop in investment, preserving aggregate demand. In this story, if the system moves out of equilibrium, the price level should increase and decrease *simultaneously*. A decrease in aggregate demand triggers a corresponding decrease in the price level, while the decrease in production results in an increase in price level. It means that *disequilibrium states are only hypothetical*. Consumption function poses another problem: the emergence of equilibrium value of output, assuming decreased investment, is only possible with increased consumption demand. However, it is not explained by the shape of the consumption function: it is evident that the autonomous component cannot change and the level of the income dependent part is also given at income Q_e (bQ_e), and, moreover, the real balance dependent part cannot be higher than it was in the previous equilibrium, since, if the quantity and velocity of money and parameter f are all unchanged, the same price level P_e appertains to output Q_e. Accordingly, in a (hypothetical) new equilibrium with decreased investment, real balance remains the same as before the change, and, hence, consumption demand seems to be incapable of compensating the drop in investment. It is doubted whether disturbed equilibrium triggers a mechanism that actually restores (or re-forms) this equilibrium.

It is also an interesting situation if we assume an interest rate dependent investment function. Considering the details discussed above, it is not necessary to specify the whole system again—it is enough to draw the segments of investment and savings (part C and D of Fig. 3.5) in Fig. 3.6.

Let us assume an economy to be in equilibrium with savings and investment functions s' and i', respectively, and with output Q_e and interest rate r_e (point E in Fig. 3.6). If, now, investment function moves (shifting from i' to i''), then excess supply in the commodity market dislodges labour market from its previous equilibrium state and output cannot be maximal (Q_1). The drop in the output forces the price level to decrease and, simultaneously, the interest rate rises (through the IS curve). *Therefore, the economy cannot stay in F*. This higher rate of interest reduces investment again—i.e. there is a shift to G along i''. In the meantime, of course, savings function cannot stay in its original position, since the higher real balance triggers its shift to s''. However, the new investment demand decreases output further (Q_2) through the new savings function, triggering an additional drop in the rate of interest—i.e. the economy cannot stabilize in G. *A tendency towards continuous contraction may emerge in the economy without having anything in the model to hinder it*. Introducing interest rate dependent investment demand is nothing else but making the downwards spiral explicit that emerges after the economy is dislodged from equilibrium. Let us consider what would happen if, in the dichotomy model, the interest rate dependent investment function autonomously changed in the full employment equilibrium through decreasing i^* in Eq. (3.15), which is equivalent to a drop in investment demand. Although employment and output would have dropped: autonomous investment stands in the numerator of Eq. (3.17); this new equilibrium can be regarded as unique after all, that is, the system could stabilize there within that framework. Assuming an interest rate dependent investment demand, this characteristic feature is missing from the real balance effect model.

The automatic occurrence of full employment is not ensured either in the neoclassical model of Keynesian approach that includes an interest-elastic demand function of money (i.e. the LM curve). The simultaneous solution of IS and LM curves that contain both the rate of interest and real output does not necessarily result in the full employment equilibrium of money and commodity markets. Its explanation corresponds to the case considered above in connection with the dichotomy model. Of course, in case of functions that are consistent with the state of full employment (in other words, it is exactly full employment equilibrium that accidentally emerges as a result), unemployment is voluntary, caused by inflexible wages, and, eventually, by workers themselves.

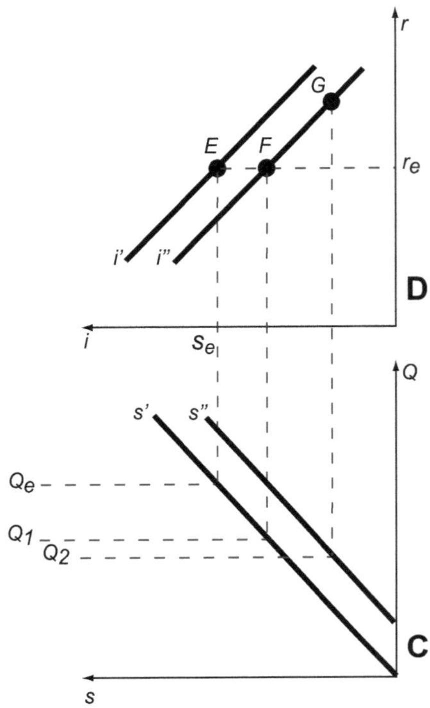

Fig. 3.6 Real balance effect—assuming an interest rate dependent investment function

Weeks (1989) calls the attention to the fact that full employment may also be hindered by the liquidity trap. [→ **liquidity trap**] However, the case of the liquidity trap can also emerge with a rigid money supply (in the widely known parable of Keynes, the central bank enhances money supply through open market operations, that is, the quantity of money is not exogenously given). If the labour market converges through decreasing in wages towards the equilibrium from a non-full employment situation, then prices (drop because of dropping wages) together with the rigid nominal money supply (M^*) increases real money supply (M^*/P), that is, an excess supply of money, and the resulting inclination to buy bonds leads to a decrease in the rate of interest. However, if the demand for money tends to infinity, the drop in prices will not generate a further decrease in the rate of interest. In the interest-elastic money market model, the case of a liquidity trap may hinder full employment even if the equality of investment and savings, resulting in the equilibrium of labour market, is possible at a positive rate of interest, but the needed interest rate is below a level that leads to an infinitely elastic demand for money. Of course, full employment cannot be achieved under these circumstances.

Liquidity trap. The theory of the liquidity trap is the theory of the demand for money. The view on the possibility of the liquidity trap means one of the major distinctions between the Keynesian theory and monetarism.

In the neoclassical-monetarist system, the views on the demand for money are incorporated in the quantity theory of money. In the form of the quantity equation

$$MV = PQ$$

the demand for money is expressed by velocity (V). Here, there is a reciprocal relation between velocity of and demand for money: the greater the demand for money is, the lower the velocity is (Mátyás 2003). One of the most important lessons from the neoclassical-monetarist theory is that the close relationship between nominal income and quantity of money can only be drawn by assuming *rigid* velocity (or that only changes in long-run and in a calculable manner). Since the neoclassical paradigm postulates the persistent equilibrium of real economy, it follows that a stabile relation between the quantity of money and the price level develops.[55]

The stability of velocity and, hence, the demand for money is expressed through the demand function for money[56]

$$M = f\left(P_1 r_b - \frac{1}{r_b}\frac{dr_b}{dt}; r_e + \frac{1}{P}\frac{dP}{dt} - \frac{1}{r_e}\frac{dr_e}{dt}; \frac{1}{P}\frac{dP}{dt}; w; \frac{Y}{r}; u\right)$$

given by Friedman (1956). It assumes the demand for money to be interest-elastic (interest rate dependent), though, for Friedman, the rate of interest was presupposed to be rigid (Friedman only accepted small changes with long-term horizons). The most important characteristic feature of this theory is that the demand for money is in a stabile (function-like) relation with the variables and the variables themselves appear as stable or stably (foreseeably) changing conditions. It is relevant for operative monetary policy, because if V is stable or, at least, changes in a foreseeable way, then nominal income (PQ) can be certainly influenced through controlling the quantity of money (M).

The theory of the liquidity trap given by Keynes (1936) attacked the stability of the demand for money.[57] The direct relation between the demand for money and velocity is eliminated in his theory of the demand for money and of the rate of interest.[58] The various

[55] It can be regarded as quite a rough form and interpretation of the quantity theory, and monetarism (even its orthodox school) can hardly be charged with the formulation of this conception in such an oversimplifying way. Quantity equation is actually the basis of the inflation theory of monetarism, but it was especially the Friedmanian Phillips curve that could demonstrate changes in the quantity of money to have real effects in the short-run at least, though velocity could be actually assumed to be constant. Friedman (1956) stresses the importance of distinguishing the vulgarized and sophisticated-elaborated forms of the quantity theory. He was obviously right.

[56] Independent variables in the demand function for money are the following: (1) real return on bonds, (2) real return on equities, (3) real return on wealth held in the form of physical goods, (4) the ratio of non-human to human wealth, (5) total wealth and (6) individual tastes and preferences as to holding wealth.

[57] Although it has to be stressed in terms of the history of economic thought that Friedman's theory of the demand for money emerged *after* the times of Keynes, the line our discussion followed is not wrong after all, since Friedman renewed the pre-Keynesian theory of the quantity equation (e.g. Hume 1752; Fisher 1928).

[58] Although it was Keynes who laid the foundations for the theory of the liquidity trap in his *General theory*, the label did not come from him.

motives for holding money, as Keynes identified them,[59] underlying the demand for money (especially the transaction and the speculative purposes) affect the velocity of money in different ways, since an increase in the transactions demand enhances, while an increase in the speculative demand lowers the velocity of money (since the latter manifests itself as holding idle balances). The speculative motive is crucial, since Keynes could only prove the interest rate dependency of the demand for money and the limited potential for monetary policy through it. However, speculative demand for money does not simply depend on the rate of interest, but on the relation between the current rate of interest and the safe level of the rate of interest: as the interest rate starts dropping due to an open market bond-purchasing action by the central bank, the public shows increasing reluctance to hold their wealth in bonds. However, the central bank is successful in decreasing the interest rate for a while—the infinite demand of the central bank for bonds meets a finite and limited supply of bonds: though many sell their securities to the monetary authority, there are always market participants who bide their time, speculating for further rises (and, equivalently, for further drops in the rate of interest). If the process of lowering the rate of interest does not halt, interest rate will be below the safe level sooner or later, so no one can expect the rise to be continued. This is the psychological moment when even the toughest and most patient speculators sell their bonds to the central bank—and the infinite demand meet an unlimited supply of bonds now. The rate of interest cannot drop further—the central bank fell into a liquidity trap. Here, speculative demand for money is infinite and funds held for the purpose of speculation are withdrawn from circulation as idle balances—while quantity of money used in transactions keeps circulating.[60] So, for Keynes, velocity unpredictably responds to the increase in the money supply. Liquidity trap is a serious concern and constraint for monetary policy, since, because of it, lowering the rate of interest may turn out to be insufficient to boost real economy in an unfavourable conjunctural situation. Keynes also utilized this line of reasoning when arguing for the high potential for fiscal policy.

However, the followers of Keynes cannot agree on the exact level of the rate of interest at which we should talk about a liquidity trap. Some authors take the view that liquidity trap emerges at a zero or near-zero level of the interest rate (Ahearne et al. 2002; Krugman 1998; Ozsvald and Pete 2003). However, this interpretation seems to be false, since, at zero rate of interest, these are not the expectations of agents and their propensity to hold money but an uncrossable "physical" threshold that prevent the interest rate from being lowered

[59] The wide (even if not general) acceptance of precautionary-, speculative-, and transactions-motives described by Keynes (1936) is out of the question. Authors of highly differing theoretical views naturally cite them (e.g. Samuelson and Nordhaus 1998; Szalkai 1995).

[60] For Keynes (1936), velocity of the total quantity of money is difficult to predict. What happens, for instance, when the rate of interest drops due the central bank open market operations? This development, firstly, affects real economy to *some* extent, so there is a rise in the demand for money for transactions purposes; and, secondly (dependent on the relation to the safe level of the rate of interest), may induce speculative demand for money to increase to *some* extent. The higher transactions demand for money increases, while the higher speculative demand reduces velocity—and the impacts of precautionary-motive that responds to changes both in the rate of interest and income should be also kept in mind. Therefore, when talking about the theory of Keynes, it is not enough to say that there is a rise in the demand for money—the motive also has to be added to which the higher demand appertains. Further problems come from the circumstance that transactions demand may withdraw cash from the (idle) balances held for speculative purposes—a rise in the money supply (generated by open market operations) induces developments that are very difficult to foresee. Velocity (V), real output (Q) and the price level (P) can equally (simultaneously) change. Quantity theory of money is not capable of more than a simple illustration of equilibrium.

further. Once zero rate of interest has been achieved, it is not the liquidity trap that hinders the reduction in the interest rate again.

The possibility of the liquidity trap can be ruled out by introducing the wealth effect. According to this, aggregate real wealth represented by the real value of funds and bonds is included in savings/consumption and investment equations. If the case of the liquidity trap emerges when nominal money supply is rigid (that is, due to a decrease in the price level), then both consumption and investment demand would intensify thanks to the increasing wealth (where this increase is caused by the drop in the price level). If the extent of the decrease in the price level is large enough, there will be full employment. So, full employment may emerge if rigid money wages do not impede it (when, of course, money cannot be regarded as neutral). However, it is a different problem, mentioned by Weeks (1989), that introducing the wealth effect is an effectual refutation of the liquidity trap and of the inconsistency between investment and savings[61] only within the synthesis model. We have seen that the lowering of the prices triggered the wealth effect that automatically took the whole system towards full employment—however, under life-world conditions, such a process (viz. a persistent deflation) may be disastrous for the economy. So, wealth effect is only a comfortable presupposition. In logical terms, it can help the neoclassical reasoning get through the difficult situation created by the possibility of non-full employment—but the realm of neoclassical economists did not become less abstract even after the introduction of this circumstance.

References

Ahearne A, Gagnon J, Haltmaier J, Kamin S (2002) Preventing deflation: lessons from Japan's experience in the 1990s. Board of Governors of the Federal Reserve System, Washington

Akyeampong E (1997) A statistical portrait of the trade union movement. Perspect Labour Income 9(4):45–54

Archer-Hind RD (ed and trans) (1888) The Timaeus of Plato. Macmillan, London

Barcsi T (2011) Az emberi méltóság filozófiája. Dissertation, University of Pécs

Barr N (2004) Economics of the welfare state, 4th edn. Oxford University Press, Oxford

Bayer J (1998) A politikai gondolkodás története. Osiris, Budapest

Begg DKH (1982) The rational expectations revolution in macroeconomics – theories and evidence. Philip Allan, Oxford

Bene L (2003) Akarat és szabadság a sztoikus és a platonikus hagyományban. Epiktétosz és Plótinosz Világosság 44(9–10):107–121

Berlin I (1958) Two concepts of liberty. In: Berlin I (eds) (1969) Four essays on liberty. Oxford University Press, Oxford, pp 118–172

Bolt K, Matete M, Clemens M (2002) Manual for calculating adjusted net savings. The World Bank, Washington, DC

[61] As we have seen, the latter can be eliminated in the real balance effect model if investment demand is assumed to be interest rate independent.

Bourke P (2012) Google Earth fractals. http://paulbourke.net/fractals/googleearth. Accessed 15 Nov 2012
Branson WH (1972) Macroeconomic theory and policy. Harper & Row, New York
Bretter Z (2004) Politika a határon: a Devlin–Hart vita. Kalligram, Pozsony
Buza M, Vas J (1980) A munkahelyi demokrácia és a szakszervezetek. Táncsics, Budapest
Cameron R (1993) A concise economic history of the world from paleolithic times to the present. Oxford University Press, New York, Hungarian edition: Cameron R (1998) A világgazdaság rövid története a kőkorszaktól napjainkig (trans: Kállai T). Maecenas, Budapest
Carter I (2012) Positive and negative liberty. In: Zalta EN (ed) The Stanford encyclopedia of philosophy, Spring 2012 edition. http://plato.stanford.edu/entries/liberty-positive–negative. Accessed 9 April 2012
Chikán A (2003) Vállalatgazdaságtan. Aula, Budapest
Csákó M (1992) A magyarországi munkaügyi viszonyok átalakításának egyes akadályairól. Szociológiai Szemle 2(4):61–67
Daly HE (1972) In defense of a steady-state economy. Am J Agric Econ 54(5):945–954
Daly HE (1974) The economics of the steady state. Am Econ Rev 64(2):15–21
Davies N (1996) Europe: a history. Oxford University Press, Oxford
Dickens C (1854) Hard times – for these times. Bradbury & Evans, London
Diederiks HA, Lindblad JT, Noordam DJ, Quispel GC, de Vries BMA, Vries PHH (1994) Van agrarische samenleving naar verzorgingsstaat. Martinus Nijhoff, Groningen, Hungarian edition: Diederiks HA, Lindblad JT, Noordam DJ, Quispel GC, de Vries BMA, Vries PHH (1995) Nyugat-európai gazdaság- és társadalomtörténet: a rurális társadalomtól a gondoskodó államig (trans: Kocsor D). Osiris, Budapest
Durkheim É (1895) Les règles de la méthode sociologique. Préface à la 2ème edition. Presses Universitaires de France, Paris, p XI–XXIV
Durkheim É (1897) Le suicide: Étude de sociologie. Alcan, Paris, Hungarian edition: Durkheim E (2003) Az öngyilkosság: szociológiai tanulmány (trans: Józsa P). Osiris, Budapest
Engels F (1845) Die Lage der arbeitenden Klasse in England. Otto Wigand, Leipzig
Erdős M, Mérő K (2010) Pénzügyi közvetítő intézmények – bankok és intézményi befektetők. Akadémiai, Budapest
Erdős T (2006) Növekedési potenciál és gazdaságpolitika. Akadémiai, Budapest
Ficino M (1576) Opera omnia. Heinrich Petri, Basel
Fisher I (1928) The money illusion. Adelphi, New York
Friedman M (1953) The methodology of positive economics. In: Friedman M (ed) Essays in positive economics. University of Chicago Press, Chicago, pp 3–43
Friedman M (1956) The quantity theory of money – a restatement. In: Friedman M (ed) Studies in the quantity theory of money. University of Chicago Press, Chicago, pp 3–21
Friedman M (1962) Capitalism and freedom. University of Chicago Press, Chicago, Hungarian edition: Friedman M (1996) Kapitalizmus és szabadság (trans: Bősze SM). Akadémiai, Budapest
Friedman M (1968) The role of monetary policy. Am Econ Rev 58(1):1–17
Gál Z (2010) Pénzügyi piacok a globális térben – a válság szabdalta pénzügyi tér. Akadémiai, Budapest
Galbraith JK (1967) The new industrial state. Houghton Mifflin, Boston, Hungarian edition: Galbraith JK (1970) Az új ipari állam (trans: Hantos É). KJK, Budapest
Galbraith JK (1973) Economics and the public purpose. Houghton Mifflin, Boston, Hungarian edition: Galbraith JK (1978) Közgazdaságtan és a társadalmi cél (trans: Szalai S). Kossuth, Budapest
Gerson L (2012) Plotinus. In: Zalta EN (ed) The Stanford encyclopedia of philosophy, Fall 2012 edition. http://plato.stanford.edu/entries/liberty-positive–negative. Accessed 8 Oct 2012
Giddens A (1993) Sociology, 2nd edn. Polity Press, Cambridge, Hungarian edition: Giddens A (1997) Szociológia (trans: Acsádi J, Babarczy E, Nagy M, Nagy Z, Tóth L). Osiris, Budapest

Greenwald B, Stiglitz JE (1987) Keynesian, new Keynesian, and new classical economics. NBER working paper No. 2160. National Bureau of Economic Research, Cambridge

Habermas J (1973) Krisentendenzen im Spätkapitalismus. In: Habermas J (ed) Legitimationsprobleme im Spätkapitalismus. Suhrkamp, Frankfurt, pp 50–131

Hahn FH (1980) Monetarism and economic theory. Economica 47(185):1–17

Hamilton K (2000) Genuine saving as a sustainability indicator. The World Bank, Washington, DC

Hámori B (1986) Politikai gazdaságtan – a szocialista gazdaság elmélete. KJK, Budapest

Hart HLA (1989) Law, liberty and morality. Oxford University Press, Oxford, Hungarian edition: Hart HLA (1999) Jog, szabadság, erkölcs (trans: Krokovay Z). Osiris, Budapest

Hartwick JM (1978) Substitution among exhaustible resources and intergenerational equity. Rev Econ Stud 45(2):347–354

Hawking SW (1988) A brief history of time from the big bang to black holes. Bantam, New York, Hungarian edition: Hawking SW (2003) Az idő rövid története (trans: Molnár I, Egri G). Akkord, Budapest

Hayek FA (1988) The fatal conceit–the errors of socialism. University of Chicago Press, Chicago, Hungarian edition: Hayek FA (1992) A végzetes önhittség–a szocializmus tévedései (trans: Pásztor E). Tankönyvkiadó, Budapest

Horváth ÁB, Szilágyi K (2004) Konszenzusból nyugvópontra – elmélettörténeti áttekintés a makroökonómia viharos évtizedeiről. Külgazdaság 48(11):4–19

Hume D (1752) Of money. In: Hume D (1994) Political essays. Cambridge University Press, Cambridge, pp 115–125

Kalecki M (1935) The mechanism of business upswing. In: Kalecki M (ed) (1969) Studies in the theory of business cycles 1933–1939. Basil Blackwell, Oxford, pp 26–33. Hungarian edition: Kalecki M (1980) A tőkés gazdaság működéséről – válogatott tanulmányok 1933–1970 (trans: Balogh A). KJK, Budapest, pp 64–73

Kerekes S (2007) A környezetgazdaságtan alapjai. Aula, Budapest

Keynes JM (1936) The general theory of employment, interest and money. Palgrave Macmillan, London, Hungarian edition: Keynes JM (1965) A foglalkoztatás, a kamat és a pénz általános elmélete (trans: Erdős P). KJK, Budapest

Király J (2009) A tornádó és a hurrikán: a 2007. év válságos hatásai. In: Muraközy L (ed) A jelen a jövő múltja Járatlan utak – járt úttalanságok. Akadémiai, Budapest, pp 295–332

Kopányi M (2003) Mikroökonómia. KJK–KERSZÖV, Budapest

Kotler P, Keller KL (2006) Marketing management, 12th edn. Prentice Hall, New Jersey, Hungarian edition: Kotler P, Keller KL (2006) Marketing menedzsment (trans: Németh Á, Dóczy B, Doubravszky J). Akadémiai, Budapest

Krugman P (1998) It's baaack! Japan's slump and the return of the liquidity trap. Brookings Pap Eco Ac 29(2):137–206

Kuhn TS (1962) The structure of scientific revolutions. University of Chicago Press, Chicago, Hungarian edition: Kuhn TS (2002) A tudományos forradalmak szerkezete (trans: Bíró D). Osiris, Budapest

Lucas RE (1980) Two illustrations of the quantity theory of money. Am Econ Rev 70 (5):1005–1014

Mankiw NG (1997) Macroeconomics, 3rd edn. Worth Publishers, New York, Hungarian edition: Mankiw NG (2002) Makroökonómia (trans: Vojnits T). Osiris, Budapest

Maróth M (2002) A görög filozófia története. PPKE, Budapest

Mátyás A (1979) A polgári közgazdaságtan története az 1870-es évektől napjainkig. KJK, Budapest

Mátyás A (1984) A polgári közgazdaságtan napjainkban. Kossuth, Budapest

Mátyás A (2002) A korai közgazdaságtan története. Aula, Budapest

Mátyás A (2003) A modern közgazdaságtan története. Aula, Budapest

Mill JS (1859) On liberty, 2nd edn. John W. Parker and Son, London

Muraközy L (2009) Állam és gazdasági szabadság. In: Muraközy L (ed) A jelen a jövő múltja. Járatlan utak – járt úttalanságok. Akadémiai, Budapest, pp 99–161
Ozsvald É, Pete P (2003) A japán gazdasági válság – likviditási csapda az új évezredben? Közgazdasági Szemle 50(7–8):571–589
Patinkin D (1989) Money, interest, and prices. MIT Press, Cambridge
Perrow C (1986) Complex organizations, 3rd edn. McGraw-Hill, New York, Hungarian edition: Perrow C (1997) Szervezetszociológia (trans: Hercsik A, Kósa K, Szalay I). Osiris, Budapest
Plotinus (250 AD) The six enneads (trans: Mackenna S, Page BS). MIT. http://classics.mit.edu/Plotinus/enneads.html. Accessed 27 Dec 2014
Pula G (2005) Az euro bevezetésével járó strukturális politikai kihívások – munkapiac. Central Bank of Hungary, Budapest
Samuelson PA, Nordhaus WD (1998) Economics, 16th edn. McGraw-Hill, New York
Schütz A, Luckmann T (1975) Strukturen der Lebenswelt. Luchterhand, Neuwied
Shaw GK (1984) Rational expectations – an elementary exposition. Harvester Press, Brighton
Simmel G (1900) Philosophie des Geldes. Duncker und Humblot Verlag, Berlin
Singh S (2004) Big bang: the most important scientific discovery of all time and why you need to know about it. Fourth Estate, London, Hungarian edition: Singh S (2007) A nagy bumm – minden idők legfontosabb tudományos felfedezésének története (trans: Szécsényi-Nagy G). Park, Budapest
Sinkovics A (1989) Állam – szocializmus – szakszervezet. Akadémiai, Budapest
Slings SR (ed) (2003) Platonis Rempublicam. Oxford University Press, Oxford
Smith VL (1991) Rational choice – the contrast between economics and psychology. J Polit Econ 99(4):877–897
Smith VL (1994) Economics in the laboratory. J Econ Perspect 8(1):113–131
Solow RM (1974) Intergenerational equity and exhaustible resources. Rev Econ Stud 41(5):29–45
Solow RM (1986) On the intergenerational allocation of natural resources. Scand J Econ 88 (1):141–149
Stead C (1998) Philosophy in Christian antiquity. Cambridge University Press, Cambridge
Stiglitz JE (1988) Economics of the public sector, 2nd edn. W.W. Norton, New York, Hungarian edition: Stiglitz JE (2000) A kormányzati szektor gazdaságtana (Mezei G). KJK–KERSZÖV, Budapest
Sydsæter K, Hammond P (1995) Mathematics for economic analysis. Prentice Hall, New Jersey
Szalkai I (1995) A monetáris irányítás. Bankárképző, Budapest
Szokolay K (2001) Szakszervezet, kormány, hatalomváltás. História 23(9–10):34–37
Taxner-Tóth E (1972) Dickens világa. Európa, Budapest
Tengelyi L (1995) Kant. Áron, Budapest
Thoma L (2002) A baloldal dilemmája: a szakszervezet. Mozgó Világ 28(1):107–114
Tóth IJ, Vincze J (1998) Magyar vállalatok árképzési rendszere. Central Bank of Hungary, Budapest
Trinkaus C (1983) The scope of renaissance humanism. University of Michigan Press, Ann Arbor
Vígh L (1997) Keynes, keynesianizmus, monetarizmus. Dissertation, Hungarian Academy of Sciences
Vigvári A (2008a) Positions and standpoints – or what does the "invisible hand" arrange? Public Finance Q 53(3):545–551
Vigvári A (2008b) Pénzügy(rendszer)tan – alapvető ismeretek rendhagyó megközelítésben. Akadémiai, Budapest
Wallis RT (1972) Neoplatonism. Duckworth, London
Walras L (1874) Éléments d'économie politique pure ou théorie de la richesse sociale. L. Corbaz & Cie, Lausanne, English edition: Walras L (1954) Elements of pure economics or the theory of social wealth (trans: Jaffé W). Allen & Unwin, London
Weber M (1923) Wirtschaftsgeschichte: Abriss der universalen Sozial- und Wirtschaftsgeschichte. Duncker & Humblot, Munich, English edition: Weber M (1961) General economic history (trans: Knight FH). Collier, New York

Weber M (1934) Die protestantische Ethik und der Geist des Kapitalismus. J.C.B, Mohr, Tübingen, Hungarian edition: Weber M (1982) A protestáns etika és a kapitalizmus szelleme (trans: Gelléri A, Józsa P, Somlai P, Tatár G). Gondolat, Budapest

Weber M (1964) Wirtschaft und Gesellschaft. Kiepenheuer & Witsch, Köln/Berlin, Hungarian edition: Weber M (1967) Gazdaság és társadalom (trans: Józsa P). KJK, Budapest

Weeks J (1989) A critique of neoclassical macroeconomics. Macmillan, London, Hungarian edition: Weeks J (1998) A neoklasszikus közgazdaságtan kritikája (trans: Szalai Z). Aula, Budapest

Zalai E (2000) Matematikai közgazdaságtan. KJK–KERSZÖV, Budapest

Chapter 4
Monetary Policy in the New Classical Framework

Pursuit of the image then prevents pursuit of the reality
(John Kenneth Galbraith: The new industrial state)
The world loves to be deceived
(Petronius)

4.1 Preliminary Remarks

Having drawn the most important conclusions from neoclassical models, we can continue our work here. We have seen that the notions on the formation of labour market equilibrium play a key role in this theoretical system and that the emergence of the equilibrium is accidental—that is, this mechanism can be proved only through maintaining strong assumptions. The existence of an equilibrium depends on strong presumptions. We have focused on (1) the automatisms that direct macroeconomic systems towards the state of full employment and (2) on the theorem of neutrality of money (i.e. its manifestation) and we have seen that these two crucial conclusions of neoclassical economics are *assumptions* in the strictest sense of the word. Therefore no proof is needed, since both appear as *axioms* in the system. Moreover, we should not forget about the logical inconsistency (or the oddities) of some of the models we investigated above, which can lead to serious malfunctions within these theoretical constructs, and the proving of the intended conclusions is fraught with serious difficulties.

After this short summary, it is time to mention briefly the content of our schedule. Having realized that the state of full employment is not so natural and evident as monetarist views on the natural level of output or employment suggest, we should review the economic policy recommendations of new classical macroeconomics.[1] We can do this in two steps. It was mentioned above (in Chap. 2) that monetarism phrased its theorems by assuming an intermediate central bank target specified in terms of the quantity of money—therefore, the first step in the investigation of these economic policy conclusions and advice is going to be conducted by staying within the original environment. Moreover, it was also stated that the

[1] It has to be stressed again that REH and new classical macroeconomics do not constitute an inseparable unit.

P. Galbács, *The Theory of New Classical Macroeconomics*, Contributions to Economics, DOI 10.1007/978-3-319-17578-2_4

REH plays a crucial role in the mechanism of inflation targeting, but this regime creates entirely new conditions for the manifestation of the rationality of expectations. Therefore, as a second step, we have to investigate how inflation targeting as a conceptual and economic policy framework modifies the functioning and functionality of the tactical elements established in the first step.

However, it is not necessary to confine our questions to the problems above, since we can come to numerous conclusions as the by-products of our investigation. The most important item among these additional problems is that a new classically oriented economic policy has a balefully limited scope due to the activity of commentators (i.e. literature referring to new classicals attributes much *less* potential to an economic policy, thanks to its unforgivable superficiality or inadvertence, *than* what the authors referred to expected). Phrasings are *acceptable* but, at the same time, *incomplete*. For example, Antal Mátyás (2003) makes the following remark: "Representatives [. . .] of the new classical macroeconomics [. . .], attacking countercyclical policy and, through it, Keynes provides even less leeway for countercyclical policy than [Milton] Friedman for whom a rule-following economic policy [. . .] could be effective at least temporarily [. . .]." This statement is undoubtedly true, since new classicals have actually voiced this message—the only problem is that they, following different logical lines (to be more precise: by assuming different transmission mechanisms), actually believed in the potential and effectiveness of *systematic* economic policies (it will be a recurrent issue, for example, when investigating supply side economics). Actually, the REH and, through it, new classical macroeconomics regularly appear in the relevant literature as the denial of the potential of all kinds of activist economic policy (e.g. Benczes and Kutasi 2010).

So, according to the commentaries, a *nihilism of economic policy* emerges from the review of new classical texts: an activist policy, based on unanticipated-surprise actions, only leads to malfunctions, therefore it is to be avoided—while systematic (anticipated) interventions are completely ineffective.[2] Elsewhere Prof. Mátyás uses a much more rigorous language by attributing the alleged theorem of the complete ineffectiveness of economic policy to new classicals.[3] This confusion is probably raised by commentators interpreting new classicals as if they had assumed the *only* channel of economic policy effectiveness to be *delusion* (exerted by monetary policy).[4] As we will soon discover, it is only one aspect of the story.

---

[2] We have to bear in mind that the new classical conception of labour market (according to which, both sides, i.e. both employers and labour unions, make all-out efforts to achieve the labour market equilibrium and, through this, maximal output) already included the signs to present the economic policy passivity as a behavioural norm.

[3] "What is the consequence of a government intervention in economy if it enhances government spending financed by money issuance to boost employment and to lower unemployment below its natural rate? [. . .] this intervention is predestined to fail." (Mátyás 1984)

[4] It seems to be very probable if we recall the remarks of Prof. Mátyás on the surprise supply function given by Lucas: "Government [. . .] could exert an influence on the level of employment

Undoubtedly, though, there is a constructive interpretation of the new classical rejection of activist economic policies, according to which *new classical macroeconomics systematically circumscribed the necessary conditions for the ineffectiveness of countercyclical interventions*. In other words, any kind of activist-countercyclical economic policy is ineffective *only if* agents form rational expectations, *and if* they anticipate every economic policy action, *and if* all the conditions are there to guarantee an immediate and persistent clearing of markets—in short: if we (can) maintain the fundamental assumptions of the new classical system. *If* all these conditions are met, the activist economic policy will turn out to be ineffective. However, without them, that is, in the absence of complete and perfect rationality and flexible markets, even systematic actions can affect real economic variables (output and unemployment) as well. This notion will be of crucial importance in our analysis, especially when the new classical system of fiscal policy, the Ricardo-Barro equivalence, will be considered (Chap. 5). *Therefore, when reviewing new classical theory of economic policy, attention should be paid to the issue of whether the findings according to which these authors branded several instruments of conventional economic policy as useless once and for all, are right.*

4.2 Concept of the Phillips Curve

4.2.1 Elementary Considerations

The doctrine of the Phillips curve is the most useful to start the discussion with, despite the fact (or because of it) that it is widely known, since it was the analytical tool by which Friedmanian[5] (orthodox) and Lucasian monetarism tried to exemplify their theorems. The original concept of the Phillips curve, although very illuminating, is outside of the scope of our interest: it is not necessary to consider the thoughts of the enthusiastic New Zealand statistician and the methodology he applied in order to review the relating new classical doctrines. In case of a historical view taken of economic thought, it would be of some use to scrutinize the theoretical and practical consequences of the Phillips curve, since the neoclassical recrudescence after Keynes (i.e. the monetarist counterrevolution) attacked the system of Keynes mainly through the notions of Phillips. However, even for Phillips (1958), the postulated stable relation between unemployment and inflation (or the growth rate of wages) was not clear-cut and definite. Keynes (1936) himself talked about a tendency at most (Hansen 1953), although there is no doubt that this tendency was assumed by Keynes as a strong proneness, which was accentuated accordingly by the plain methodology applied by Phillips. The concept was simple: the higher the

and production only if it was capable of moving the actual price level from the level expected by the public." (Mátyás 1984)

[5] Mentioning his surname refers to Milton Friedman in the followings.

effective demand (partly through countercyclical economic policy), the higher the *resulting* inflation. [→ **inflation and unemployment**]

The honoured importance attributed to the Phillips curve is that part of the Friedmanian tradition, which was inherited by new classical macroeconomics. Friedman started reconsidering the concept of the original Phillips curve as early as the 1960s in the course of which, while trying to preserve the relation between unemployment (as a real variable) and the price level (as a nominal variable) strongly resembling the original idea, the direction and the nature of the relationship fundamentally changed. Phillips, citing Keynes, presumed a steady trade-off between the growth of wages and the level of unemployment (later, his followers replaced the wage index by the rate of inflation). Here, the basic rapport was that a decrease at any level of unemployment can be done only through (strictly speaking: *by* and *at the expense of*) an increase in wages (and, eventually, in prices) due to the emerging bottlenecks in employment.[6] *Through this, a rule, easily translatable into the language of operative economic policy, emerged offering at least as painful a choice between the level of unemployment and inflation for (economic) policy makers as it was useful.* Subsequent interpretations of the Phillips curve, did in a dubious manner, highlight the discrepancy between the system of Keynes and the statistical observations summarized by Phillips (cf. e.g. Phelps 1967b, or, based on it, KVA 2006). Keynes was reproached mostly for an alleged conclusion that full employment can be maintained *without* any inflationary costs—quite simply, what needs to be done is to reach the highest possible level of employment reconcilable with a stable price level (by using mostly fiscal and partly monetary policy instruments). [→ **inflation and unemployment**] For Keynes (1936), reaching this level of full employment was an issue reduced, in fact, to an empirical problem, since, above this level, any increase in aggregate demand triggers a rise in price level, keeping real output unchanged (Keynes revealed nothing more about this particular level of unemployment). *However, Keynes never stated that prices and wages reveal complete stability before full employment is achieved*, thanks to the bottlenecks in employment which hinder a flexible growth in employment. According to these interpretations, Phillips' most exceptional merit was to call both public attention and that of policy makers to the painful trade-off between inflation and unemployment (though the theoretical foundations for this relationship were definitely provided by Keynes). Only the most superficial interpretations of the system of Keynes may venture to claim that the coexistence of full employment and perfect price stability depends on the fine-tuning of an economic policy alone (and, eventually, overreaction from an economic policy will lead to a situation in which full employment will be maintained through an increased price level). Prices start rising more or less before full employment is reached (while

[6] It is exactly the characteristic that was highlighted by one of the most cunning definitions of the Phillips curve. According to it, "In some cases prices and wages were assumed to be mechanically adjusted to the level of excess supply in each market [viz. in commodity and labour market], such that price and wage inflation was a decreasing function of the rate of unemployment: the so-called Phillips curve." (KVA 1995)

beyond that only prices can increase); however, *it follows from the analyses offered by both Keynes and Phillips that a specific rate of unemployment can be preserved at a specific rate of inflation* (that is, at a stable rate of devaluation of money) *even in the long-run.*

An impartial historical review of economic thought would need to mention that, after Phillips but before the time of Friedman, further (also empirically based) concepts of the mechanism investigated by Phillips emerged. For example, Bowen and Berry (1963) arrived at the conclusion that a *change* in the rate of unemployment determines wage dynamics much more forcibly than the *level* (i.e. the rate) of unemployment. Thus the two distinct stages of the evolution are doubtlessly associated with Phillips and Friedman, even if they are not even adjacent. Also, a few years prior, Richard Lipsey (1960) made an effort to offer a theoretical foundation consistent with the results of Phillips, according to which changes in wages do not simply relate to the level of unemployment but rather to the conditions of the labour market determined by excess demand and excess supply. It is still important to detect the level of unemployment, given that labour market demand and supply cannot be directly witnessed, while the level of unemployment provides us with reliable information about labour market dynamics. Unemployment (rate) is, therefore, an instrumental variable here.

4.2.2 The Orthodox Monetarist Interpretation of the Phillips Curve

Friedman adamantly denied the existence of the trade-off stressed by Phillips and substituted it with another mechanism. Figure 4.1 supports the discussion.

This figure made its way to the literature as the expectation-augmented (sometimes: expectation-adjusted) Phillips curve. The basic mechanism is very simple, even if a deeper analysis does tend to raise a few problems (which even Friedman found hard to tackle). If an *equilibrium* state, depicted by point E (that indicates the natural rate of unemployment as well) is the economy's kick-off position, and inflation (π) rises from A to B for some reasons, the system will temporarily shift towards F, reducing unemployment below its natural rate. However, this is only a provisional state (i.e. the economic system cannot be fixed below the natural rate of unemployment over the long haul as this requires special conditions), the Phillips curve will shift from P' to P'', and the economy returns to its natural rate of unemployment (G) albeit at a higher price level. It is important to note that the economy is in equilibrium at the unemployment rate U_N where it will return later, at a higher rate of inflation, following a temporary fluctuation.

The language we use here is simple and concise, avoiding any interpretation bound to throw light on the underlying elements of this mechanism—not even Friedman fell short of offering a consequent explanation for the economy's perceived trajectory. Differently put (which brings us closer to reality), *the explanation*

Fig. 4.1 The Phillips curve as modified by Friedman (1977)

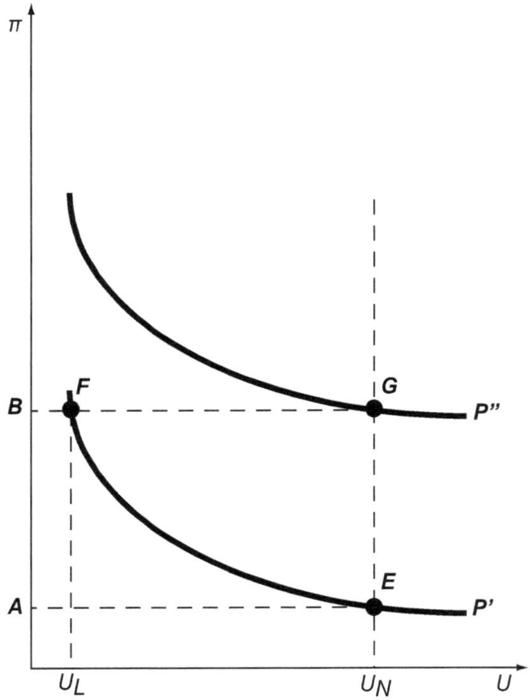

offered by Friedman was exposed to continuous modification. Empirical analyses revealed it to Friedman for the first time that the actual (statistically estimated) Phillips curve *failed* to behave in the manner predicted by the theoretical construct: there was no trace of the steady relationship between unemployment and inflation—it was possible for *several* (according to the theory derived from observed data: *infinite*) rates of inflation to appertain to *any specific* level of unemployment. In other words, a particular level of unemployment had implied ever-increasing rates of inflation.

Initially, Friedman came forward with a story closely resembling the original explication of the Phillips curve. The conception was very simple: employees take their expectations about price (and, hence, wage) dynamics into consideration when deciding on labour supply.[7] If the authority controlling the economy is set to achieve a level of unemployment below its natural rate, it needs to enhance the *growth rate* of money, through which aggregate demand will increase (it is

[7] As we will see, the only problem is that expectation-based decisions are attributed only to employees, while employers are assumed to determine employment (i.e. their demand for labour) considering actual real wage. However, it has to be stressed that considering expectations does not mean that employees enhance their labour supply expecting an increase in their real wage–the mechanism works in the opposite direction. It is a deviation from the expected dynamics that triggers the effects.

supposed to be a twofold expansion: demand on the investment and consumption side would rise, due to the decrease in the rate of interest and, in a direct manner, due to the higher growth rate of money, respectively). Here, economic policy does not aim for a countercyclical control since the economy is in equilibrium—the purpose of intervention is to set the economy to (and fix it permanently at) levels of production and unemployment more favourable than their natural levels. Thus the Phillips curve of Friedman was definitely *not* built on a countercyclical parable, since the fundamental problem stemmed from the potential of monetary policy in the *persistent* controlling and *prolonged* boosting of a real economic growth.

In the above case of increased money growth rate, market participants (employers and employees) enhance their supply as a response to expanded demand. Therefore the rise in prices is the *consequence* of a pulling effect triggered off from the demand side of the economy.[8] In order for the initially favourable employment effect to weaken, the experience of employees on prices needed to be corrected, through which employment will eventually return to its natural level (it is completely indifferent whether it happens through employees learning to correctly perceive the dynamics of their salary and, as it remained unchanged, reduce their labour supply to its initial level accordingly; or they claim higher salaries, also to correctly perceive real wages in order to offset the higher price level, resulting in a decrease in the demand for labour). Inflation plays a twofold role in this explanation (cf. Friedman 1968). The *higher increase* (sic!) in prices is the result of the pulling effect coming from the commodity market. However, the disappointment of inflation expectations (i.e. excess inflation) appears as the *cause* of the expansion in employment (and, as far as the breaking with the Keynesian theory is concerned, the latter role is crucial, since, for Keynes, inflation was the *consequence* of commodity and labour market dynamics[9]).

Inflation and unemployment As a part of his general theory of employment and inflation, Keynes himself laid the foundations of the Phillips curve becoming widely known and accepted through the ardour of the enthusiastic New Zealand statistician, Alban William Phillips. First of all, Keynes focussed on the relationship between aggregate demand and unemployment–inflation, for him, was a cost considered acceptable (which was in fact to be accepted) incurred by employment problems being taken care of. Keynes did not explicitly mention in his *General theory* that unemployment can be eased through additional inflation, although this trade-off can undoubtedly be derived from his system.[10]

[8] Even if we lost the explanation that traces the rise in prices back to wage (and cost) dynamics, which was a characteristic element in the theory of Keynes. For Keynes, inflation, after all, was cost-push inflation, meaning that the increase in the price level triggered by the expanded demand manifests itself through a rise in (labour) costs. For the sake of trustworthiness of the history of economic thought, it should also be mentioned that distinguishing demand-pull and cost-push inflation is not a relevant aspect as far as the theory of Keynes is considered.

[9] The contrast, however, is not completely appropriate, since Keynes never investigated the macroeconomic effects of (disappointment of) inflation expectations.

[10] The *General theory* of Keynes was published in 1936, while the (only) Hungarian edition in 1965. If an economist takes the trouble to skim through the home papers in economics from these times onwards, he will realize that Keynes was almost completely ignored by the Hungarian

For Keynes (1936), the expansion of employment is the response to the stimulation of demand–and the price level changes in accordance with the way marginal costs react to changes in level of employment. It is exactly the essence of the concept of involuntary unemployment that if there are involuntarily unemployed people in a national economy, employment can expand thanks to a rise in effective demand, without any increase in (wage) costs, since those involuntarily unemployed are ready to work even at the prevailing wage level. However, they do not have the opportunity to enter into employment under the given conjunctural conditions. Reaching the saturation point is the state from which the intensity of labour force utilization can be enhanced further only if the increase in salaries exceeds inflation.

As a macroeconomic system approaches full employment, that is, as unemployment declines, intensifying pay rises are needed to persuade employees to perform additional work units, and there are considerable salary claims from those who were previously unemployed on a voluntary basis. The Keynesian theory explains on these grounds why the Phillips curve is steeper at lower levels of unemployment. Of course, employment decreases as a response to a drop in aggregate demand–however, money wages are sticky downwards, i.e. at higher levels of unemployment, the decrease in nominal wages is slower than the decline in the rate of employment. The Phillips curve becomes increasingly flat in these sections.

The relation between the level of employment and price dynamics appears as a different story. The diminishing marginal product of labour may be capable on its own of raising the prices when employment increases. This is something that is not too hard to realize. When unemployment decreases, the marginal product of labour also declines necessarily, though this process is accompanied by an increase in wages. Therefore, an increase in prices, simultaneously with the expansion in employment, will unavoidably be triggered. Moreover, there are several resources exploited in production, and, as aggregate demand expands, some of them can respond flexibly but others are rigid–so certain factors of production, due to their intensified exploitation, can trigger tendencies pointing towards inflation. This is also true for wage costs. Keynes made the observation that nominal wages

authors. However, this confusing situation is easy to explain: Keynes could add only a few novel theses on the theory of centrally planned economies to those already known in Eastern Europe thanks to Marx. It is so very true that the history of economic thought is segmented in Marxian terms so that Prof. Mátyás (1963) distinguishes only two phases, a pre- and a post-Marxian stage (the latter embraces the theory of Keynes). Moreover, Keynes was highly criticized for abandoning the labour theory of value. Sardoni (1986) suggests various aspects of similarity between Keynes and Marx (while deploying a vast literature). He mentions, for example, that Keynes explicitly cited Marx in the previous drafts of his *General theory* (the final version contains only three, highly unimportant citations). The resemblance and parallelisms these two theories bear impel some authors to draw rather bizarre conclusions sometimes. For instance, Brandis (1985) highlights that Marx died in the year (1883) in which Keynes was born and, moreover, mentions the fact that both Keynes and Adam Smith were born on the same day of the year (5th July). On these grounds, he infers that the faith in reincarnation may not be an ungrounded assumption. Perhaps Alexander (1940) summarizes the most clearly. Although Keynes (1936) underlined that followers of (neo) classical economics were expected by him to fluctuate between a belief that the *General theory* was simply wrong and another belief that it had nothing new to say, he completely ignored those with a background pertaining to Marxian theory. They were pretty sure, since Keynes promised nothing new *to them*. Undoubtedly, the theory of Keynes was radically new to the British academic circles of the 1930s. "The Keynesian system is not only thoroughly consistent with the Marxian but also supplements it at certain critical points." (Alexander 1940) Of course, Keynes did not copy Marx–it is interesting, for example, that, over and above Keynes (and nearly at the same time) Michał Kalecki (1935) could also reveal the need for countercyclical economic policies (for further details, see Vigvári 2008).

tend to rise even before full employment is achieved: employers show more inclination to accept wage claims under conditions of expanding aggregate demand. It is also important that people with the same skills get the same salary when employed, and employers do not (or cannot) take individual differences of productivity into consideration–which also contributes to a rise in wage costs. Enhancing employment (and reducing unemployment and under-utilization of productive capacities) therefore implies an increase in inflation characterized by the fact that the growth of inflation exceeds the growth of the cost-unit, due to a diminishing productivity.

Yet again, we should take the role and importance of expectations into consideration. In case of certain commodities, producers cannot adjust their supply instantly, nor can they respond immediately and flexibly to *unanticipated* increases in demand. It is a fundamentally different situation if entrepreneurs are prepared for those changes in advance, that is, if their expectations turn out to be in accordance with later market developments. One of the consequences will be that the flexibility of employment will be low in specific sectors of the economy. Therefore a tendency pointing towards inflation will emerge. Although this was not elucidated by Keynes, it would be hard not to realize that *it is far from a steady and prevailing trade-off between unemployment and inflation.* Any rate of inflation (as determined by other characteristics of the macro-economy) can appertain to any given level of unemployment.[11]

It goes without saying that the theoretical grounds summarized by the Phillips curve were created by Keynes himself. However, it has to be noted that the system of Keynes leaves greater leeway for simultaneous changes in employment and the price level. Multiple Phillips curves may also emerge in the theory of Keynes depending on the sectoral distribution of the aggregate demand. If employment responds flexibly, the negatively sloped curve presents itself as a basic situation–but if demand is directed towards sectors which have a low flexibility of employment, we have a steeper Phillips curve. If employment is completely inflexible, the curve may even be vertical. Consequently, there is not one *single* Phillips curve given that consumer taste and the sectoral distribution of aggregate demand are always changing.

Actually, money illusion (that is, regarding nominal changes as real changes—cf. Patinkin 1989) is not enough to explain this mechanism: it requires two additional assumptions. According to one, prices respond differently to modified demand conditions: an increased aggregate demand exerts its influence on commodity prices sooner than it does on labour market prices (in other words: wages are more rigid). Therefore, *the drop in unemployment is, after all, the result of decreasing real wages* and an accurate judgement of the situation by employees is the only reason for the return to an initial (natural) rate of unemployment (i.e. the end of the money illusion, when they finally recognize the actual dynamics of prices and wages). The other (arbitrary) assumption refers to a special informational asymmetry: whatever employees are unaware of in connection with the changes in (real and nominal) wages and prices can be clearly observed by employers.

---

[11] It is more complicated when the expansion of productive capacities is not precluded either. The relation between inflation and unemployment is more sophisticated in a case like this. It should not be forgotten either that Keynes, on longer time-horizons, attributed higher importance to technological change in the determination of price dynamics and did not preclude even a drop in prices. The link between prices and demand becomes clearer if it is also considered that Keynes' (1936) analyses were performed assuming a given (i.e. unchanging) set of equipments.

Friedman explained the rigidity of nominal factors through the existence of contracts designed in line with former expectations (as we will see, it became a common argument in disequilibrium REH models—for further details see Sect. 4.3.4. below). Which set of prices shows greater extent of flexibility as a response to demand is an empirical problem on the one hand and, on the other hand, it requires more than some regressions, that is, a theoretical foundation (in other words, a precise basis and explanation of what stickiness means) if we intend to turn it into a pillar upon which a theory rests as an underlying assumption. In such cases, the monetarist literature usually refers to long-term contractual obligations that prevent prices from flexibly adjusting to changing demand conditions. However, an explanation like this is insufficient in this case, since *different extents of flexibility* of different sets of prices need to be theoretically founded. If there is no such explanation in sight (Friedman is known to have failed to come up with one), we are free to assume a contrary system of flexibilities where the response from money wages is faster.[12] However, the question is whether the trajectory of the economy described by Friedman is still a possibility. The *initial* increase in money wages (the increase that starts earlier) accompanied by a slower (later) reaction of commodity prices results in an *increase* in real wages, leading to an increase in labour supply but, simultaneously, a decrease in the demand for labour services (since, in the meantime, employers correctly perceive the changes in prices—this circumstance is crucial in Friedman's original conception). Of course, we have to take the notion of money illusion into consideration as well, since workers lack the knowledge to assess the actual dynamics of their real wage—an increase in money wages is, for them, the equivalent of an increase in real wages (even if, in this particular situation, this seems to be the case). Therefore, the initial expansion in employment would fail to take place under these circumstances—what is more, a contrary chain of events would unfold (in the labour market it is always the demand that sets the level of employment in case of an excess supply), and only a balancing out of changes in these two sets of prices (i.e. the restoring of the original real wage level) would make the economy return to the natural level of unemployment. The lessons from all this are perceived to be really tangible if the flexibilities of those two sets of prices are assumed to be non-delayed (to be the same); in other words, if our starting point is an equal extent (and, more importantly, equal timing) of changes in prices and wages driven by excess demand. Employees, according to the original presumptions of Friedman, expect a previous trend of price changes to continue, so higher money wages are perceived to be an increase in real wages, i.e. they fall victim to money illusion. However, employers are aware of the fact that real wages did not actually change, so the economy does not move away from the natural rate of unemployment, not even temporarily—all it does is move vertically upward (say, from E directly towards G). Despite all efforts by Friedman, money illusion *alone* cannot account for the EFG shift of the economy along the expectation-augmented

[12] To do this, we are required only to phrase the arbitrary assumption that the contracts leading to the temporary stickiness of prices in labour and commodity markets have different maturity dates.

Phillips curve. This requires (as an implicit assumption) a special pattern of changes in prices and wages: a pattern for which no explanation, not even a loose one, was provided by Friedman.

Money illusion is not *identical* with the situation when these two requirements, i.e. price flexibility and informational asymmetry (both or one of them) are met. These two requirements need to provide an explanation for employment effects. But then, what is money illusion? *Someone erroneously* perceives nominal changes as real changes. What happens if only one of these two requirements is met? In the first case, there is only an informational asymmetry, that is, employees do not realize the increase in prices. However, wages and prices change at an equal rate *in this case*. Employees fall victim to money illusion, since they perceive (falsely) an increase in their real wages. A favourable employment effect, however, cannot emerge, since employers know that real wages do not change. In the second case, the only difference is in price dynamics and we do not have to take the information asymmetry into consideration: again, employees fall victim to money illusion through a perceived increase in real wages. The same perception is experienced by employers, thus the outcome, if anything, is a decline in employment. However, if we eliminate this information asymmetry by assuming that both sides perceive the dynamics of real wages correctly, then, due to the principle of short side, an expansion in employment will again be prevented from taking place (because of an actual decrease in real wages). *Therefore both requirements (price flexibility and information asymmetry) need to be met concurrently in order for favourable employment effects to evolve.*

In theory, we could take an additional option into account, according to which, seemingly, the assumption of information asymmetry is not in fact needed. If real wages are perceived to be *increasing* by employees and, by employers, to be *decreasing*, we can follow through Friedman's story (i.e. the story of a boost in employment). In this case, it is really true that both sides fall victim to errors, i.e. there is no information asymmetry. However, this case is not relevant as the story is about sticky wages in an inflationary environment—sensible options stand for either (1) the *perception* or (2) the *non-perception* of a decrease in real wages. While assuming the existence of sticky wages and an inflationary environment, talk about a perceived increase in real wages is not a relevant aspect.

Friedman resorts to a rather loose explanation when he says this: "Producers will tend to react to the initial expansion in aggregate demand by increasing output, employees by working longer hours, and the unemployed, by taking jobs now offered *at former nominal wages*." (Friedman 1968; italics added). Employees can respond to the expansion in aggregate demand by working longer hours at previous levels of nominal wages (which can in no way lead to an increase in real wages[13]) *only if involuntary unemployment does exist*. This leads to an undeniable

[13] Referring to money illusion does not save the reasoning of Friedman, since he emphatically talks about employment conditions prevailing at former nominal wages. If nominal wages do not change, employees might even fall victim to money illusion (that is, expect former price dynamics

inconsistency with the interpretation of the natural rate of unemployment, since an expansion in employment, while *nominal* wages remain unchanged (or, at best, while *real* wages remain unchanged), is possible only in case of a non-equilibrated labour market, in which certain groups of unemployed people are prepared to take jobs at prevailing rates of money wage (or real wage), which, due to the insufficient aggregate demand, they cannot do. Later, we will revisit this issue and highlight the fact that the natural rate of unemployment, is voluntary unemployment in the strict sense of the word. As a footnote, David Begg (1982) scrutinized the consequences of rational expectations presumed in a disequilibrium model (i.e. it is not characterized by continuous market clearing). Under these conditions, fully anticipated (systematic) economic policy actions result in traditional and well-known effects derived by the Keynesian theory. At this point all we need to do is to refer back to the models of Chap. 3 where we were able to study the effects of changes in the quantity of money conducted by the central bank, assuming rigid nominal wages: the conclusion was that demand for labour services is not independent of the money supply.

Broadly speaking, authors assuming a general equilibrium environment and a continuous clearing of the market do not usually understand why disequilibrium economists presume institutional arrangements (such as long-term wage contracts) causing some nominal variables to become rigid and being blamed for the fail of market clearing and, in this way, for the eventual failure of full employment. They argue that these restrictions would lead to negative consequences that would prevent their emergence right at the outset—since utility maximizing agents would be interested in their elimination in order to achieve, say, full employment. Therefore these compulsions will turn out to be irreconcilable with certain microeconomic principles. However, the reasoning of Frank Hahn (1980) was motivated by other considerations. For him, transaction costs and information deficiencies might provide an explanation for the lack of continuous market clearing, therefore disequilibrium models can still be underpinned in terms of microeconomics. Thus everything depends on the microeconomic foundations chosen and the assumed characteristics of agents and of the economic environment. [→ **microeconomics]** For example, a cost associated with training the labour force may pose an obstacle to continuous market clearing.

The implicit information asymmetry underlying the mechanism provided by Friedman is an additional oddity. The logic here is that naive (adaptive) expectations of employees coexist with realistic perception of employers (revealing the actual price dynamics). In other words, while employers do in fact perceive an increase in prices (which erodes nominal wages through inflation, making an expansion in labour demand possible), employees functioning as buyers in the same markets see (by relying on their expectations) the continuation of a previously perceived dynamics of prices—which, of course, is an error. We can rightfully

to proceed), real wage as they perceive does not exceed the former level. *Unemployment, therefore, does in fact exist and, in this case, its existence is brought about by, insufficient aggregate demand.*

assume that, as far as the setting of real wage is concerned, the price level of goods in the market basket of employees is relevant. Although employees may be in the dark as to the actual (national level) inflation dynamics, they should be able to correctly perceive the prices of the commodities they actually purchase (or, at least, to spot current changes as early as possible, if an unanticipated rise occurs). *Money illusion will thus become one of the most controversial and unintelligible (not only in comparison to life-world conditions) elements of the story told by Friedman (which is still necessary to preserve the functionality of his mechanism).*

Shaw (1984) tries to play down the significance of this information asymmetry by offering explicit and cute assumptions arguing that, as far as the perceived-by-employees real wage is concerned, expected inflation is relevant, while, simultaneously, employers make their decisions on the demand for labour services by taking actual real wage (*actual*: determined by *actual* inflation) into account. Freidman (1968) refers to this strong assumption as an unimportant (and evident) marginal condition that does not need to be undergirded. It should be stressed again that *a reference to money illusion is problematic as stripping employees of their basic situation awareness seems to be arbitrary, whereas this is not the case with employers*. It is not supported by any intention to abstract; it looks as though Friedman had the intention of forming his basic assumptions underlying this mechanism (i.e. the expectation-augmented Phillips curve) with the aim of the *desired* results being seen as a consequence (*EFG* path). It seems that introducing this presumed difference in knowledge (along with the notion of money illusion) *was necessary to keep the views on the natural rate of unemployment defendable*. If we abandon this mechanism and look at the simple Marshallian cross of the labour market, we can spot a discrepancy in the way that employees and employers respond to the changes in real wages: what is good for one is bad for the other. Friedman was forced to tamper with assumptions because the employment effects he 'concluded' could not have followed from a pure market mechanism. He had to abandon the conventional set of abstractions (i.e. the set that neoclassical paradigm traditionally applies). The only question is whether his new assumptions stem from an idealization or if, yet again, this is about an arbitrary alchemy (F-twist). *The expectations-augmented Phillips curve rests on unfounded assumptions which Friedman devised from pre-set conclusions (natural rate, EFG path) by moving backwards, but he did not bother to base his assumptions on abstraction*. Hardly anywhere else in monetarist literature is there a better example for applying the methodology of positive economics.

In the paragraph above, we made the claim that introducing an assumed information asymmetry and, simultaneously, money illusion was necessary to support and defend the notion of the natural rate of unemployment. Since if involuntary unemployment is not ruled out, these dubious assumptions will become unnecessary and an increased demand will be capable of generating positive employment effects without them. However, there is the issue of a change in the growth rate of the quantity of money and then the stabilization of this new rate, which would bring about protracted effects (in a Keynesian sense) under these circumstances. Weeks (1989) expressed the view that the temporary nature of those effects also opened the door to questions regarding the very existence of governmental demand management. It is precisely this reasoning which gained prominence among new classical economists with the aim of making the conclusions drawn look even more universal, i.e. they argue that deviations from a full employment equilibrium become entirely insignificant, although real business cycles may still emerge alongside other mechanisms. This emergence of business cycles, which are partly independent of the new classical mechanism of Phillips curve and, eventually, of the fulfilment of expectations, is an undeniable benefit and unquestionable merit of the new classical theory. It is a topic to be dealt with in more detail when the analysis of the Lucasian production function will be conducted.

Shaw (1984) makes an additional mistake by relating the adjustment of expectations to the maturity structure of labour market agreements, arguing that the return to the natural rate might occur following the expiry of contracts at the earliest. Although these agreements can provide an explanation for the different dynamics of prices and wages, Shaw's interpretation suggests that these wage contracts are responsible for workers being unable to perceive the actual dynamics of the price level and their real wage. It seems to be a deliberate interfusion of two different effects. We will see later that the constrained flexibility of certain nominal variables can trigger real adjustment—however, this effect is entirely independent of money illusion. *Although wage contracts are concluded in the context of a specific rate of expected inflation, to infer from this that the actual path of inflation (deviating from expectations) is not (and cannot be) perceived by employees, would be an error.* Nothing prevents employees, disappointed in their real wage expectations, from withdrawing their (unduly) increased labour supply during the expiry period of wage agreements responsible for wage rigidity. This would only be possible if wage and labour contracts were concluded for the same period and employees were stripped of the opportunity of terminating these agreements during the maturity period, i.e. the reduction in labour supply was prevented via formal obstacles. However, classic texts do not stipulate this assumption. Bearing this in mind, the beneficial employment effect generated by excess inflation can only be a short-lived one, and *even in this case* it will only last until employees start to perceive a *de facto* (albeit unexpected) increase in prices. *The expansion in employment triggered by money illusion is a short-lived process* (it lasts as long as it takes to realize the de facto consumer price dynamics) *which may not even occur*. Therefore *reference to money illusion is not an ab ovo misconception, what needs to be done is to find its adequate scope*. A persistent employment effect is hardly realistic to assume on grounds of money illusion. Not even Friedman (1977) denied this. However, the information asymmetry eliminated in the final version of his Phillips curve was significantly more confusing than the application of money illusion.

Friedman (1977), almost certainly under the influence of Lucas, tried to dissolve the problem, raised by information asymmetry, in later versions of the model by causing both sides of the labour market to fall victim to an error which, eventually, eliminated the hitherto assumed information asymmetry. Employers face a signal extraction problem, i.e. although they perceive the increase in prices (and, simultaneously, the rise in the price of their own product), they cannot decide whether it is an increase in the general price level or a change in relative prices, which is favourable for them.[14] In order to avoid error implying adverse consequences (that is, in order to take profit opportunities), they enhance output (not doubting a favourable re-distribution of relative prices) and, in order for this to happen, they recruit workers through paying higher rates of money wage. It should also be noted that this reinterpretation of the model could dissolve the previous, highly problematic stipulation on price dynamics: the rise in prices, here, starts earlier in a natural manner, since it is this effect that triggers labour market adjustment and the expansion in production after all. *Through introducing the signal extraction problem, money illusion, which is responsible for labour market real effects, can remain in the game, making information asymmetry unnecessary.* Employers, when making a decision regarding their demand for labour services, consider only the factual increase in the prices of their own products (and, on these grounds, real wage as they perceive actually drops), and since they are not sufficiently informed on changes in the price level (that is, prices of products other than theirs), an information deficiency similar to that of employees also emerges (although the expansion in the demand for labour services is, after all, not triggered by money illusion). However, other parts of the story do not change: *employees continue to realize the increase in prices with some delay and interpret higher money wage rates as a rise in their real wage* (without this, i.e. in absence of an expansion in labour supply, we cannot talk about a shift from the natural rate of unemployment) and in order for that swing to end, the actual dynamics of real wages will need to be recognized.

However, it is important that, in the final version of the model, involuntary unemployment is not a necessarily implication anymore, since a rise in nominal wages (perceived by employees, suffering from money illusion, as a rise in real wages) triggers changes in labour market conditions. [→ **Phelps and Phillips curve**]

Phelps and the Phillips curve. An ambiguous phrasing in his prize lecture by Friedman (1977) suggests that the re-interpretation of the Phillips curve and the notion of natural rate of unemployment was a co-work with Edmund Phelps.[15] Far from it. Although the theory of

[14] It is a special characteristic that, in the original Friedmanian conception of the Phillips curve, the central bank was able to generate additional inflation so that it was known only to producers, while it could remain unknown to the mass of employees for a long time. A further oddity replaces it in this new classically inspired version of the Phillips curve, since, now, central bank can boost inflation with unchanged relative prices. One can hardly imagine an inflationary process that keeps relative prices intact (of course, this is mostly an empirical question).

[15] Edmund S. Phelps, professor of Columbia University (New York, NY), Nobel laureate of 2006. Júlia Király (1998) labelled Lucas' Nobel prize as "overdue", that is highly true for Phelps, since his grandiose theories, in spite of the Prize, lost much of their prior economic relevance. It is also interesting that Phelps has undertaken a significant role in prompting Lucas, and, in turn, Lucas received his Nobel prize in 1995. Phelps was awarded the highest scientific award (undeservedly) late with regard not only to the evolution of economic theory but his own career. Perhaps it is the most interesting question as to the theory of Phelps whether he should be classified as a representative of mainstream economics. On the scientific program of strengthening the microfoundations and on the influence of Phelps on the attitude of mainstream economics, see Howitt (2007). It has

Phelps was important primarily from the point of view of New Keynesian developments (cf. Vincze and Madarász 2005), stressing the discrepancies is highly illuminating in theoretical terms.

It goes without saying that both Phelps and Friedman assumed a negatively sloped short-run Phillips curve and that both denied a stabile long-term trade-off between inflation and unemployment (for Phelps, any adjustment in expectations on the basis of actual data shifts the short-run Phillips curve accordingly; cf. Phelps 1967a). In other words, none of them believed in the potential of fiscal and monetary policy in output stimulation in the long run *through demand management*.[16] It is an additional similarity that both authors stressed the existence of a level of unemployment (or, rather, employment) at which actual and expected inflation is equal. However, it is still true of the models of Phelps that higher employment leads to higher inflation (and not vice versa, as for Friedman). A rise in inflation caused by an unexpected expansion in aggregate demand increases the level of expected inflation as well and expectations cannot be fulfilled (as expectations also follow the adaptive scheme in the theory of Phelps, which means that agents anticipate previous experiences to continue) unless employment reaches its equilibrium level. Significant, too, is the fact that since expectations follow factual inflation, the higher the inflation, the higher the employment sacrifice of curbing inflation and inflationary expectations. Phelps' special equilibrium concept of employment, *the equilibrium rate of unemployment* (sic!), is, after all, also consistent with multiple rates of inflation and its special character is assigned only by the equilibrium of expected and actual inflation. The fact that, in Phelps' view, there is a low likelihood of full employment even at this equilibrium level of unemployment, is another problem, which means that involuntary unemployment should be taken into consideration. Phelps did not contest the fact that countercyclical economic policies can contribute to considerable reductions of involuntary unemployment (through unanticipated interventions), while, for Friedman, such an action did in fact have negative and adverse effects, since the economy moved away from the assumed equilibrium state (Phelps 2006).[17]

---

been pointed out that Phelps, just like Friedman, contributed to achievements in economic theory through reforming the central doctrines of mainstream economics (Grey 2002; KVA 2006). However, there is talk about a challenge to mainstream theory and about serious contradictions (e.g. Heckman 2008).

[16] There was consensus in the evolution of high theory as to the fundamental impotency of monetary policy in enhancing real economic performance. Keynes demonstrated it via the case of liquidity trap, while monetarists argued for this inefficiency on various theoretical grounds. For Keynes, the central bank can exert an influence on the interest-rate-dependent part of aggregate demand through affecting the rate of interest–however, this potential may be constrained by a liquidity trap and, moreover, the marginal efficiency of capital may occasionally collapse to such an extent that it cannot be compensated by any drops in the rate of interest. We can hardly find another thesis in which the antagonistic economic theories could similarly agree. Theorem of (persistent) impotency of monetary policy can, with reason, be regarded as an element (and manifestation) of an economic policy consensus. The significance of this common point cannot be overstated, however, literature still ignores this problem. For a bracing exemption, see Fischer (1977). It is obvious that arguments of Keynes with regard to monetary policy had an important role in stressing the efficiency of fiscal policy.

[17] That's why it cannot be said that any countercyclical monetary policy intervention would have been needed in the theory of Friedman. For him, the ultimate question was what happens when monetary policy tries to dislocate real economy from the natural level of unemployment and output. Strictly speaking, it is growth-enhancement and not following of countercyclical purposes. Since, as Friedman argued, monetary policy is not capable of stimulating real economy in the long-run, dislodging it from the natural level, the outcome of such an attempt will be a policy-generated

Phelps based the explanation of his augmented Phillips curve primarily on microeconomic foundations, highlighting the role of relative wages, pointing out how firms try to influence quit rates through changing these wages and, by doing so, their levels of employment. The well-established, familiar relationship between the change in wages and the level of unemployment was seen being formed again, albeit through a more complicated mechanism (Phelps 1967b). Setting the equilibrium rate of unemployment was based on these microfoundations as at the equilibrium rate of unemployment an average firm will raise its wage offer at a rate equal to the expected rate of increase in the average wage rate (KVA 2006). Under the influence of an unexpected economic policy action, a specific macroeconomic system follows the same trajectory in Phelps' models as in Friedman's theory (a short-lived drift from the equilibrium followed by a return to it), but along completely different mechanisms. For Phelps, expectations and errors are also crucial: individual producers consider unanticipated increases in demand as being particularly beneficiary for them, therefore they believe that their own increases in prices and wages will exceed similar actions taken by other firms–while at the same time output and employment also expand. These errors are eliminated by the end of this process and firms lose interest in maintaining an elevated level of employment. Price level, real wage as well as employment are driven towards their new rest-point values and, simultaneously, expectations of a cumulative increase in wage and price level are brought into line with actual increases (cf. Phelps 2006).

On second thoughts, the assumption of information asymmetry in Friedman's original conception[18] on Phillips curve is a *necessary but not sufficient condition* for real effects (that is, for a temporary reduction of unemployment).

Friedman, as we have seen, explains the inflexibility of wages by a rigidity caused by labour contracts. This in itself is problematic, if we think of Friedman's methodological confession in which he refers to the accidental nature of underlying assumptions (Friedman 1953). This is not more than an analogy of the principle of "the end justifies the means" cloaked in the guise of academic jargon. One may construct a model in whatever way if its predictions perform relatively well in terms of comparisons to observed facts—this is the view that Friedman holds. However, if we are of the view that we are allowed to construct our underlying assumptions as freely as possible (by using "construct" in a literal sense here again, incorporating procedures of arbitrary *alchemy*), it is not fair to justify one of these assumptions and to argue for its application by asserting that the given assumption is in accordance with directly observed reality, in other words with life-world conditions. If we cite *actual* effects of wage agreements as ones leading to the rigidity of nominal wages when we want to use this fact as an assumption, how can, at the same time, we argue that the empirical testability of assumptions is not a relevant aspect in judging our models?

---

business cycle. The need for a countercyclical monetary policy appears only in new classical macroeconomics for the first time through postulating cycles defined as white noises. The problem to be investigated considerably cuts the relevancy and merits of Friedman's theory, since he scrutinized an issue ("Is monetary policy capable of following long-term growth purposes?") that had been already answered consistently by high theory even before the time of Friedman ("No, it is not.") and he could only underline this answer. Is it a scientific achievement if someone demonstrates again (true, along different lines) that the dinosaurs died off?

[18] In which employers perceive price dynamics to be true all along.

It seems that Friedman found himself backed into a corner since he had ready-made conclusions prior to the establishment of his model (i.e. expectations-augmented Phillips curve). In other words, Friedman intended to underpin his view (that there is no steady trade-off between inflation and unemployment) denying the Keynesian theory by a compelling framework. Such a relation may not in fact exist—however, it is argued here that with Friedman's conclusions resting on shaky ground, his Phillips curve fell short of confirming this.

This theory (and its interpretation) seems to be problematic because it is unclear how a temporary fall in unemployment follows from the mechanism underpinned *somehow* (well or not so well) by assumptions. Employers' reactions are easy to understand. Money wages stiffened by agreements and increasing commodity prices naturally inflate nominal wages (reduce real wages), i.e. additional employment will be profitable and possible under these circumstances. However, the behaviour of employees is unclear. Theoretically, they are also aware of the rigidity of their money wage rates as they are supposed to be familiar with wage contracts, a crucial element in the creation of this stiffness. Their perception of the increase in price level is also a natural phenomenon, given that the story is precisely about the prices they are faced with, day in day out. It follows from these two circumstances that employees *should* perceive a decrease in real wages and, accordingly, it would seem nonsensical to talk about increasing their labour supply, which *is* a necessary precondition for a higher level of employment. It is natural that decreasing real wages mentioned in the story trigger an increase in the demand for labour services: it is nothing more than moving along the labour demand curve (the lower the real wage, the higher the demand for labour—this is precisely what is expressed by a negatively sloped labour demand curve). However, a drop in labour supply *should* follow from a decrease in real wages. In one way or another, Friedman had to prevent this mechanism from taking shape in order to avoid having to arrive at this conclusion. To do it, he needed to introduce, albeit implicitly, the information asymmetry mentioned above. Employees can only avoid reducing their labour supply if, quite simply, they *are unaware* that they should do it. *This is* the weakest element in the theoretical framework provided by Friedman.

One might say that employees are prevented from acting by labour agreements that lead to nominal wage rigidity, and that labour supply evolves under the duress of an inflexibility associated with wage contracts. However, our real problem is that employees do, in fact, act—by increasing their labour supply as a response to new circumstances altered by additional inflation. This can in no way be explained by a wage rigidity. For additional employment, an increasing labour supply is needed because, as a result of the short side rule, employment is always determined by the lower quantity of labour market demand and supply. If we still want to talk about an expansion in employment with decreasing real wages, we should assume employees to be completely blind to everything around them as has been mentioned above—or firms to have the power to set the level of employment in accordance with the demand for labour services. It would be equivalent to implicitly assuming slavery, but Friedman stopped short of stipulating this. Thus with real wages on the decline, it is the fundamental flaws in the situation awareness, attributed to employees, which is to be blamed for the expansion in labour supply.

Quite simply, this theory lacks a driving force that might explain the aforementioned expansion in employment. The rigidity of wages precludes increasing money wage rates and even if this was the case, it would rule out the explanation for a rise in the demand for labour services: increasing prices *and* wages do not put anybody in a mood for increasing labour demand, since real wages remain unchanged. If wages and prices rise in a manner that it will lead to the reduction of real wages, a rise in the demand for labour services can thus be explained. But what about the rigidity of nominal wages referred to above? And why (on what grounds, that is) can it be argued that prices increase at a higher rate than money wages, resulting in a decrease in real wages? This is nothing but an arbitrary assumption that reveals Friedman's real "Friedmanian" face spectacularly. We can hardly lean on empirical facts—at best we might say that, *in our opinion*, wages are *a little bit* more rigid in reality than commodity prices. This brings to mind Friedman's famous metaphor of helicopter money (Friedman 1970): bank notes are scattered evenly from a helicopter flying above agents. How can an inflation occurring at varying rates of increase in different sets of prices be consistent with this parable? Although it is to some extent a digression from our topic, but it might still be worthwhile to remember how wisely, in the case of David Hume (1752) or Irving Fisher (1928), quantity equation gave leeway for changes in relative prices. Even if we were ready to accept that Friedman, constructing his Phillips curve, assumed various dynamics of prices, this tends to reflect badly on his theory if we look at the big picture, given the contradictory nature of the things he had said at other times and places. Despite our efforts to eliminate these contradictions, we are faced with others. Friedman's Phillips curve makes a strong attempt to lean on price dynamics which was assumed by him axiomatically (that is, without accurate theoretical foundations) and which, at the same time, contradicts his views put forth in his various publications on quantity equation. A Phillips curve with reliable foundations might have been capable of producing a more sophisticated and vivid version of the quantity equation.

4.2.3 Relations to the Natural Rate of Unemployment

What do we know now? Conception of the Phillips curve underwent fundamental transformation during the monetarist counterrevolution. The stabile relation (as Keynesians assumed) between (the rate of) inflation and the level of unemployment was replaced by another consistent trade-off between unemployment and *the change* in (the rate of) inflation[19] (while the notion of the absolute price level had

[19] One could state that there is a relation between unemployment and the error of inflation expectations–but this option is preserved for the new classical Phillips curve (though,

already paled into insignificance[20]). As only surprises or unanticipated impulses matter, and as expectations become adjusted over time (in other words, we should not expect market agents to be incapable of realizing actual price dynamics in the long run), any favourable effect on unemployment can only be temporary. Therefore, a short-term and a long-run Phillips curve will need to be distinguished with the former (negatively sloped) curve describing the path of temporary moves and the latter, which indicates the natural rate of unemployment, determining the centre of fluctuations. Thus the long-run Phillips curve tends to be a vertical line, or it is much steeper than short-run curve(s) that make leeway for a temporary trade-off. However, we also had to take note of the fact that the political-operative exploitability of this relation depends on several assumptions, too, in the context of both the nature of price structure and of the information asymmetry of agents left unspecified. Of course, the problem of operative exploitability raises additional questions, given the question mark behind the issue of what kind of economy-wide structural effects follow from a recurrent misinformation to which the public is exposed—this topic will be dealt with at length when considering monetary policy.

At the same time, the notion of the natural rate of unemployment also changed—or at least it became more sophisticated. According to the initial interpretation on the natural rate, this natural rate is a variable determined by the structure of real wages and emerges from the equilibration of labour market demand and supply, as a manifestation of a general equilibrium. There are only temporary deviations from it, since the fluctuations of real wages, i.e. their appropriate changes (following the rules of Walrasian markets on excess demand and excess supply) always restore the equilibrium. Under these circumstances, an equilibrium real wage is uniquely determined by the technological level of the economy and the composition and volume of investment (Friedman 1968). However, it is evident that willingness to work is also affected by other economic conditions, therefore its level is far from constant—on the contrary, it is continuously changing under the influence of these causes. Friedman (1977) referred to them as "non-monetary" factors, including, among others, unemployment insurance and other forms of assistance offered to the unemployed, the effectiveness of labour markets plus the barriers or encouragement and various incentives to working. In his summary, Shaw (1984) accepts the same definition, although his phrasing is less accurate when he briefly refers to the natural rate by mentioning that it is "determined by real forces operating in the economy" which are independent, as it were, of inflation which is, in turn, a pure monetary effect. Therefore it seems to be of fundamental importance to distinguish between monetary and non-monetary (real) factors when talking about the determination of the natural level of unemployment and output.

---

undoubtedly, that could be applied properly just like in this case). Later, the difference between the change in the rate of inflation and the disappointment (i.e. the error) in inflation expectations will be highlighted (with particular attention to some cases when they are equivalent).

[20] In the more sophisticated words of Shaw, even if an absolute price level has a role in determining unemployment, the relation (that, hence, indicates a direct relationship with the original concept of the Phillips curve) is at best temporary (Shaw 1984).

This concept also triggered the objection of John Weeks (1989), whose criticism is of a terminological-semantic nature, rather than one based on conceptual considerations. He stressed that a variable the actual (and highly unsteady) level of which is affected by several economic factors can hardly be regarded as "natural". However, interpretations of and investigations into the concept itself sometimes use careful wording. It is a thought-provoking fact that concepts of the natural level of unemployment and of output are often mixed up in the texts as synonyms (which, having studied the single-commodity neoclassical world, should not strike us by surprise). For example, Shaw (1984) says that "the economy would exhibit an inherent tendency to gravitate towards its 'natural' (near full employment) level of output if left to its own devices [...]". On the other hand, this natural rate of unemployment does not imply that *measured unemployment* in an economic system staying at this level *would be zero*—what this means is that everyone willing to find work will eventually do so at the prevailing level of real wages. Therefore, unemployment is strictly of a voluntary nature (cf. Begg 1982). Moreover, one aspect of using the adjective "natural" seems to be justified as unemployment, for Friedman, has only one level (rate) that is consistent with an unchanging rate of inflation, *ceteris paribus*. This special numerical value can be referred to as the natural rate of unemployment, since a shift from that necessitates the difference, in one way or another, of factual inflation from inflation expectations. We can reveal the conceptual importance of the natural rate of unemployment and of the natural level of output without having the ability to measure them. Perhaps the most important characteristic of the natural level of output is that, on its basis national nominal income can be divided into price level and real output in such a manner that national nominal income will be determined by aggregate demand but the price level will, at the end of the day, hinge on the behaviour of employees and the suppliers of products (cf. Lucas 1973a).

4.2.4 New Classical Macroeconomics and the Phillips Curve

The Friedmanian concept of the Phillips curve was only a transitory form, since, as we have seen, its mechanism was determined by several disturbing assumptions and by multiple, unfounded clauses meant to eliminate those presumptions. Among them, information asymmetry was our primary topic: it turned out to have crucial importance in triggering favourable (albeit temporary) real effects. Attention was also paid to the assumed presence of money illusion.

The new classical interpretation of the Phillips curve was aimed at removing this puzzling inequality of knowledge, but its mechanism still requires money illusion. Shaw (1984) recounts a story in the introduction of which he stresses that labour market effects are definitely (and exclusively) the consequences of the signal extraction problem; however, the details fail to convince us. Shaw talks about (similar and already known) difficulties of employees in perceiving relative prices, but, for him, the result of this effect is still that workers (although they correctly perceive a change in their nominal income) fall victim to errors (or uncertainties) in judging real wage—and all of this is the result of their inability to perceive an increase in prices of the commodities they purchase. On second thoughts, this is nothing more than money illusion under a different guise. To understand this, all we need to be aware of is that the fact that on each side of the labour market a different story is told—despite the need for an identical story in order that both the employees and the employers fall victim to the signal extraction problem. The supply side mechanism is a familiar one—as we have seen, even Friedman himself referred to it in a later version of the Phillips curve. *In this case*, our independent variable (which is of a psychological nature) is the uncertainty surrounding the dynamics of relative prices—the analogy of which on the employee side would be an erroneous (or uncertain) judgement on relative real wages. Judging by the manner of their wording, Blinder and Fischer (1979) are more sophisticated but unable to get rid of money illusion. What happens here is that *they introduce the signal extraction problem affecting the employees as a consequence of money illusion*. In their story, employees work in separated segments of the labour market, so the information they have on other markets is imperfect and is obtained with considerable time lags. Moreover, employees can perceive price level dynamics correctly only one period later than their labour supply decisions are made. Therefore, they can make these supply-related decisions while being familiar with money wages alone—the actual dynamics of real wages, as the assumptions dictate, will only be known one period later. Anyway, Blinder and Fischer try to turn this story into a signal extraction problem by assuming that money wages (hectically) change from period to period due to both relative and aggregate (absolute) price changes and, under these circumstances, employees infer changes in relative (real) wages favourable to them. However, money illusion continues to be the underlying component of the story: nominal wages, unlike the price level, are correctly perceived by workers.

Thus, in a consistent version of the story, employees would decide to enhance their labour supply because they might think they are better off than other clusters of workers as a result of changes in relative real wages. However, this narration of the story requires a complete restructuring of the labour market—at least, the tradition that derives labour market effects *only* from real wages should be abandoned. As long as the independent variable of market for labour services is real wage, money illusion remains an unneglectable prerequisite for explaining the effects of unanticipated demand shocks generated by central banks (as long as we want to separate *actual* and *perceived* sides of effects). Moreover, if we continue to emphasize the signal extraction problem rather than that of money illusion—to make employees believe in a favourable rearrangement of relative real wages—this also requires an actual increase in real wages (note that producers, as an analogy of the story, decide to boost production because a general increase in prices is thought to be an increase in the prices of their own products due to excess demand—that is, an increase in each price did occur). The first event in the story pertaining to the signal extraction problem of employees in a logically consistent way must be an actual increase in real wages[21] (implying a general increase in real wages that is *falsely* interpreted by each employee only as increases in their own real wages, respectively); in this case, however, it would be dubious why employers perceive this increase in real wages falsely and, under these conditions, why they decide to enhance their demand for labour services.[22] Some kind of information asymmetry would again be needed, in which, still arbitrarily, employers would now be inadequately informed. In this way, however, the story would go on by assuming that firms decide to expand employment in spite of the increase in real wages, because now *they* misinterpret the rise in prices (similarly to employees in the original story) and, hence, the dynamics of real wages, say, because of their expectations. However, a *de facto* increase in real wages will be perceived by employers on the basis of their expectations as unchanged real wages only if they expect an inflation *higher than the actual inflation*. In this version of the story, in order to boost output, a central bank would be in a position to create a surprise situation by applying a growth rate of the quantity of money *lower* than expected. However, this is far from expanding aggregate demand, therefore, in this story, the drive to trigger the expansion of employment is, after all, eliminated. Despite the fact that in logical terms this story does make sense (the signal extraction problem of employees was told in a logically consistent way), it is still untenable, given the latter circumstance. We seem to have failed to get the problematic information asymmetry and money illusion out of the way.

Despite all of this, the interpretation of the Phillips curve underwent a change at the hands of new classicals. The interpretation of the natural rate definitely needs further specification, since the content element of the concept was perceived to be problematic as early as the appearance of the model provided by Friedman. The natural rate of unemployment was, in that model, said to be that same level (or rate) of unemployment which, *ceteris paribus* (i.e. assuming that additional economic factors that affect natural rate constant remain unchanged), prevails at an unchanging rate of inflation. This definition cannot be preserved anymore in the new classical approach, since *a disappointment of inflation expectations* (i.e. expectational error) or *a change in the rate of inflation* can be regarded as synonyms in Friedman's model. In other words, market participants always err when the central bank, through generating excess (additional) demand, sets a rate of inflation differing from the previous level. New classical macroeconomics replaced the adaptive expectation mechanism postulated by Friedman by rational expectations and, in the case of rational expectations, a change in the rate of inflation and expectational errors are no longer considered synonyms—consequently, the explanation of the natural rate of unemployment equally needs to be modified. *If expectations are rational, a change in price dynamics* (*i.e. in the path of inflation*)

[21] If this line of reasoning was followed, we would not be allowed to refer to money illusion, that is, to the evaluation problems from perceiving changes in price level with an undue extent of correctness.

[22] Moreover, we ignored the circumstance up to now how a central bank could generate prices that change in a completely unexpected way.

does not necessarily lead to erroneous expectations, i.e. a shift from the natural rate of unemployment (whatever it is—as not specified by new classicals in details) *is not considered necessarily despite a changing rate of inflation.* However, errors are still essential—*therefore, in the new classical theory, the natural rate of unemployment is a particular level of unemployment which prevails when inflation expectations are met* (with other factors kept unchanged).

It has to be stressed that it is here that the sharpest contrast between the economic policy consequences, drawn from these two theories, becomes visible. Orthodox monetarism assumed a steady (and, hence, systematic) acceleration of inflation[23] to be capable of keeping macroeconomic systems below the natural rate of unemployment (quite) permanently. New classicals abandoned this idea. Literature usually summarizes the discrepancy between the interpretations given by Friedman and those provided by the new classicals, by saying that the difference between short- and long-run Phillips curves are eliminated in the new classical concept, in other words any economy can move only vertically, even in the short run. Therefore, our model macro-economy is, after all, permanently at the natural rate of unemployment and only the rate of inflation accompanying any given state of employment can change (cf. e.g. Shaw 1984). Despite its popularity, this interpretation seems to be superficial and sketchy. *New classical macroeconomics never claimed that macroeconomic systems are permanently in a state of maximum output*—rather than that, *new classical authors narrowed the scope of economic policy to deviate economies from equilibrium*, therefore only a limited set of surprises (errors) could trigger non-equilibrium states of employment. As Eq. (4.27) will later help us understand that the difference between expected and actual inflation is not necessarily zero even with rational expectations, thereby opening the door to the assumption of a random walk around the natural rate of employment.

Before moving ahead, it might be worthwhile to summarize our conclusions drawn so far on the Phillips curve. Friedman, in his first interpretation of the trade-off between inflation and unemployment, assumed an arbitrary information asymmetry between employers and employees, since an increased demand for labour services is caused by *de facto decreasing real wages*, while an increase in labour supply is triggered by *real wages falsely perceived as rising*. This latter phenomenon is called money illusion. A more recent interpretation, provided under the influence of Lucas, eliminated this annoying informational difference by making producers fall victim to the signal extraction problem, (also) implying that both employers and employees need time to perceive the actual economy-wide price dynamics—despite the fact that making a decision on the demand for labour services is unrelated to producers' uncertainty about price level dynamics (since the real wage which they perceive does in fact fall). Employers make the mistake of inferring a favourable change in relative prices from the general price dynamics, but

[23] Considering this characteristic, the theory was labelled with the ever-recurrent adjective "accelerationist" (cf. Taylor 1979).

in the new story, when making their decisions on the demand for labour services, the accompanying assumption by employers is based on the increase in the prices of their own products rather than the general price level. Therefore, the need to perceive the actual drop in real wage does not require a knowledge on the general price level anymore. Thus money illusion was accompanied by no circumstance causing an inconvenience (an informational difference) and the whole story became more credible: employees keep their labour supply higher until they realize that the increase in nominal wages did not imply a rise in real wages. Temporary real effects of money illusion are short-lived ones in all versions of the story. It was also highlighted that money illusion was needed probably in order that the (temporary) expansion in employment could be attributed to misperceived real effects, as with involuntary unemployment, the boost in aggregate demand could still have enhanced employment. Also, it has to be mentioned that money illusion can hardly be regarded as a consistently deployable argument for the effectiveness of (surprise) economic policy actions due to the ever-faster perception of actual price dynamics (that is, macro-economies can shift from the natural level of output and unemployment for periods that become shorter with the elapse of time). The difference between the Phillips story, offered by Friedman under the influence of Lucas, and the new classical story, take us to the introduction of the next section.

Microeconomics. The intention to build macroeconomics on microfoundations is a direct neoclassical heritage of mainstream theory. General equilibrium analysis investigating the interactions of multiple markets of an economy through applying mathematical devices is not a problem of microeconomics.[24] The notion of a stabile market system emerged from the interaction of rational and completely informed agents and the auctioneer controlling their relations. Sometimes, mainstream economics modified the axioms of the Walrasian model significantly. Lucas and new classical macroeconomics adhered the ideal of rational market participants, however, replaced complete information by rational expectations and incomplete information sets. Islands models illustrate this effort particularly well: information deficiencies underlain the mechanism that can upset the equilibrium of macrosystems from time to time (albeit not systematically).

[24] The scope of general equilibrium analysis still seems to be unclarified in the system of economic theories. On technical grounds, it is reckoned to microeconomics, though the purport is not to analyze the dynamics of single markets or the behaviour of individual consumers or producers–microeconomics can deal with this problem without general equilibrium theory. Rather, general equilibrium analysis could be the prolegomena for a systematic neoclassical-mainstream macrotheory (as we have seen, these models sometimes quite directly rely on the Walrasian order of things); however, mainstream economics tries not to reveal this relationship. Stressing the general equilibrium analysis deserves attention for didactical purposes as well (either as settling of courses in microeconomics or as an introduction to macroeconomics), since it is always the missing link that could place macroeconomics to microfoundations. At the start of macroeconomic courses, the Keynesian invention is often mentioned, according to which macro-level investigations need special aspect and specific concepts, and even the parable is told stressing that micro- and macro-level rationality is something radically different (economic decisions being rational in micro-terms are not necessarily expected to be rational at an aggregate level–if all the firms cut wages, it can hardly resolve but deepens a recession). Then, after all, we use models that root in the neoclassical tradition.

In terms of microfoundations, the views of Friedman make a special case. For him as well, the swing from equilibrium is governed by an informational mechanism. Agents do not have enough information to evaluate their status correctly–however, postulating adaptive expectations is a direct break with former rational traditions. Agents exploit only information from the past when forming their expectations. On second thoughts, this can be seen not only as a break with previous traditions of mainstream economics, but postulating the adaptive scheme raises the suspicion of removing axioms from the notion of abstraction. Information on future outcomes is available, but agents completely disregard it. The success of the REH over this adaptive scheme is due to the fact that adaptive expectations could hardly be consistent with the traditional and common assumption of utility maximization (cf. Shaw 1984).[25] While Walras assumed rationality and complete information, Friedman's point of departure was non-rationality and incomplete information–later, new classical macroeconomics and the REH, assuming rationality and incomplete information, became part of this evolution.

On the face of it, Friedman's Phillips curve, based on adaptive expectations, (and the macro-mechanism stemming from that) is an abortive effort, but to actually interpret it like that would be tantamount to an ill-advised statement. It is more important than it seems on the basis of this analysis. Macro-theories based on rational foundations were often criticized and even if the difficulties revealed by the DSM theorem are swept away, the alleged failure of mainstream economics may be attributed to microfoundations postulated to be rational.[26] Rational microeconomics can underlie rational macroeconomics at best–however, if we do not intend to design a pure theory but a system that approximates life-world conditions and outcomes, it is worth considering to apply *irrational* (i.e. non-rational) *microeconomics* as a basis on which to rely. Friedman's system can be regarded as an effort to create irrational microeconomics; only the manner of irrationality can be questioned here. Irrationality *as Friedman assumed* is not from abstraction, it can hardly be proved by the acts of actual market participants. However, it should not be forgotten that behavioural economics has already contributed to mapping the irrationality of agents through innumerable results (see e.g. Tversky and Kahneman 1974, 1986, 1992; Kahneman and Tversky 1979). If our aim is to establish irrational microeconomics in order to lay the foundations of an alternative non-pure theory that can approximate actual socio-economic processes better, we might choose to establish the axioms of this irrational microeconomics on the basis of results of behavioural economics. If nothing else, the outcome is going to be interesting. Our only problem is that we still do not know how irrational we are. [→ **experimental economics, paradigm**]

Finally, mention must be made of the explanatory variable underlying orthodox monetarist and new classical Phillips curves. It was argued earlier on that a fluctuation in employment is triggered by *a change in inflation* in the orthodox monetarist case and by *a disappointment of expectations* in new classical macroeconomics. Orthodox monetarist and new classical Phillips curves fundamentally modified the nature of the trade-off between inflation and employment (output) available for economic (more specifically: monetary) policy. For Friedman, any change in the rate of inflation led to errors on the part of market participants,

[25] If we attach considerable importance to the microfoundations in the evolution of mainstream economics, labelling Phelps as a mainstream theorist does not seem to be complicated anymore. The Phillips curve he elaborated has definite microfoundations.

[26] Whether building a macro-theory on microeconomic axioms is necessarily undermined by the critique DSM-theorem exerted is another important issue. In other words: *is the validity of all micro-founded macro-theory constrained by the DSM-theorem*?

therefore a specific macroeconomic system could be dislocated from the natural level *even in the long run*—moreover, this statement is identical with the assumption regarding the efficiency of systematic growth-enhancing economic policies as well, since a *persistent* acceleration of the rate of inflation (disregarding all the evidently destructive effects of such an action) made it possible for unemployment to remain permanently below its natural level (of course, the pragmatic benefits of an effort like this can be questioned). In Friedman's system, a long-term reduction of unemployment is the result of a persistent accumulation of similar short-term effects (i.e. ones that reduce unemployment). However, it has to be noted that *countercyclical monetary policy actions are not needed in this system, given the permanent presence of a natural level (rate) of unemployment.* For Friedman, the dilemma of monetary policy is not a choice between an effort toward eliminating business cycles or allowing them to run their own free paths. The business cycles that could make this question relevant are absent. The problem is whether monetary policy is capable of a long-term spurring of the performance of the real economy—which is equivalent to the stimulation of real economic growth by monetary policy instruments. Friedman's answer is that the ability of this long-term growth enhancement is available for monetary policy after all—however, policy makers should not exploit it because of its destructive inflationary consequences.

For new classicals, an inflation-accelerating systematic economic policy action results only in a rise in prices without having any real effects, since market participants are in a position to adapt themselves to the increased quantity of money and to higher prices—it is, therefore, not possible here to trigger systematic effects by tampering with certain nominal variables. The discrepancy between *a change in inflation* and *a disappointment (i.e. error) of expectations* (which was only necessary to stress the distinction between factors capable of triggering real effects) will be eliminated soon.

4.3 Monetary Policy and New Classical Macroeconomics

4.3.1 Rational Expectations and Surprises

Although both the Friedmanian and the new classical interpretations of the Phillips curve stressed the sole efficacy of *surprise in terms of deviations from the natural levels*, the scope of surprising (unexpected) actions was drastically limited by new classical economists—it was the main lesson to learn from our analysis of the Phillips curve (beyond studying and revealing several, logically inconsistent circumstances). First, the emphasis should be put on considering the nature of surprises mentioned in each theory. To do it, once again, we have to take up the thread of the discussion in Chap. 2 on the differences between adaptive and rational expectations.

Generally speaking, adaptive expectations can be characterized most spectacularly through agents forming their expectations on the basis of time series models. A classic Box-Jenkins time series model is built on lagged values of its dependent variable and, at most, tries to exploit pieces of information hidden in forecast errors (cf. e.g. Franses 1998). In formal terms, it means that such a model makes predictions about the future path of a given time series on the basis of its past behaviour alone. By contrast, (if we temporarily disregard our reservations regarding the subject) for REH, agents forming rational expectations use complex, multivariable models, implying that they exploit information not only in past values of their dependent variable but also from other sources in order to make predictions about the future dynamics of this dependent variable (say, the rate of inflation). Rational expectations thus mean that not only most recent information on future outcomes but all knowledge on past events is to be exploited as Sargent and Wallace underline it (Sargent and Wallace 1975).

Agents are observing economic policy actions and the circumstances under which policy makers making efforts to control a particular macro-economy have taken them. *Measures taken as parts of a countercyclical economic policy are always situation-bound actions*. If agents experienced in the past that, say, monetary policy always responded by increasing the money supply in case of unfavourable fluctuations in unemployment, next time they will both expect it and adjust their behaviour accordingly—but exchange rate changes or changes in input prices (and innumerable other variables) may also be included among the factors. The fact that an economic policy regularly responds by taking the same actions in certain specific situations makes interventions systematic and, hence, predictable. So, perceived conditions determine expectations for the next period, since agents expect economic policy to adhere to the distilled rules when taking actions. Of course, if government communication pronounces certain (economic) policy changes or announces the outlines of the program to be followed by economic policy in the (near) future, all of this will also be taken into consideration by agents with rational expectations when making predictions about future outcomes. This learning process is nothing more than the exploitation of past experiences and the adjustment of behaviour—that is, of expectations[27] (cf. Shaw 1984).

Most criticism was directed at the economic policy consequences of the high theory of REH and new classical macroeconomics rather than the REH and the new classical macroeconomics themselves with the criticism being aimed at the thesis of inefficiency of the *systematic* countercyclical monetary and fiscal policy.[28] The basics of the relating argumentation by REH are reviewed above. According to this, agents forming rational expectations are informed about the future of economic

[27] Even if the success of this learning process may be doubted on grounds of discussion in Chap. 2. Some aspects of the difficulties of the learning process will be investigated in detail when reviewing the systematic and random components of a monetary policy rule.

[28] It has to be mentioned that inconsistencies are sometimes found in the REH literature, that is, it is not clear whether authors talk about *monetary* or a(n) (general) *economic* policy (cf. e.g. Brimmer and Sinai 1981).

policy, therefore a planned monetary (that is, purely nominal) policy shock would only affect the price level and the real economy would refute to respond even in the short run—that is, Phillips curve is a vertical line even on this time horizon, since agents (according to the assumptions of REH) are capable of making unbiased predictions about the future path of monetary policy. However, our foregoing arguments, generally speaking, were meant to throw light on the fact that if this unbiasedness cannot *realistically* be maintained, these economic policy conclusions will mostly be invalid. It is a statement of fundamental importance by new classicals that only surprising (unexpected) actions are capable of triggering real effects and by doing so, they can dislocate an economy from its natural level of output (whatever it is). Moreover, the resulting assertion is of the same importance that questions the possibility of all kinds of systematic monetary (and fiscal) policy (cf. Barro 1977). Rationally formed expectations raise high and heavy barriers around (the possibilities of) economic policy, since *any* systematic economic policy (i.e. an economic policy following *any* rule) constitutes part of the economic environment surrounding agents who are capable of learning the rules of this policy and also adjust their behaviour to future actions of economic policy in advance. Thus any effects can only be triggered by unsystematic (random) actions.

However, the picture is not as clear as it is depicted by John B. Taylor (1975). For him, if agents are capable of making unbiased predictions about the next steps of a monetary policy, there will be no real economic effects—and, in his view, all that this unbiasedness requires is the *observation* of central bank actions by agents under different macroeconomic conditions. One of the most serious problems of intermediate targets specified to manipulate the quantity of money is that the selected instrument proved to be mostly uncontrollable even for central banks, therefore assuming market agents to have the ability of observing its dynamics correctly is optimism at its worst. Not to mention a second implicit statement hidden in Taylor's comment that proclaims agents to be capable of *correctly perceiving* and *precisely interpreting* economic constellations (i.e. they can thoroughly perceive and analyse not only central bank actions but also the situations justifying these actions). We cannot turn a blind eye to the fact that *perceiving* something is not the same—and, in the present context, it is far more than a simple semantic play—as *accurately perceiving* or *understanding* something. It is true that if these requirements were met, predictions would be unbiased. However, the real problem is that new classical authors refrain from finishing sentences such as the one above, and by doing so, fail to make note of the fact that it is hard to build a theory on plausible grounds via these assumptions. Naturally, the option according to which unbiased predictions are made possible by economic policy announcements on future actions sets considerably less stringent requirements—this case has already been mentioned in connection with the problem of inflation targeting. This reasoning often emerges in the REH literature as a kind of easing of assumptions in the theory of Muth (e.g. Fischer 1977).

The new classical interpretation outlined above highlighted not just the importance of surprises (Friedman had previously made efforts in that direction) but tried to stress the efficacy of *completely unpredictable* monetary (and general economic) policy actions in terms of generating deliberate deviations from the natural level of unemployment/output. Differently put, *new classical macroeconomics accurately defined the (model) environment in the dominance of which only unexpected economic policy actions can be efficient.* In Friedman's theory, systematic economic policy actions could exert a temporary influence on the real economy, given the fact that for him, the only criterion for efficacy was a (deliberate) deviation of certain target variables (inflation was chosen for this special role) from their previous trends. Within the realm of new classical macroeconomics, systematic,[29] i.e. predictable actions are not capable of exerting such an influence.[30] The explanation is simple: as a shift from the natural level of output and employment requires unanticipated (i.e. unpredicted) inflation dynamics and since systematic economic policy steps are no longer capable of bringing into play the king of surprise required for this to happen, only unexpected (random or, if you like, non-systematic) actions can trigger the disappointments of inflation expectations.

4.3.2 The Island Models

Before moving ahead toward further investigations of economic policy recommendations of the new classical macroeconomics, we need to take a short but very useful detour. This detour, aimed at studying the island models given by Lucas, is expected to be fruitful for various aspects. On the one hand, these models can help us understand the nature of the frictional mechanism that was supposed by Lucas to underlie monetary shocks. Most comments on the oeuvre of Lucas highlight that one of his major achievements was to establish the macro-theory on micro-foundations (cf. e.g. Király 1998). His island models served this very purpose, even if he was neither the only nor the first scholar to move economic thinking toward that direction—the earliest works of Phelps provide ample evidence of that fact. On the other hand, we can scrutinize the characteristics of this micro-founded macro-theory (in this context, the aggregate supply function of Lucas is a focal point to be examined) and see whether it is marred by some kind of logical inconsistencies. However, our primary purpose is to review the evolution of the Lucasian aggregate supply function.

Lucas laid the foundations of new classical macroeconomics in two papers (Lucas 1972, 1973a) followed by the concept of an *equilibrium* model of the

---

[29] Weeks (1989) stresses that new classical macroeconomics, by emphasizing the inefficacy of systematic changes of the money supply, raised the banner of neutrality of money as well.

[30] The debate indeed focused on whether the view is correct that only unexpected effects are capable of dislocating the equilibrium temporarily (this view is labelled as LSW-theory after its elaborators, Lucas, Sargent and Wallace) or the short-run potency of anticipated monetary policy steps must be admitted as well, occasioned by the limited flexibility of prices–as a footnote, this view is equivalent to assuming price dynamics to be dependent on past path of changes in prices (cf. Gordon 1979).

business cycle (Lucas 1975). It is by no means an exaggeration to claim that these three short treatises summarize the theoretical and empirical message for which Lucas was awarded the Nobel Prize by the Royal Swedish Academy of Sciences in 1995 (cf. Király 2005).

Lucas integrated his seminal papers into currents that had been already dominating economic thought for some time. In 1969, Phelps envisioned (and urged to start researching towards) a new theory on inflation and unemployment that was to be characterized by the assumption of incomplete information (Phelps et al. 1969). The other source of Lucas' theory was constituted by the island models. Prior to Lucas, Phelps also made efforts to apply macro-structures built from isolated economic units (Phelps 1969) but Lucas regards Samuelson's work (Samuelson 1958) as his immediate precursor[31] (cf. Lucas 1972, 1995).

The purpose of the first Lucasian island model was to establish an analyzing framework to support the understanding of the nature of the relationship between inflation and real economic performance by assuming that this relation offers no trade-off exploitable by economic policy. Lucas' intention was to prove that the *Phillips curve exists without existing*. In every period, N agents are born (where N is a positive integer) living for two periods. That is, there are always N young and N old individuals living together (so, there is a constant population of $2N$) inhabiting the two islands (i.e. two isolated markets) of the world. In the first period of their lives, each agent supplies and consumes one part of the amount of the (unstorable) product. The other part is up for sale, offered to the older generation. *There are no differences in productivity among the young, all of them achieve the same output*[32] and when making their production and consumption-related decisions, it is not only current price level but the price expected for the next period that they take into account. Beyond labour force and the commodity it is producing there is fiat money in the setting as a third factor to be taken into consideration. This is issued by the supra-islands government and enters the economy by means of a beginning-of-period transfer paid to each member of the older generation. The stock of money and, hence, monetary demand is allocated across these two islands equally, since *members of the older generation live in equal numbers on each island* and each member holds an equal amount of money. This condition prevails even after the new transfers are paid, so the growth of money is proportional. The older

[31] Samuelson's model proved to be flexibly extended, since a model specified on two overlapping generations can be easily expanded to include a third age group (it is so true that Samuelson's model had already been built from three overlapping generations and only Lucas simplified the setting to two clusters) making it possible to investigate and interpret investment in human capital, which seems to be its most beneficial characteristic for Júlia Király (2005). Moreover, one can easily introduce capital in addition to labour force, so capital investment cycles can be also studied in this modified version of the initial model. Of course, the original model of Samuelson was directed at purposes other than the goals at which it was aimed later by Lucas: Samuelson's main interest was modelling interest rate dynamics.

[32] The single-commodity character of the economy is completely evident here as well, hinted slightly by Lucas.

are not allowed to leave an inheritance, so unspent money balances revert to the government. Inter-generation transfers are precluded (cf. Lucas 1995). An exchange process occurs on these two isolated markets where the commodity supplied by the young is traded for the money owned by the old. Members of the older generation are distributed evenly, therefore equal numbers of the old live on each island, so money supply is also distributed evenly across the two markets. *However, members of the younger generation are allocated stochastically to one island or the other*: an improbable albeit viable option would be that the young will live on one of the islands only. After populations are established, communication between islands is ruled out (accordingly, exchange is allowed only between partners living on the same island). Under these circumstances, trading is performed by means of an auction that guarantees the emergence of the market clearing (equilibrium) price.

Agent-based models. Agent-based models (in other words: agent-based simulations) were brought about with the express purpose of getting the better of traditional modelling techniques (and philosophy) of mainstream economics. This school is still in its infancy and the bulk of the seminal papers are between 5–10 years old.

After all, the intellectual content underlying these models is the same as the one that has determined the approach of mainstream economics since the time of Adam Smith. According to it, the economy is a complex adaptive system (the expectations of agents, as in reality, are based on beliefs about expectations formed by other agents), in which production, consumption and trading as organized activities emerge as results of uncoordinated actions of market agents. This is a direct manifestation of the notion of the invisible hand. Agent-based models scrutinize the characteristics of the order evolving from the interactions of the agents and the way in which interdependence among interrelating agents modifies their behaviour (Oeffner 2008). Agent-based models are *simulation models*, meaning that a researcher launches the agents defined according to his own interests and purposes in an artificial (computer) environment and observes how these agents behave in the course of the successive steps of interaction. By means of these models, economists often study certain typical markets (e.g. financial market–for further markets see Nicolaisen et al. 2001; Tesfatsion 2000) but macroeconomic extensions form yet another burgeoning topic.

These models follow a bottom-up perspective, so it is a brand new approach to building macroeconomics on microfoundations. While even new classical models smuggled the Walrasian auctioneer back in their systems, the "phantom" who coordinates transactions in order to achieve equilibrium (after all, he presumes equilibrium *by* and *in* his activity), by contrast, agent-based models are interested in the outcomes of the interactions of a large number of autonomous agents without any global top-down controller (Gaffeo et al. 2012). These models abandon the rationality assumption of mainstream economics, too, thus economists presume bounded rationality (often an explicit adaptive scheme) instead of rational expectations, and, simultaneously, postulate incomplete decision-making abilities on grounds of the results from behavioural economics. This in itself is capable of approximating infinitely abstract mainstream models to life-world circumstances (agent-based modellers, brimming with self-confidence, tend to regard mainstream models simply as unrealistic and contradictory to life-world experiences) which paves the way for clarifying the conditions that lead to phenomena such as sudden and non-rational changes in investor opinions leading to dramatic fluctuations in asset prices. As an additional innovation, heterogeneous market agents who are usually capable of learning from each other are put on the stage. In this way, these models make it possible to investigate, the economic effects of this learning process on the one hand and, the interferences of actions on the other. These learning processes are sometimes depicted with quite a bit of sophistication: since agents are forced to base their expectations on beliefs about the expectations of others, these beliefs are uncertain and hypothetic. Agents are continuously testing these hypotheses, i.e. their expectation models (which are, after all, theories describing markets) and replace the poorly performing models from time to time with better models to underlie their trading behaviour (e.g. Westerhoff 2009). Thus models that describe markets and dynamics of certain market variables are in competition with each other (homogenous and converging expectations are out of the question), while the best performing model will be selected (Arthur et al. 1996; Chan and Steiglitz 2008). Behaviour develops following an evolutionary logic.[33] This setting is radically different from the world of rational expectations, and even if the homogenous structure of expectations presupposed by the REH emerges, the question of what circumstances lead to this situation or to heterogeneous expectations[34] is still an intriguing problem.

However, it should be realized that agent-based models and mainstream economics have the same roots, although the differences are, of course, significant. Mainstream models are pure mathematical constructs in which theorems emerge as the results of rigorous mathematical deductions. When these models became the focus of attention in the 1970s and 1980s, mathematical knowledge was quite remarkable, accompanied by limited computer capacities to support researchers. This pressure explains why mainstream research was built on axiomatic-deductive methodology so strongly and on the improving techniques of econometrics (which, by then, were helped by a computer technology capable of providing adequate support). Compared to these puritan mainstream theories, agent-based models are far too complex for consequences and results to be drawn on pure mathematical grounds. Economists put the emphasis on simulation, instead, so what we have here is an *empirical* investigation of interactions. In spite of the robust mathematical foundations, these models resemble the methods of experimental economics rather than the imposing

[33] Manifestations of the evolutionary logic in economic theories sometimes quite directly adhere to the Darwinian theory of evolution. There are models in which entities (e.g. trading rules; Hirabayashi et al. 2009; Lin et al. 2007) create new generations. The so-called fitness-function describes the viability of these entities: those who are the fittest have the highest chance to create successors. A successor inherits the genes from its parents, but mutation is in the game, i.e. any new generation is going to differ from the ancestors somehow. Researchers use this technique based on the so-called genetic algorithms even to solve highly interesting and specific searching and optimization problems (e.g. Gruca and Klemz 2003).

[34] Of course, it has to be admitted that these agent-based models help us take a critical view on mainstream theory and models. As new classical macroeconomics also emerged with an explicit intention to build solid microfoundations, it may arouse suspicion if a school with a strong desire to differ fundamentally from new classicals talks about the need for microfoundations as well. However, when the underlying philosophy of agent-based models is thoroughly scrutinized, the difference between bottom-up (agent-based models) and top-down (mainstream economics) conceptualizing aspects becomes clear itself immediately. Agent-based models put light on the essence of micro-founding efforts of new classical macroeconomics, according to which economists are building top-down, i.e. they postulate a macro-structure in the beginning (define the system as a system), and try to trace some components of the macro-structure back to certain (abstract and presumed) characteristics of agents (cf. Oeffner 2008). The monetarist way of micro-founding is confined only to creating a market from a large number of agents that are completely identical. The Phillips curve (of Friedman and Lucas) is an outstanding example, where market mechanisms are deduced from the postulated behaviour of a multiplied average individual. In such a case, it is in fact indifferent whether the behaviour of only one agent or that of many is scrutinized. It is not hard to realize that the shape of the Phillips curve can be explained well on the basis of the behaviour of only one agent.

mainstream models. However, *the results distilled from agent-based models are the consequences of the basic assumptions just like in the case of mainstream models.* A researcher cannot arrive at conclusions other than those which he himself formed in his assumptions regarding the system and the agents (behavioural patterns to be followed by the agents put in a computer application or the way they learn from each other, etc.–this is all defined by the modeller himself; cf. Howitt 2007). Because of the modified methodology, however, these assumptions are capable of approximating reality more efficiently. The reason why a simulation technique is preferred to mathematical deductions is that these assumptions are applied in greater numbers and in more complex forms than in mainstream models, so the resulting picture is much more complicated. This similarity also occurs in the context of problems under scrutiny as there exist agent-based simulation models that scrutinize the prerequisites for prolonged crises (Dosi et al. 2008). Answers are different of course, since agent-based models, given their structure and methodology, are a suitable instrument to focus more efficiently on minor details than pure mainstream models investigating fundamental and comprehensive laws. Therefore agent-based models that follow conceptual lines different from our present analysis highlight the occasional and fragile (unstable) nature of general equilibrium. [→ **microeconomics**]

In the world of Lucas, information deficiencies come from the fact that relative and absolute changes in prices cannot, with certainty, be distinguished accurately, in other words it is unclear for a producer whether the price of his own product changes due to a real or a nominal shock (it is evident that a nominal-monetary shock would affect the price level alone—while a real shock, not accompanied by a monetary expansion, would only modify the price ratios). Each market participant knows the pre-transfer holdings of the older generation but not their post-transfer balances (up to the next period), members of the younger generation will, therefore, only obtain information on this based on the evidence provided by demand dynamics (and by price level reactions). If m denotes the pre-transfer money stock per member of the older generation and m' indicates post-transfer balances, then the rule that governs the relation between stocks can be written as

$$m' = mx, \tag{4.1}$$

where x is a random variable. Thus, dynamics of the nominal money supply is not known to the suppliers.

So, it has to be noted that possibilities of two kinds of shock is in the offing. One is the allocation of the young across the islands. Since the money supply (whatever its extent) is distributed evenly across the markets (that is, members of the older generation represent the same monetary demand on each market), only the allocation of goods between the markets (islands) can trigger changes in relative prices (one should expect higher relative prices in markets where lower quantities of commodity are produced). *However, suppliers are unaware of this since all forms of communication between individual islands are excluded from this framework*— that is, if the populations of these islands undergo a change from one period to the next (this being the only source for real shocks), that information will only be available to agents by drawing relevant conclusions from the changes in prices. If this were to be the only factor capable of affecting prices, there would be no question of information deficiencies, since changing prices would clearly indicate that a real shock has occurred. However, a clear and unambiguous inference is impossible because monetary expansions are unforeseeable, since the growth of money supply is not known to producers. *The root of confusion is that an awareness of nominal shocks, based on information available to members of the younger generation, appears via price dynamics.* An increase in price in each market may indicate a rearrangement of demand across the islands (a real shock) as well as a monetary (or, if you like, nominal) shock that emerged due to a change in the money supply. However, producers have to make their supply decisions on grounds of information sets conveyed by the price level, while they are supported, at best, by the knowledge of the probability density functions of the stochastic variables governing the monetary expansion and the allocation of new generations across the islands. And since their inferences are not accurate, that is, the consequences of real and nominal shocks cannot be separated entirely (even if the information available is exploited as comprehensively as possible), monetary changes trigger real adjustments: a monetary expansion leads to an increase in real output. On these grounds, however, the converse statement can also be justified: a fully anticipated monetary change would not trigger any real effects (so, consequences of a simultaneous real shock, if any, could be separated).

Lucas assumed agents forming rational expectations. This is exactly the circumstance by which he could outperform the island model of Phelps, since Phelps had postulated an adaptive scheme (cf. KVA 2006). Market participants utilize *all* available information, but this is not enough to perfectly foresee future dynamics—that is, true predictions cannot be made. For Lucas, imperfect information was the basis of the frictional mechanism that guaranteed *non*-neutrality of money in his framework after all. It took quite some time for Lucas to make an assumption regarding this friction: initially, his attempts were focused on multi-generation models in which, similarly, money was exogenously given. But in this framework, in other words, under friction-free conditions, the relation between inflation and real output could not take shape: money was strictly neutral, changes in the money supply triggered only proportional shifts in the price level (Király 1998). For Lucas, *non-neutrality of money was implied by the assumption of isolated markets.* So, Lucas could preserve the quantity equation without having to cope with the inconsistency of false dichotomy. Quantity equation guaranteed the potential of monetary authority to affect prices, but, from this, Lucas did not infer the neutrality of money but a definite causal relation which, because of the characteristics of the system he created, did not guarantee the potential of economic policy with which to systematically influence real output. This completes the details outlined about dichotomy in Chap. 3. However, the temporary non-neutrality of money had already proved valid in Friedman's Phillips model because of money illusion. *If money illusion is present, dichotomy cannot exist.* What does money illusion and classic dichotomy mean? The point about money illusion is that market participants perceive certain nominal changes (erroneously) as shifts in real quantities. What is most important here is its *consequence*, since, after all, money illusion is the phenomenon that transforms nominal changes into real adjustments. In this way, money illusion interlinks the nominal and real sphere. By contrast, dichotomy means that the dynamics of nominal and real variables are isolated (and, moreover, independent of each other). *Money illusion and classical dichotomy cannot coexist.*

The frictional mechanism created by Lucas is based on the *signal extraction problem* mentioned earlier on: due to incomplete information, each market participant can only perceive the price in his own market along with the relation between this price and its expected value, agents thus draw conclusions and make supply decisions on these grounds. In the course of this decision-making process, agents should distinguish between unanticipated changes in the price level and equally unanticipated shifts in relative prices[35] (Ellison 2003). And there is a further characteristic to highlight. Expectations, as we have seen, are *not true*, in other words they are marred by errors; *however, these errors do not include systematic biases, i.e. there is no trade-off between inflation and real output available for economic policy to exploit.*

Utopia. Max Weber (1904) regards the realm of the models of neoclassical economics (using his own words: abstract economic theory) as utopian for the ideal(−typical) character of the description they gave on economy. Weber's traditionally rigorous and severe argumentation precludes it to be only a lofty rhetorical flower. Where does the conceptual similarity mentioned by Weber stem from?

The history of utopia[36] as a literary genre can be traced back to ancient times. The value underlying literary works is a special longing to get away that was fed by a displeasure at the present walk of life. The dissolution of this tension was manifested in creating ideal countries, eras and peoples–with the purpose of literary-artistic delighting. Utopias depict space and time where and when everything was/is/will be alright: there is no ponderous physical working, nature abounds in its fruits, peace and justice prevail. The imagined perfect world was placed in a diversity of times (distant past, present or future) and space (e.g. the country of the Phaeacians, Sparta, Persia etc.). We can even find utopian ideas in which the advancement of convenience and felicity of the human race hinges on technological improvements and the spreading of automotive machines. However, the most important ancient utopia that highly influenced the later development of this genre was Plato's Republic in political science. Contrary to the preceding legislators of the Greek traditions, Plato outlined laws, structure and functioning *not* of a concrete state; instead, he came up with the idea of a perfect country, unfettered both by time and space (Németh 1995).

---

[35] The problem Lucas scrutinized had been something different from that identified by Júlia Király as the key question. The author writes that "[Lucas] regarded markets of an economy as isolated islands that can get pieces of information available for all the islands (say, dynamics of money supply) but do not have detailed knowledge on each other. This assumption seems to be acceptable even today, in the age of »perfect information« if we consider the fact that while aggregate data are available almost free, one must pay such a high price for special data that it is not worth buying or it cannot be purchased at all." (Király 1998). Rather, a precise phrasing of the problem can be given by saying that agents, although they *know* the price dynamics in their own markets, are not able to *interpret* them; and it is the less true that they posses more information about the aggregate level, since it is exactly the events happening in other markets (on other islands) about which agents know nothing. So, the analogy does not go on all fours.

[36] The word "utopia" is used in three senses at least. The compound of Greek origin was applied for the very first time by Thomas More as the name of the country he imagined. Its meaning grew richer afterwards as started appearing as the name of all kinds of envisaged countries and, simultaneously, became the generic term for literary works about such countries (Morton 1969).

Utopia by Thomas More[37] was the most important stage in the modern history of utopias. The utopian state-vision of More is based directly on the thoughts of Plato. As a matter of fact, More considers all Plato's arguments as presumptions and, while relying on these principles, he gives the description of a fictitious perfect-ideal (but envisaged-in-operation) state–it is Utopia[38] (Morton 1969). Our most important inference is that utopias always depict worlds that definitely differ from the economic-social-political conditions of the world experienced in reality. There is a direct socio-critical intention which remains true for negative utopias as well (Orwell).

However, Daniel Defoe's *Robinson* is the utopia that bears the most direct relation to the economic thought. Of course, the criticism of contemporary society is tangible here, since the sailor who finds himself on a solitary island becomes the symbol of the lonely struggle of the outcasts of London. More importantly, though, Robinson, a man facing natural forces also becomes the symbol of lonely individuals having to brace themselves for competition in free market capitalism. Despite the huge distance between Robinson's island and an ideal world, it is still a utopian vision, since the personal character and abilities of a human being, in his loneliness, are revealed freely, displaying a flawless purity. Robinson and his island eventually emerge as the utopia of the bourgeoisie, that of a man engaged in husbandry and capable of mastering nature (Ungvári 1956). Robinson's success in economics was, by now, only one step away. Robinson is the prototype of Man having his share of the economy as a lonely individual and, although as the protagonist of a novel he was not a *homo oeconomicus* but a moral creature, he could be easily stripped of these characteristics. Working on his small plot of land, our lonely hero is confronted by nothing else but a series of decision-making situations of an economic kind (Varian 1999), while his island-sized world offered the perfect setting for a framework where the models of economic theory are scrutinized (Mérő 2004). Having rid Robinson, without too much difficulty, of his uniquely human characteristics, economists were now in a position to put the economic man, whose only concern is to make rational decisions, into the shoes of quite a directly experienced individual.[39] [→ **homo oeconomicus, labour union**]

The world of neoclassical economics is, in fact, a utopia, since it illustrates the pure and perfect nature of the kind of work that the economic man performs. However, when these economists start attributing normative meaning to the consequences drawn from their models created under laboratory pure conditions, as it were, this utopian character emerges much more powerfully. It is not only a pure display of certain features that are of crucial importance for an economic analysis, but the definition of an ideal world and socio-economic status to be achieved, as opposed to the directly observed and experienced spatiotemporal reality. In this situation economic models are seen as 'the Promised Land,

---

[37] More, sharing in the current of Christian humanism, lived under the joint spell of Greek-Latin antiquity and Christianity at the same time. He was the son of an interesting era: there were vast debates, say, on if an angel could be at multiple places at the same time or she was bounded by the spatiotemporal constraints similarly to us. More perished on the block in requital for his consistency of principle and faith (for further details, see Chadwick 1993).

[38] For scientific socialism, More was one of the very first communist thinkers (cf. Mihalik and Szigeti 1984). The idea is that perhaps Plato was the only one preceding him in these efforts (Morton 1969). More's communism is highly controversial. Some authors deny that More can be discredited by communism even because of his Christianity and his views charged to be pre-communist (e.g. the lack of private property, strict rules as to dressing, interpreting society as a huge family) rhyme rather to cloistered life. Some of his assertions were quite simply meant to be jokes (Nigg 1979).

[39] It is interesting that Weber (1934) relates the highest level of (actually manifested) economic rationality also to bourgeoisie.

unlike the current conditions characterized by inefficiency, wasted resources and failure to guarantee the highest level of welfare.

A more precise explication of the signal extraction problem was given by Lucas in a paper one year later (Lucas 1973a). This paper was directed at an empirical verification of the relationship between real output and inflation: according to his hypothesis, real output is invariant to changes in inflation dynamics, implying that there exists a natural level of output. The most important invention of the paper is an analysis of the supply side. The previous island model was preserved in many ways, although a large number of separated competitive markets (islands) were introduced as a new feature. The relation of demand and supply still reflects stochastic effects, but it no longer comes from a random allocation of suppliers across markets. Here, changes in demand become unexpected, that is, demand in each period is distributed unevenly across the markets. Relative and absolute (general) price changes, yet again, emerge together and simultaneously.

The dynamics of (real) output is fraught with fluctuations. We can separate a trend y_{nt} common in all markets and a cyclical component $y_{ct}(z)$ following different paths on different islands. Individual supply is written as

$$y_t(z) = y_{nt} + y_{ct}(z). \tag{4.2}$$

Originally, Lucas described the cyclical component by using the following equation:

$$y_{ct}(z) = \gamma\left[P_t(z) - E\left(P_t \middle| I_t(z)\right)\right] + \lambda y_{c,t-1}(z), \tag{4.3}$$

that is to say, the cyclical component is dependent on perceived relative prices and its own lagged values. Since specifying a distributed lag model is always an empirical problem, authors usually include only relative prices as explanatory variables (cf. e.g. Ellison 2003):

$$y_{ct}(z) = \gamma\left[P_t(z) - E\left(P_t \middle| I_t(z)\right)\right]. \tag{4.4}$$

Whatever form is taken into consideration, $P_t(z)$ is the actual price of market z in period t while $E\left(P_t \middle| I_t(z)\right)$ is the expectation about the general price level, formed on grounds of an information set, available in same market for the same period.[40] Since $y_{ct}(z)$ indicates the deviations from the trend, it can be either positive or negative. Accordingly, a supplier of market z decides to enhance his production when he perceives an increase in the price of his own product to a larger extent than that experienced in the general price level—in other words, when relative prices change favourably to him. The frictional mechanism in the system remains the same: suppliers do not have sufficient information about the aggregate (general) price level. In case of a complete and perfect knowledge, all the agents on the supply side

[40] It is wise to refer to information set I_t as the information underlying expectations for period t.

could make competent production decisions as they would correctly perceive the relative price changes affecting their own products. However, since aggregate data is not available, they are forced to make decisions on grounds of their (rationally formed) expectations about the general price level. In other words, although the relation of the local price $P_t(z)$ and the general price level P_t is relevant as far as a production decision is concerned, due to incomplete information, suppliers are forced to replace P_t by expectations formed about the latter.

Suppliers could learn the probability density function of the aggregate price level dynamics from their previous experience, since knowing the previously experienced demand shocks, the (common) trend component of output and of cyclical fluctuations offer this information, even if an *accurate* prediction *cannot* be based on this. For Lucas, the price level follows a normal distribution with constant variance, which means, that $P_t \sim (\overline{P}_t; \sigma)$. Thus the general price level can be presented as

$$P_t = \overline{P}_t + \varepsilon_t, \tag{4.5}$$

where $\varepsilon_t \sim (0; \sigma)$. Accordingly, it also holds that

$$E(P_t|I_t) = \overline{P}_t, \tag{4.6}$$

therefore, what follows from this is that

$$P_t = E(P_t|I_t) + \varepsilon_t. \tag{4.7}$$

To put it shortly, information set $I_t(z)$ applied in the expectation formation process consists of the actually perceived price $P_t(z)$ and historical data pertaining to the aggregate price level—the latter is incorporated in the expected value $\overline{P}_t$. In this model, the price in each market is moved from the general price level by a normally distributed stochastic component with constant variance, that is

$$P_t(z) = P_t + z_t, \tag{4.8}$$

where $z_t \sim (0; \tau)$. Rational expectations for the price level in period t (with the structure of the information set described above being borne in mind):

$$E(P_t|I_t(z)) = E(P_t|P_t(z); \overline{P}_t). \tag{4.9}$$

What this points to is that a supplier in market z forms his expectations on grounds of the directly observable price and the historical data pertaining to the aggregate price level. If the information set did not consist of multiple components, agents would be in a relatively favourable situation, because

$$E(P_t|P_t(z)) = P_t(z), \text{ since } E(z_t) = 0, \tag{4.10}$$

and

$$E(P_t|\overline{P}_t) = \overline{P}_t, \text{ since } E(\varepsilon_t) = 0. \tag{4.11}$$

Expectations about the general price level must be some weighted average of the price in market z and the expected value of the price level. But what grounds are available for agents on the supply side to underlie the weighing of these two values? Bearing in mind the details described above, it might be said that

$$P_t(z) = P_t + z_t = \overline{P}_t + \varepsilon_t + z_t = E(P_t|I_t) + \varepsilon_t + z_t, \tag{4.12}$$

that is, the price in market z deviates from the expected value of the general price level under the *joint* influence of two shocks. However, since a supplier in market z forms his expectations about the general price level on grounds of the local price observable in his market, we could also write that

$$P_t(z) - \varepsilon_t - z_t = E(P_t|I_t). \tag{4.13}$$

These shocks *cannot* be observed by market participants in period t, so their numerical values need to be obtained through some kind of guesswork. The economic interpretation of Eqs. (4.12) and (4.13) can be provided in a different way as well: suppliers of market z are capable of observing only the relation between (1) their expectations about the general price level and (2) the directly observable price in their own market (as we have seen it above); however, no specifics about the difference ($\varepsilon_t + z_t$) between (1) and (2) are known to the suppliers. The decision on weighing, in formal terms, means realizing the factor that moved the price observed in each market in the past from the expected value of the general price level.[41] If (as we have seen) σ^2 is the variance of shocks moving the general price level from its expected value, and τ^2 is the variance of shocks moving local prices from the general price level, the ratio

$$\theta = \frac{\tau^2}{\sigma^2 + \tau^2} \tag{4.14}$$

shows the extent to which local prices were affected in the past by relative price shocks ($0 \leq \theta \leq 1$). Bearing all this in mind, we can return to the equation

[41] At this point in the discussion, it has to be recalled that local prices can be deviated from the expected value of the general price level by *both* a global price shock *and* a relative price shock *in any combination*. A genuine problem, therefore, is that the actual effect is a mixture of these factors.

$$E(P_t|I_t(z)) = E(P_t|P_t(z);\overline{P}_t) \tag{4.15}$$

and, on grounds of the details above (particularly $P_t(z) = P_t + z_t$), we can continue by saying that

$$E(P_t|I_t(z)) = (1-\theta)P_t(z) + \theta\overline{P}_t. \tag{4.16}$$

It is worth summarizing our previous results. These results can be acknowledged without any difficulty in the course of a verbal discussion if two extreme cases are considered. First, there were only global price shocks in the past, and, naturally, the relative prices remained unchanged at the same time. It is evident that $\theta = 0$ here. If this is the case, global price level dynamics can easily be inferred from local price changes, since these changes (as there were no shifts in relative prices) reflect increases and decreases in the aggregate price level. Therefore, if $\theta = 0$, then $P_t(z)$ is weighed $(1-\theta) = 1$ in forming expectations (thus the expected value of the price level is disregarded, and its weight is $\theta = 0$). By contrast, if $\theta = 1$, that is, the global price level was not affected, only relative prices were, by shocks in the past (which means that the price level remained constant or changed at a constant rate), the price in market z cannot in this case be applied to underlie expectations on the aggregate price level; the resulting weights work well again, since we form our expectations on the basis of the expected value of the global price level (since $\theta = 1$), while the directly observable price is disregarded, since it does not contain any useful information (that is, $1-\theta = 0$). Of course, there are innumerable stages in between, when both sources of information are in the game.

The relationship between these two shocks, one might add, has some significance in setting the slope of a *short-run* Phillips curve. If τ^2 is relatively small, that is, relative price shocks in the past were unimportant, the Phillips curve will be almost vertical as changes in local prices indicate shifts in the aggregate price level (obeying nominal shocks). If that is the case, suppliers will either hardly respond or not respond at all to demand shocks generated by the monetary authority. This is exactly where the efficiency of central bank surprises stems from, since a change in prices triggered by a deviation from the previous (known and expected) monetary policy (that is, a pure nominal shock) is regarded as a real shock, to which the reaction is to adjust output.[42] The effect, of course, cannot be long-lasting, since the *long-run* Phillips curve is a vertical line. This mechanism is of great importance. As a reminder, let us summarise the situation: the more effort a monetary authority makes to mislead market participants, the more limited its potential becomes due to the past experience of agents. Still, *situations in which money is not neutral might present themselves within the realm of island models.*

[42] Begg (1982) phrases rather loosely by saying this: "[. . .] no matter how we define the rest of the model and no matter which systematic parts of policy rules are altered, the effect on the path of real output will be nil." This is not explained by the restoration of PFH. *This statement is completely wrong since it precludes even the temporary effectiveness of surprises.*

Based on these findings, we can phrase the (individual) production function prevailing in market z (while disregarding the lagged values of the cyclical component):

$$y_t(z) = y_{nt} + \theta\gamma\left[P_t(z) - \overline{P}_t\right]. \tag{4.17}$$

The sum of these local production functions will take us to the aggregate supply function:

$$y_t = y_{nt} + \theta\gamma\left[P_t - \overline{P}_t\right], \text{ or, in other form } y_t = y_{nt} + \theta\gamma[P_t - E(P_t)]. \tag{4.18}$$

It was not useless to spend some time on the islands of Lucas (we will return to the topic when discussing his business cycle model), since we could scrutinize the role Lucas attributed to expectations in the explanation of fluctuations in real output. Now, we can move on and drop these models, since the dynamics of the supply side will soon be in the focus of our attention. As for the full model, it does not provide us with additional knowledge as its key message is that only unanticipated demand shocks can cause real output to move away from the trend. Both local and aggregate supply functions illustrate that in the world of Lucas, the dynamics of expectations constitutes the only shock that affects real output. The relation between inflation and unemployment (or real output) comes from the confusion (error) that mars the judgement of producers when perceiving relative and absolute price changes. If expectations are *accurate* (i.e. *true*), the economy continues to be in the state of equilibrium output. However, the accuracy of expectations can be distorted by real and nominal shocks affecting prices. Expectations are only channels through specific factors (real and nominal shocks) actually govern real economic performance, but these factors exert their influence on real output and employment only through expectations and the relation between expected and actual price dynamics. We need to emphasize this as a few years later key new classical authors[43] Sargent and Wallace attributed a completely different role to noises affecting the real economy—this is one of the key topics of the next section. For this reason, we have to revisit the issue of the Phillips curve.

4.3.3 Explication of Aggregate Supply

Although it is true that a shift from the theory of Friedman towards new classical macroeconomics does not necessitate the reshaping of the Phillips curve (since surprises are capable of diverting the economy from its natural level of

[43] Strange as it is, new classical macroeconomics was launched to its conquest by the seminal paper of Sargent and Wallace (Sargent and Wallace 1975; cf. Király 1998), however, both of them had come out of Lucas' overcoat.

unemployment and output in the short run—in other words, there *could be* a leeway for short-run, negatively sloped graphs in the new classical Phillips curve as well). However, differences are crucial. In Friedman's model, systematic demand management can lead to short-run effects, but for new classicals this option is altogether absent. *We have an additional neutrality theorem to complete the notion of the long-run neutrality of money as money is neutral for new classicals even in the short run* (Weeks 1989; Gordon 1979). However, we have to take this statement with a pinch of salt as it is as suspicious as it is curt. For the orthodox monetarist (Friedmanian) line of thought, the non-neutrality of money was, simply put, caused by the disappointments of expectations, i.e. the fact that actual and expected outcomes were different. It was unnecessary in the discussion about efficiency to divide monetary policy actions into *systematic* and *unexpected* sets. In the long run this error (if further shocks did not emerge, thereby adding to the previous disturbance) faded away and the system was readjusted to a state of stable equilibrium. In the new classical case, the short-run neutrality of money is caused by the fact that agents now are capable of anticipating systematic economic policy actions, therefore these interventions will no longer produce an effect, not even in the short run. *Short-run neutrality is a matter of interpretation, i.e. it pertains to the efficiency of actions that can be anticipated under rational expectations but that are unexpected in an adaptive framework.* However, if we generalize the criterion of short-run efficiency (seen purely as a surprise) maintained by orthodox monetarism to the system of new classical theory (disregarding the content of this surprise, i.e. the technical details of actions capable of making surprises), the conclusion we come to is the same: *money is non-neutral in the short run, not even in this case*; unproclaimed (random) political actions *have* temporary real effects. *Money, therefore, can only be regarded as neutral both in orthodox monetarism and in the new classical macroeconomics if its changes are anticipated or correctly perceived.* Viewed from a different angle, this is what the issue is about: the flexibility of prices ensures the short-run neutrality of money, while the empirically supported case of incomplete price flexibility (or, in other words: gradual adjustment of prices) is compatible only with the hypothesis of long-run neutrality (for an explication regarding price flexibility see Gordon 1979).

Sargent and Wallace (1975), in reference to a 1973 paper by Lucas, analysed above, introduce a 'surprise' supply function:

$$y_t = a_1 k_{t-1} + a_2\left(p_t - {}_{t-1}p_t^e\right) + \varepsilon_t, \tag{4.19}$$

where real output in period t (y_t) is dependent on the capital stock available at the end of the previous period (k), the price level in period t (p_t), an expectation about this price level formed in the previous period (${}_{t-1}p_t^e$) and a white noise error term (ε_t) that cannot be predicted in any way. The first component on the right, determined by the capital stock, should, without any difficulty, be regarded as the natural level of output, which takes us to the function referred to in literature (e.g. Begg 1982) as the Lucasian production/supply function:

$$Q_t - Q_t^* = \alpha(p_t - p_t^e) + \varepsilon_t. \tag{4.20}$$

If the natural level of output (Q_t^*) on the left is *exogenous*,[44] the left side depends on the joint effect of the two terms on the right. The main problem is that *a systematic countercyclical economic policy cannot be based on the efficiency of surprises that can in no way be anticipated*. Differently put, although the new classical theory does to some extent provide leeway for contingent departures from the natural level, these contingencies *cannot* be exploited by economic policy. Upon scrutinizing the Lucasian production function in Eq. (4.20), it becomes obvious that *the equilibrium state is accidental even in the case of accurate expectations* as a result of the error term (ε_t) on the right. Earlier on, this was referred to in an argument that the emergence of real business cycles is *partly* independent of the new classical Phillips curve, since even if expectations are fulfilled, the white noise term in Eq. (4.20) generates (random and unexpected) cycles. According to the production function formulated in Eq. (4.20), the source of cycles is, on the one hand, the non-fulfilment of expectations and, on the other hand, a white noise completely independent of the fulfilment of expectations (and, consequently, of the Phillips curve). Equation (4.27) will shed light on the nature of the disappointment of inflation expectations. Even if these expectations are fulfilled, that is, actual price dynamics correspond to the path that was expected by agents, an equilibrium state (i.e. an output equal to the natural level) could emerge only if the error term in period t was zero. These errors are unpredictable, since it follows from the rationality of expectations that *all* information available before the end of period $t - 1$ was used while forming those expectations—thus prediction errors can only be random.[45] (However, it also needs to be stressed that there is a difference between expecting something and acting in accordance with that expectation—the assumption of price flexibility is, therefore, equally a prerequisite for the inefficiency of systematic monetary policy as the central bank would be in a position to trigger real adjustments in case of sticky prices.) Therefore, monetary policy is unable to come up with a preliminary intervention program in order to respond to these unexpected shocks.

Under these circumstances, monetary authority might decide on a policy to be followed by throwing a dice, as it were, which would possibly worsen a negative shock or eliminate a positive (favourable) effect. Random nominal shocks *generated by monetary policy* make expectation errors (unexpected inflation) reflecting these shocks turn into white noises[46] the distribution of which is independent of the

[44] Stressing of exogenity will become really important apropos of supply side applications of fiscal policy, where we are going to realize that the exogenous nature of natural level of output is not an axiomatic element of the theory.

[45] Moreover, it naturally follows from all of these (that is, from the rationality of expectations) that systematic, therefore completely predictable monetary policy steps are also ineffective, since known rules cannot generate surprise-inflation.

[46] Following Lucas, Blinder and Fischer (1979) also stress that surprise-inflation is a white noise.

error term ε_t on the right. Thus the dynamics of real output is the resultant of the joint effect of two stochastic variables, the sum total of which *might even be* zero (in formal terms, this would be interpreted as output reaching its natural level). Since shocks are unpredictable, an operative instrument of monetary policy (the quantity of money) and the change in prices it brings about might neutralize the exogenous shocks as an accidental effect at most—worse still, it may aggravate them. One of the most important economic policy conclusions of new classical macroeconomics relates precisely to these details: *a positive shift from the natural level of output is, at best, temporary, while the efforts to generate negative differences would be completely nonsense*. Monetary policy is at its best when it tries to stabilize the economic environment. *We can create a situation in which money is non-neutral in the short run, but economic policy will no longer be capable of building the potential of a systematic control over real output* (not even in the short run).

The significant differences between the aggregate supply function Lucas originally presented and the production function in Eq. (4.20) derived from that need to be underlined. Lucas introduced the effects of nominal and real shocks affecting a macro-economy into his system through price expectations: *if expectations are true, output in period t is at its natural level* (*that is, at its own trend*). However, the well-known and widely accepted aggregate production function described by Sargent and Wallace (1975) also provides leeway for shocks (ε_t) *independent* of price expectations—resulting in the accidental nature of equilibrium and in the inefficacy of countercyclical efforts of monetary policy.[47] *Suppliers living on the islands of Lucas are fundamentally capable of adjustment*: if unexpected shocks generated by monetary policy do not confuse them, *production on each island will instantly adjust to stable* (constant, or steadily growing nominal) demand, *thus real output will, after all, remain at its natural level. Monetary policy cannot contribute to the neutralization of real fluctuations, since these fluctuations are, after all, generated by the monetary policy itself*; such fluctuations cannot occur while a predictable monetary policy is being implemented (it has to be stressed, of course, that the efficacy of monetary policy was judged in new classical macromodels by its potential to eliminate cycles and *not* by its capability in long-run growth enhancement, which is a huge step forward compared to Friedman's framework). *If price expectations are true, real fluctuations will be absent from the picture*. Under these circumstances, moving real output from its natural level could be the only alternative goal of monetary policy (to fix output at its natural level, all the central bank needs to do is to act today just as it did yesterday). However, the room for

[47] It is obviously not independent of the debate that is, in fact, about faiths. According to it, some economists evidently regard the economy as an equilibrium system, even if a given macroeconomic unit departs from this state again and again; others regard these swings as the evident lack of equilibrium (cf. Király 2000). However, it seems to be a different case, since Sargent and Wallace do not talk about macroeconomies moving away from the state of (general) equilibrium from time to time but rather that equilibrium can emerge and can be achieved only accidentally, thanks to random effects.

manoeuvre is limited here because of the rationality of expectations. Moreover, factor θ, described above, provides a reasonable argument to support the fact that with an aggregate price level of high volatility (indicating the purpose of the central bank to deceive), suppliers are more likely to infer an increase in the general price level (that is, a nominal shock) from increases in local prices—therefore real effects, if any, will be insignificant.

The world of Sargent and Wallace is radically different: here, real shocks are possible even independently of price changes—economic policy makers may, therefore, consider the elimination of cycles. The above reasoning makes us realize that equilibrium is accidental because of these shocks—and nothing will change even if a monetary authority is actively involved in trying to eliminate fluctuations, since information to underlie the planning of adequate actions is not available (ε_t is not foreseeable). *Within the realm of Sargent and Wallace, market participants are exposed to the negative effects of real shocks even without the assistance of monetary policy*—and the effects of a monetary policy that does activist intervention are entirely unpredictable (due to the fact that central bank actions pertaining to the shock that needs to be eliminated cannot be planned in advance, since even monetary policy will fail to foresee the effects it tries to neutralize).

Hahn (1980) drops a very sophisticated hint on the issue that the neutrality of money (whatever the timeframe) is not independent of expectations. He argues that the real variables of a macroeconomic system are independent of the quantity of money (and, hence, of price level) only if these variables are *completely* predictable. The neutrality theorem could, therefore, be rephrased by saying that money is, as it were, *neutral if its dynamics is perfectly foreseen by agents*—or when (or on the time horizon over which) agents are able to adjust to a change in the quantity of money (note that this statement is equivalent to the former one if flexible prices can be assumed). The long-run neutrality of money means that agents can adjust their behaviour to the new state generated by the higher quantity of money in the long run alone—in this case money is non-neutral in the short run because of the failure of expectations (recall the orthodox monetarist Phillips curve provided by Friedman). The short-run (that is, immediate) neutrality of money requires agents to perfectly foresee changes in the quantity of money—however, it should be clear by now that *it is not possible* (not even Lucas and his group would accept this). The case of short-run neutrality (like the long-run case) is conditional: it requires *expectations and behaviour to be in accordance with actual processes*. This is that very situation which the new classical literature refers to as the signal extraction problem: if these crucial variables (quantity of money and price level) were perfectly foreseen, a change in the quantity of money would *not* trigger *any* real effect, only a proportional shift in the price level would follow from the change in the money supply.[48]

[48] Note that the neutrality of money was established in the false dichotomy model so that money got thrown on the system only after real equilibrium had already been found–in other words, money had no role in setting the equilibrium.

It is not surprising if, on this basis, we argue that *the foundations of orthodox monetarist and the new classical Phillips curve are identical*, since both describe what happens to real output when expectations are not fulfilled. This becomes quite evident when we read the words of Frank Hahn (1980) who, when tackling the problem of the neoclassical foundations of monetarism, sometimes seems to confuse the reader, since his findings blend in with the orthodox monetarist as well as the new classical doctrines.[49] *The only* (although significant) *difference is in the way agents form their expectations and the factors* (processes, changes, etc.) *that can surprise them.*

It should also be noted that in the new classical system (just like in orthodox monetarism) deviations from the equilibrium level of both employment and output was thought of as something originally *harmful*: the inevitable consequences of certain shocks (possibly caused by economic policy itself). If an economy was free from such shocks, it could permanently stay at the salutary, natural level of output. Here Phelps' theory should be mentioned, who also applied an equilibrium concept regarding unemployment, but, for him, demand management contributed to reducing involuntary unemployment, even if only temporarily.

However, *a countercyclical monetary policy is not necessarily directed at the long-term stimulation of demand and, through it, output; and temporary effects could be sufficient to dampen the consequences of certain negative shocks.* Whether or not an economic policy (or, more specifically, monetary policy) is capable of this is the very problem that has established one of the most important dividing lines between Keynesian-oriented[50] and monetarist theories. **[→ Phelps and Phillips curve]** It must be stressed that new classical macroeconomics highlighted the fact that *systematic countercyclical monetary policy* is inefficient *under specific conditions* (conditional inefficacy), although they generalized these findings to cover a broader set of economic policy instruments with a degree of questionable success. For example, Hahn (1980) points out that as for portfolios of debtors and creditors, changes in the price level are nothing to sneeze at and, he adds, this effect could only be ignored if debtors and creditors were *sufficiently* (sic!) alike. However, a rearrangement of macro-social income conditions could hardly be regarded as nothing more than a nominal change. A progressive tax system, for example, has real effects if prices change proportionally to the quantity of money, even if its mechanism is well-known to market participants and surprise-effects are entirely absent. At the same time, mention should be made here of the emergence of a crowding-out effect cited often in literature: private investments that fail to

---

[49] As an instance, this phrase can be referred to both schools: "[...] real variables can diverge from their natural values only if agents make mistakes [when forming expectations]." (Hahn 1980).

[50] It must be noted that Keynes himself was also sceptical about the efficacy of monetary policy. For him, central bank stimulates the interest-rate-dependent part of aggregate demand through controlling the rate of interest. However, this potency could be limited by the liquidity trap and by the fact that the marginal efficiency of capital sometimes collapses to an extent that may not be offset by drops in the rate of interest. This is the reason why Keynes put his main emphasis on fiscal policy.

materialize due to public spending also indicate malfunctions of the real sphere (cf. e.g. Erdős 2003). As we approach real-life conditions while moving away from the vantage point of new classical models, it will be increasingly difficult to deny that even if precisely defined conditions (rational expectations, perfect price flexibility, etc.) are met, transmission mechanisms can still be found through which even changes in the price level, known to have occurred in response to completely anticipated monetary policy actions, can have real effects. However, this is a failure of new classical macroeconomics only inasmuch as they failed to offer a realistic description of actual economies (in other words: new classical macroeconomics remained a pure theory). The message is clear and unambiguous: *the (model-like) circumstances under which anticipated monetary policy is inefficient can be properly outlined*. If reality differs from the picture portrayed by these models (of course it differs!), the accompanying effects will also change.

Naturally, the Lucasian production function can be consistent with adaptive expectations as well, as Friedman himself is of the view that surprises alone were capable of inducing positive (or negative) deviations of output from its natural level. This view is confirmed by Gordon's formal arguments where it was proved that the expectations-augmented Phillips curve can in fact be easily produced by inverting the Lucasian production function, because

$$p_t = \frac{1}{\alpha}H_t + p_t^e + \zeta_t, \tag{4.21}$$

where $H_t = Q_t - Q_t^*$ and ζ_t is the stochastic error term (Gordon 1979). On second thoughts, this (formal) similarity is not extraordinary, since surprises in both systems are capable of diverting output from its equilibrium (natural) level. After all, the only difference lies in the nature of these surprises.

As it was pointed out earlier on, only random effects can trigger deviations from the natural level/rate in the new classical system (however, these surprises appertain only to the current period[51]). Now, the question is, what sources can these unexpected effects be traced back to. Shaw offers a twofold separation:

1. Exogenous unpredictable shocks: a natural disaster, for instance, unexpected weather conditions or even a war.[52]
2. Unexpected shocks which the monetary authority generates by pursuing an erratic and unpredictable policy (Shaw 1984). In this case, a monetary policy has an *active* role in misleading agents.

[51] Easing these requirements led to a modified version of LSW-theory, discussed below.

[52] It is very important to stress that these shocks are outside of the territory of policy. It turned out to be significant, first of all, in relation to the new classical views on the equilibrium of macro-systems. Considering these questions, we might realize that even a completely systematic monetary policy rule (one with no random elements) cannot guarantee a permanent macroeconomic equilibrium. *Under stochastic conditions*, where contingencies also contribute to the disappointment in expectations, *equilibrium becomes completely accidental*.

Thus an economic policy is still in a position to induce shifts from the natural rate—however, there is a question mark behind how the benefits of a policy aimed at such fluctuations relate to its costs. In other words, is it worthwhile to pursue a policy like this? However, it is to be reconsidered whether a disappointment of expectations can be caused by *deliberate* actions of the monetary authority (that is to say, can a central bank actually *make a decision* to mislead agents), or is it far more true that the central bank plays a *passive* role in generating noises coming from the monetary policy? Consider a simple form of a central bank operative target, specified in terms of the quantity of money:

$$m_t = \gamma_1 m_{t-1} + \zeta_t, \tag{4.22}$$

where m_t is the logarithm of the quantity of money in period t and ζ_t is a white noise (Begg 1982). The fact that this monetary policy rule does not declare the monetary aggregate on which it is phrased is not negligible—anyhow, quantities of money cannot be fully controlled via monetary policy. If the price level is determined by the quantity of money, price level dynamics is not completely predictable, not even in case of a known monetary policy rule γ_1. Therefore, in this case the disappointment of inflation expectations in the Lucasian production function (which, as a matter of fact, was established by Sargent and Wallace) is not generated by the central bank through its deliberate steps but is caused by the fact that the future path of the quantity of money, therefore the price level, can be predicted only loosely (only the expected values of relevant variables are available to actors, even if a precisely defined and credibly followed monetary policy rule is implemented). The regulatory options of the central bank, having a target specified in terms of the quantity of money, are limited. If the error term on the right were to be disregarded, it would be obvious that the central bank makes efforts to achieve a tight control and can do it since it gives a well-defined rule for the dynamics of the quantity of money—these efforts, however, are doomed to failure as a result of the random effects embodied by the error term. Our conclusions are similar if we take into consideration the fact that central banks are capable only of regulating the monetary base, while it is the broader aggregates that act as mediums of exchange—that is, money multiplication should not be disregarded. Under model-like circumstances, the multiplier determined by the reserve requirement ratio can offer accurate predictions for the paths of the broad aggregates (Samuelson and Nordhaus 1998). However, reality is much more complicated, since if such a stable and close relationship between the monetary base and the broad aggregates (that is, money in circulation) were to exist, the problem of monetary policy control would be resolved. 'Clear case' is a notion which designates the upper threshold of multiplication; the actual dynamics of the money creation process is determined by the money demand and credit needs of non-monetary agents (Vigvári 2004). Based on this, a quantity limit for aggregate M_1 can be described as

$$M_0 \leq M_1 \leq \frac{1}{R} M_0, \tag{4.23}$$

where M_1 is money in circulation (including demand deposits) and R is the reserve requirement ratio ($0 < R < 1$). Since the dynamics of M_1 is not clear-cut, it can be assumed that there is only an empirical rule of thumb available for the central bank and market participants, one which describes how aggregate M_1 responded to changes in the monetary base in the past—so, this empirical approximation gives the average changes in M_1 as reactions to monetary policy actions. Let this relation be denoted by ϕ. Now, the rule of changes in the quantity of money available for the central bank is

$$M_{1t} = \phi M_{0t} + \eta_t, \tag{4.24}$$

where η_t is a normally distributed and serially uncorrelated error term with zero mean. The central bank cannot control money in circulation tightly.[53] The monetary authority is only capable of a direct control over the monetary base, while broader aggregates are linked to high-powered money only loosely (Barabás and Hamecz 1997). Consequently, however, even rationally expecting agents are unable to foresee the future path of M_1. If a close (direct) relationship between M_1 and the price level is assumed, *price expectation errors emerge naturally but still independently of the central bank*, and they can be unbiased (expectation errors do not contain systematic biasedness, so their mean is zero) at best (i.e. when expectations are rational). Although the monetary authority might try to eliminate these disturbances through unexpected changes in the monetary base or reserve requirement ratio, it cannot do it, since error terms η_t are unpredictable, even for the central bank. Hence, the problem of true inflation expectations was reduced to the predictability of the quantity of money, but accurate predictions are *not* possible under stochastic conditions. Inflation expectations are, therefore, fraught with persistent (and, given their contingency, unforeseeable) errors. Now, we can formulize the relation between actual and expected price level as

$$p_t = {}_{t-1}p_t^e + \eta_t', \tag{4.25}$$

or, in another form as

$$p_t - p_t^e = \eta_t'. \tag{4.26}$$

Since there are two white noise terms in the Lucasian production function on the right, because

[53] It is equally true for open and closed economies. In case of open national economies with fully convertible currencies, the big picture becomes more complicated due to international capital flows which break money in circulation off the monetary base even farther.

$$Q_t - Q_t^* = \alpha\eta_t' + \varepsilon_t \tag{4.27}$$

(a reminder: one of them is responsible for inflation expectation errors while the other is for the deviation of real output from the natural level, even when expectations accidentally turn out to be true), one of the direct consequences of this fact is that divergences of output from its natural level are not serially correlated—that is to say, business cycles (when output gaps reveal a trend) do not (cannot) exist, since actual output takes a random walk around its natural level.

Let us make a slight detour now to consider Eq. (4.24). Such equations *as regression equations* are used to test the (extent of the) rationality of expectations. In this a case, period *t* volume of the monetary aggregate serving as the medium of exchange[54] is on the left, while its lagged value(s) and other macroeconomic variables are on the right (e.g. the rate of unemployment, monetary base, etc.), all known in period *t*. In this view, the expectation functions of agents can be found numerically through estimating Eq. (4.24). If all the explanatory variables are inputted, we then have the expectation on the relevant monetary aggregate for period *t*; and the difference between actual and this expected data is the expectation error. For a detailed review of this methodology, see Barro (1977) or Barro and Rush (1980), where dividing the changes in the quantity of money into expected and unexpected components is done precisely on these grounds. Of course, it is also possible to estimate the expectation function on, say, inflation: in this case, period *t* inflation is on the left and lagged values of that and other macro-variables are on the right. Expectation errors should be interpreted just like in the case of the monetary aggregate, since they are the differences between predicted and actual paths of inflation. However, this modelling technique is fraught with methodological and conceptual difficulties.

Errors emerge because, according to this methodology, we utilize pieces of information when estimating our equation that are not necessarily available for agents *when they form their own expectations*. What is this all about? Let us assume that a market participant intends to make a prediction about the January 2005 volume of monetary aggregate M_1 (or the rate of inflation) in December 2004, and that we want to reproduce his expectations on the basis of data between 2000 and 2010. If his only information source is the previous dynamics of M_0 (i.e. the time series of M_0), in forming his expectations for January 2005, he in fact will only have the information that becomes available by December 2004, whereas we make our estimations on the basis of the entire length of time series M_0 (in other words: we fit our regression model to data from the *entire* period of 2000–2010). Indeed, it would be a serious oversimplification to assume that market participants infer the future dynamics of broad aggregates only from the volume of M_0 (this reasoning can be extended to cover time-series modelling of aggregates M_1–M_3 as well). However, it cannot be assumed (particularly in light of our findings above) that agents in

[54] Each monetary aggregate fulfils different functions of money. Hence, M_1 serves transactions primarily, while M_2 and M_3 tend to accentuate the treasuring function (Vigvári 2008).

reality disregard specific qualitative information (e.g. official declarations of the future path of monetary policy, expert predictions) when forming their expectations. It is obvious that aggregate M_0 is only one (albeit important) source of information that can be utilized, but this should not be handled in the leisurely manner referred to above (even though it has become something of an approved habit), since we have to guarantee the continuous incorporation of new (emerging) pieces of information (in other words: information not known to our agents should be precluded from underlying expectations in our model). The procedure of expectation formation by market participants should be portrayed as a continuous activity in the course of which more recently obtained pieces of information are incorporated as they re-estimate their models (or even change the functional form).

That's why part of literature tries to abandon this traditional methodology as on the one hand it is not obvious what kind of information is used by various agents (this problem can be avoided by assuming all agents to have the same information set), while on the other hand, certain factors are hard (or impossible) to quantify but have a trivial influence on expectations (which is either hard to model or cannot be modelled at all). It has to be stressed that it is not an implicit denial of the REH. What we argue is that none can be fully informed by observing only the dynamics of aggregate M_0 or M_1; and that (this being equivalent to the previous statement) there may be pieces of information excluded during this process in which expectations are formed. If the monetary authority announces an expansionary policy starting during the next period, any future increases in the quantity of money will be anticipated by all the agents (the announcement effect of economic policy), although its impact cannot be modelled on grounds of past values of M_0 and M_1.

As soon as it was completed, the Barro-Rush methodology faced serious criticism aimed at the underlying-implicit logic of these models (cf. e.g. Weintraub 1980). For this very reason, some authors resort to explicit expectations, i.e. they do not *generate* expectations using some (presupposed) models but obtain these data from empirical surveys. To review this alternative methodology, see e.g. Carns and Lombra (1983). However, Barro (1977) mentions the methodological problem caused by an undifferentiated usage of pieces of information from different times, stipulating the caveat that if there is available (announced) information about the future path of monetary aggregates, a distinction is not justified (in this model, Barro explains the dynamics of M_1 by, among other macroeconomic indicators, its own lagged values). This, we believe, is too lenient an attitude, since not even the central bank is capable of holding aggregates M_1–M_3 tightly on a preannounced path (cf. e.g. Szalkai 1995). The central bank may announce such a path (through declaring a band at best, admitting, to some extent, its own limits), but if the monetary authority were to fix points rather than a band, the actual dynamics of aggregates would, in all likelihood, more or less deviate from this path. However, for Barro, the actual dynamics of monetary aggregates and their preannounced path (if any) should be treated as an equivalent (which is possible only in a model-like, deterministic environment at best), that is, factual experiences, emerging *subsequently*, may be regarded as *a priori* information. There is no (and cannot be) a priori information that can ensure the *ex ante* knowledge of the dynamics of a broader monetary aggregate. Of course, it seems that such a compromise cannot be precluded in principle as far as the monetary base is concerned (a central bank can always hold the monetary base on a preannounced target path), but Barro, Carns as well as Lombra fit regressions for aggregate M_1. To sum it up, monetary authority is not capable of keeping monetary aggregates beyond M_0 on a narrow path, although some control is, naturally, possible. However, these actions exert their influence with a considerable delay (short-term interest rates, reserve requirement policy, open market operations), sometimes at the expense of serious sacrifices for the real economy (that is, sticking to a target needs to be viewed with circumspection due to its implied sacrifices). Thus actual data and a preannounced path for the broader aggregates could hardly be regarded as each other's equivalent. The reasoning behind Nicholas Kaldor's approach is rather convincing when he argues that if a monetary authority were determined to exert a very strong influence (in a restrictive manner) over the quantity of money available, these efforts would soon result in the loss of this controlling potential, since markets would resolve the problem of switching over to an alternative settlement system or of replacing the official currency (Kaldor 1970).

Similarly, it is also true that agents *would be* able to make predictions only with errors on grounds of the dynamics of M_0.[55] However, it also needs to be stressed that including broader aggregates as explanatory variables can widen the information set and enhance the fit of predictions, but there remain further pieces of information that can still improve these predictions—and they may actually improve in reality, but modelling on monetary aggregates is incapable of handling these sets.[56] After all, this methodology results in questioning the rationality of agents and Barro, through the expectation formation scheme embodied in his modelling technique, deprives his agents of exploiting a particular set of information. Barro could not successfully address the problem by incorporating various macroeconomic indicators (e.g. unemployment rate) as explanatory variables, since the additional information they provide is still unable to substitute, say, announcement effects. *Barro's agents utilize less information than is actually available.* It should also be noted that this informational bias would not be possible to be resolved, not even by a procedure according to which estimations in consecutive periods are performed only on grounds of information sets that become available by the relevant period (rolling forward). Now, estimations of expectations in December 2004 would be performed *not* on grounds of the full-length time series (whose duration ends in 2010) but only on a sub-sample that lasts not later than December 2004. This technique would help us eliminate the problem of anteriority alone, that is, we could only rule out information sets not known to agents upon the formation of their expectations in consecutive periods. At the same time, sets that

[55] Using conditional is justified here by the fact that we are free to assume this: market participants are not content with forming expectations on the basis of M_0 only.

[56] It will become clear soon that this problem cannot be resolved even by incorporating further explanatory variables over and above the monetary aggregates, as Barro (1977) did.

were probably known and actually utilized would still be precluded. Broadly speaking, the same arguments would need to be involved if we were to explain why central banks abandoned the idea in the period following the 1980s of the quantity of money being used as an intermediate target, since this target variable could help neither central banks nor market participants. Therefore, the modelling of expectation errors, executed in the manner Barro suggested, precludes some information sets in advance, while arbitrarily considers others as given and known, and, if for no other reason than this, it can hardly be regarded as a defendable conception. Moreover, there is the additional problem of an unpredictable relationship between individual, broader aggregates and the monetary base, which can only be tuned roughly, even by using central bank instruments. After this methodological detour let us return to our previous line of reasoning.

New classical macroeconomics stood no chance of expecting a wide acceptance and approval as long as its findings were not consistent with obvious life-world facts. Blinder and Fischer (1979) made an attempt to argue for the possibility of persistent business cycles by incorporating the dynamics of a storable output. *This inventory stock* declines as a reaction to an unexpected rise in aggregate demand, since producers have less difficulty having to respond to any changes in the macroeconomic environment by reducing these stocks (that is, the reaction of sales outstrips the answer given by production). Naturally, after the current period, producers start to replenish their inventories to reach previous level, that is, production "remembers" former shocks for multiple periods. As a consequence, output in the following period will be higher, irrespective of whether or not inflation expectations will be fulfilled at that time. The modified version of the Lucasian production function can be written as

$$Q_t - Q_t^* = \alpha\left(p_t - p_t^e\right) + \lambda\left(N_t^* - N_t\right) + \varepsilon_t, \tag{4.28}$$

where N_t is the stock of inventories at the beginning of period t while N_t^* is the optimal (and desired) stock. This, however, implies not only the consistency of new classical macroeconomics with long-lasting business cycles, but that the moderated version of the LSW-theory makes leeway for systematic economic policies, since if the output gap in period t is not independent of previous inflation expectation errors, it follows from this that production decisions are made under the influence of effects already known (that is, not unexpected). *In this case, fully anticipated* (known) *changes are among the determinants of production* (cf. Gordon 1979). Even if a monetary shock was unexpected in the beginning, its effects will be felt even when it is already known, monetary policy thus turns out to be definitely capable of exerting a systematic countercyclical control over output and unemployment in the short run (while a long-run control still has adverse inflationary effects). *In this case, randomness* (i.e. an unexpected nature) *is no longer a relevant criterion*. Since real business cycles are undeniably present, it might be a reasonable purpose for monetary policy to support the start of economic growth (or to mitigate boosts). Although, according to Eq. (4.28), the output gap is still affected by a white noise, it does not prevent a monetary policy from taking countercyclical actions. This white noise may take (or takes) fluctuations in the growth process, but a monetary policy is not forced anymore to respond to an unforeseen white noise. The presence of business cycles makes the monetary authority tackle these cycles. Thus its actions do not need to be random. With rational expectations underway, surprises alone can do the job. In this way a countercyclical short-run control becomes *possible* and the presence of business cycles makes effects *long-lasting*.

Now, we know what countercyclical monetary policy and its symmetry mean. Monetary policy *may* have effects on real output, but it does not imply a long-run potency to enhance real economic performance and growth. The direct consequence of such an effort would be nothing but considerable additional inflation—therefore it is to be avoided. Monetary policy should confine its controlling efforts to countercyclical management, that is, it should help macroeconomies remain operational at the level of potential output, i.e. in equilibrium.

A relating inner logical inconsistency occurs, since it is not true even with completely flexible prices and rational expectations that only unexpected actions (effects) can move output and unemployment from their natural levels. If, however, we want to sustain the exclusive effectiveness of surprises, assuming rational expectations, (prolonged) cycles will in that case be precluded. Thus the dilemma is given: we either have business cycles in the model, or we focus on the exclusive effectiveness of surprises assuming that rational expectations are voiced.

Lucas himself also wanted to contribute to investigations into the sources of cycles, so he created an *equilibrium* model of business cycles (Lucas 1975) by which he tried to address the problem that deviations from equilibrium state could not be long-lasting in his first island models.[57] That is, prolonged real business cycles cannot emerge within these frameworks. Before Lucas, business cycles were regarded as disequilibrium phenomena, where the term "disequilibrium" was a stipulation that variables relevant for the analysis (e.g. wages and prices) are not endogenous but are exogenously given. From this perspective, the 1975 cycle model and the early 1970s island models have the same root, since in the above sense of the word they were all equilibrium models (KVA 1995; Karni 1980). This characteristic was pointed out by Lucas himself in his 1975 paper, stressing that prices and quantities in his cycle model are all determined in competitive equilibrium (Lucas 1975). Initial effects in this model are also caused by unanticipated monetary shocks, and an accelerator effect is used to ensure long-lasting cycles, otherwise put, to generate serially correlated cyclical differences. Friction, similarly to the previous models of Lucas, comes from incomplete information (thus incomplete knowledge does not simply imply that future is not unpredictable but also that agents do not have perfect information about the current state of the economy—and, additionally, the past dynamics of some variables are also incompletely known), plus separated markets interpreted as islands are still in the game.

[57] In formal terms, this means that divergences of real output from the natural level are not serially correlated.

However, it remains true that *the path of real economy is affected by past, familiar, i.e. non-surprising monetary shocks through the assumed accelerator effects*. However, deviations from the natural level will be eliminated with the passage of time. These last two models (although the discussion could be brief, on grounds of our previous lessons) have considerable importance, since introducing long-lasting business cycles necessarily *provides leeway for the efficacy of known (anticipated) monetary policy actions*. Still, Lucas' model of the business cycle cannot be regarded as a satisfactory answer (as he himself pointed this out; cf. Lucas 1995) and the Nobel laudation also stressed that imperfect information about monetary aggregates can hardly explain empirically the emergence of business cycles (KVA 1995). However, it has to be noted that while deviations from the level of output appertaining to full employment could only be white noises, the monetary policy inefficiency theorem still held true. Meanwhile, the existence of prolonged, real business cycles could not be explained on these grounds. However, cycle models created to provide theoretical descriptions for these cycles invalidated the inefficiency theorem. In theoretical terms, new classical macroeconomics faced a grave dilemma: the inefficiency of anticipated monetary policy actions and the theory of business cycles could not be sustained simultaneously (i.e. consistently).

4.3.4 Further Steps on the Road to a Normative Monetary Analysis: Constraints of Monetary Policy Inefficiency

The monetarist literature soon identified the dangers of a monetary policy based on erratic (unexpected) changes of direction. From the very beginning, the monetarists mostly advocated a stable and clear monetary policy, one that provided a far more limited room for economic policy control and fine-tuning than the Keynesian theory. *Not even Friedman's theory focused on predictability aimed at stressing procyclical economic policies. What it emphasised, instead, was a passive role.* Monetary policy probably worsens the prevailing negative tendencies in the economy if it tries to respond, no matter how strong its efforts are to eliminate them. It is not an aim of monetary policy to enhance economic growth. It should, instead, make an effort to ensure a stable and reliable environment that is necessary for a healthy path of boost. Friedman had simple views on what he believed to be a right monetary policy, as for him such a policy was confined to a careful control over the quantity of money that best fits specific macroeconomic constellations. Friedman believed monetary policy to have rather limited options in eliminating the non-monetary causes of macroeconomic malfunctions (Friedman 1968).

The new classical literature stressed the importance of predictability even more forcefully. If systematic (and, in accordance with changing conditions, often abrupt) changes in the directions of economic policy had implied dangers even in Friedman's theory, this was true even more of unexpected economic policy actions. On the one hand, these unexpected actions destabilize a macroeconomic environment and, on the other hand (and this is also an important aspect), these randomized steps can hardly help economic policy achieve its goals (cf. Shaw 1984; although reference to the conclusions drawn from the island models is also an adequate argument). If a move from the natural rate can only occur due to random effects, a non-natural unemployment and output level can be maintained in the long run only with expectations that disappointed in the same direction—which is about pursuing a systematic policy, therefore market participants who form rational expectations cannot be misled over the long haul. Considering all this and being mindful of Friedman's recommendations, Lucas also regarded the stable growth of the quantity of money (meaning monetary aggregate M_1 for him) as the only right and wise monetary policy to follow (Lucas 1980). As a footnote, this requirement goes back a long way, as early as his 1972 seminal paper. In that framework, the decision on the monetary policy to follow was in fact a commitment to a probability density function[58] which can be characterized well by an expected value and a standard deviation. Lowering this standard deviation (that is, shrinking the probability density function to the expected value) results in a situation characterized by a constant money growth rate. All other policies lead to an erratic and unexpected growth of money supply (where, naturally, the extent of this unexpectedness is in direct relation to the standard deviation of growth rate x). Lucas (1972) demonstrates by formal means that following a constant money growth rate results in a Pareto-optimal allocation. To accept this, all we need to do is to realize that a commitment to a constant money growth rate helps us avoid unfavourable real adjustments owing to unexpected monetary contractions.

It has to be noted that new classicals did not share the same view regarding a desirable monetary policy rule. Sargent and Wallace (1974) showed a negative attitude toward Friedman's constant money growth rate rule. For them, a central bank rule capable of lowering the variance of real output can be introduced and, strange as it sounds, they in fact argued for the potential success of a systematic, countercyclical monetary policy (this rule and the monetary policy based on it might well be regarded as systematic, given that they regulate current money growth rate on the basis of the state of the economy during the previous period). Of course, this rule is also anticipatory, i.e. a monetary authority will not try to mislead agents in order to stimulate real output. Sargent and Wallace argue that the converse of the relationship between the optimal feedback rule and real economy is also true: if a monetary policy changes the parameters of the rule, this modification does not leave unaffected the parameters of the difference equation describing the dynamics of the real economy (cf. Lucas 1973b). Thus a real economy will adjust to policy turns and only monetary surprises can divert the system from its natural level. The only problem is that the conclusions of Sargent and Wallace cannot be drawn from the threefold equation system they specified—to demonstrate this, let us make a brief digression. The system they described looks like this:

[58] We might remember that the growth rate of idle money held by the members of the older generation was determined by a stochastic variable x; cf. Eq. (4.1).

$$y_t = \xi_0 + \xi_1(m_t - E_{t-1}m_t) + \xi_2 y_{t-1} + u_t; \tag{4.29}$$

$$m_t = g_0 + g_1 y_{t-1} + \varepsilon_t; \tag{4.30}$$

$$E_{t-1}m_t = g_0 + g_1 y_{t-1}, \tag{4.31}$$

where y_t is the deviation of real output from its natural level, m_t is the money growth rate, $E_{t-1}m_t$ is the money growth rate in period t as expected in the previous period, while u_t and ε_t are white noises independent of both each other and further variables in the system. It can be assumed that $\xi_1 > 0$ (i.e. a surprise that monetary policy generates triggers a temporary boost in output) and $(-1) < \xi_2 < 0$ (since if $|\xi_2| > 1$ the system would in this case be explosive and if $1 > \xi_2 > 0$ business cycles would emerge, that is, deviations from the natural level would be systematically positive or negative in multiple consecutive periods; with ξ_2 as assumed here, y will fluctuate around its expected value probably with a damped oscillation). Because of the natural rate hypothesis it is certainly true that $E(y) = 0$. As rational expectations are assumed, the public will form expectations on the basis of the parameters of the rule actually applied by the monetary authority. If the equations of money supply dynamics and expectations are used in the equation that describes real output fluctuations, the result will be this:

$$y_t = \xi_0 + \xi_1(g_0 + g_1 y_{t-1} + \varepsilon_t) - \xi_1(g_0 + g_1 y_{t-1}) + \xi_2 y_{t-1} + u_t, \tag{4.32}$$

that is

$$y_t = \xi_0 + \xi_2 y_{t-1} + u_t + \xi_1 \varepsilon_t. \tag{4.33}$$

Strange as it is, real output dynamics is independent of the monetary policy rule adopted. If we want to phrase our findings by taking into account the variance rule given by Sargent and Wallace, the conclusion will be that the variance of real output is

$$var(y) = \frac{\sigma_u^2 + \xi_1^2 \sigma_\varepsilon^2}{1 - \xi_2^2}, \tag{4.34}$$

which, as suspected, is independent of the parameters of the monetary policy rule if this rule is known to agents. It is worth considering what happens if the monetary authority introduces a new rule

$$m_t = g_0' + g_1' y_{t-1} + \varepsilon_t \tag{4.35}$$

after period $t-1$ which is not known to market participants. Now, our output equation takes the form

$$y_t = \xi_0 + \xi_1\left(g_0' + g_1' y_{t-1} + \varepsilon_t\right) - \xi_1(g_0 + g_1 y_{t-1}) + \xi_2 y_{t-1} + u_t, \tag{4.36}$$

that is,

$$y_t = \left(\xi_0 + \xi_1 g_0' - \xi_1 g_0\right) + \left(\xi_1 g_1' + \xi_2 - \xi_1 g_1\right) y_{t-1} + \xi_1 \varepsilon_t + u_t. \tag{4.37}$$

Based on this, the important variance relevant for us is

$$var(y) = \frac{\xi_1^2 \sigma_\varepsilon^2 + \sigma_u^2}{1 - \left(\xi_1 g_1' + \xi_2 - \xi_1 g_1\right)^2}. \tag{4.38}$$

Variance can be minimized if $\xi_1 g_1' + \xi_2 - \xi_1 g_1 = 0$ is true, i.e. the monetary authority should guarantee

$$g_1' = g_1 - \frac{\xi_2}{\xi_1} \tag{4.39}$$

to be true when setting its new rule. Since $(-1) < \xi_2 < 0$ was required in advance, therefore g_1' is higher than g_1 (note that a change in g_0 has no effects on real output variance). On these grounds, we could say that the central bank has an instrument at hand by which it can lower the fluctuations in real output—significantly, though, for this to happen the central bank is required to cause a surprise, since, as we have seen, real output is independent of the monetary policy rule in case of a known rule (and this is not unusual as in a situation like this surprises are the outcome of random shocks alone). As the new rule becomes known, i.e. as it becomes integrated into expectations, real output will follow its own variance path again. But when does a setting emerge in which the central bank has to generate a surprise as specified above? Because of the constraint on ξ_2, if $y_{t-1} > 0$, then $E(y_t) < 0$, assuming that the central bank does not change the previous rule. However, if it introduces a new rule, according to which $g_1' > g_1$, the monetary authority may be hopeful that it will be capable of generating a monetary surprise to possibly prevent real output from deviating negatively from the natural level. If, however, expectations are rational, it cannot do this, since agents anticipate the next action of the central bank on the basis of the known value of y_{t-1}. A further problem is that the residual u_t is also unpredictable, it is, thus, impossible to set an appropriate rule and cause a surprise. Our findings, with some necessary amendments, are, therefore, consistent with what has been pointed out above. And now, after this digression, back to our original topic.

The inefficiency of systematic countercyclical monetary policy will be true only under very strictly defined conditions. If the nominal variables (prices and wages) are flexible, an exogenous change in the money supply does not affect anything real (this is the case of the Walrasian general equilibrium and the neutrality of money, also referred to by Weeks). We have a completely different situation if flexibility cannot be assumed. Stanley Fischer (1977) scrutinized the possibilities of short-run output control with regards to these very adjustment limitations. In his view, real effects are still possible even *with* rational expectations and a fully anticipated monetary policy. Fischer's findings successfully complete those of the new classicals in which the latter highlighted the inefficiency of countercyclical monetary policy, since, for him, output fluctuations do not come from information deficiencies. Therefore, no surprise is needed. Fischer, as opposed to the traditional way of reasoning, underlines the significance of non-indexed nominal agreements that are concluded for periods longer than the time needed for the monetary authority to respond to changing macroeconomic conditions and that lead to some stickiness in nominal wages[59]; what it boils down to is that this element, which is analogous to the Keynesian argumentation, is directly responsible for the efficiency of a countercyclical monetary policy. In this way, monetary policy will surely pursue a policy of stabilization in cases of supposedly random shocks. While new classical models were directed at correctly specifying the conditions (assumptions) under which an anticipated countercyclical policy is ineffective, Fischer's results successfully highlighted the fact that there are situations in which this policy could still turn out to be effective. Therefore, there is no contradiction here.

In his *Infláció* [*Inflation*], Tibor Erdős (1998) writes the following: "Reality contradicts the »rational expectations« hypothesis. If this theory was right, the inflation rate could easily be reduced, without any rise in unemployment [. . .] by constraining the money supply [. . .]." Apparently, Prof. Erdős questions the entire theoretical system of the REH. If the predictions of a theory are not in accordance with factual observations, one option indeed is to be opposed to the relevance of this theory—but this option is viable only if there is a consistency between underlying assumptions and factual circumstances. What is the REH argument *on the basis of continuous market clearing*? If agents form rational expectations, i.e. they can anticipate the (systematic) actions of economic policy, they will, in this case, adjust to these changes in advance and the result will only be changes in some nominal variables, while the real system remains intact. However, it should be clear by now that the possible real effects of a systematic economic policy do not necessarily imply the lack of rationality of expectations. In other words: if disinflation can be carried through only at the expense of sacrifices in the real economy (a fact corroborated by statistical data), this fact alone is insufficient to question the validity of the REH. Rational expectations aside, the lack of continuous market clearing (that is, of flexible prices) leads to the same result. What happens is that actual life-world conditions are not in accordance with the underlying assumptions necessary for conclusions to be drawn. As for Prof. Erdős' logical dual, expectations in a macroeconomic system can be rational in spite of the fact that real effects are generated by systematic economic policy actions, and in an identical manner, inflation can only be brought down at the expense of real economic sacrifices (note

[59] In his analysis, Fischer studied only the problems of stickiness caused by wage agreements–although he mentioned that other forms of contracts can trigger similar effects.

that instantaneous adjustment is not possible if actions are perfectly anticipated and some nominal variables are sticky—cf. Fischer 1977).

Fischer's (1977) findings are encountered in the writings of other authors as well. For example, Robert Gordon (1979) highlighted the stickiness of price adjustment. If we give credence to these restricting propositions, we have to raise a new question as well and *express doubt regarding the argument that the neutrality of money follows from the rationality of expectations.* Our answer, based on the findings below, will obviously be negative, i.e. we do not have to give up rational expectations in order to invalidate the notion of the neutrality of money. *Expectations can be rational, while money is non-neutral. The inefficiency of a countercyclical monetary policy does not stem from the REH but from the assumption of a general equilibrium environment (continuous and instantaneous market clearing).* As it was mentioned earlier on, *even a fully anticipated economic policy can trigger desired real adjustments in a disequilibrium environment.*

Fischer underlines the consequences that follow from the lack of immediate and complete adjustment (note that this adjustment, if sustained, precludes any real effects) and points out the objections which this development raises against the standard argumentation of new classicals (for the view on this expressed by the new classical macroeconomics and the REH, see Carns and Lombra 1983). Fischer opposed the Lucasian surprise supply function, saying that based on that function, it looks as if only expectations formed in period $t-1$ had a significant effect on real output in period t. If, however, we take into consideration the fact that the period lasts for no longer than a year, we have to extend the set of relevant expectations, since there are contracts[60] the maturity of which exceeds this period, and it follows that as far as the current (period t) output is concerned, expectations formed before period $t-1$ are also relevant, since at the time these contracts were concluded (e.g. in period $t-2$, knowing that they will still be valid in period t), agents formed their expectations for period t as a basis, therefore not only expectations of period $t-1$ are relevant. However, we cannot lose sight of the fact that the monetary policy of period t becomes fully anticipated in (not earlier than) period $t-1$ therefore, oddly enough, the efficiency of monetary policy is caused, yet again, by the divergence of actual and expected inflation dynamics (that is, agents in period $t-2$ will form erroneous expectations about period t inflation, even if they will evaluate it correctly in period $t-1$).

With the help of his model, Fischer could perform a theoretical trick through which the disappointment of expectations can still play a key role in affecting real output, while the government will be in a position to systematically control and exploit this error. It needs to be repeated, over and over again, that the time schedule of availability of information is very important in this line of reasoning. Over and

[60] These contracts require special conditions to emerge–as Fischer also stressed. It is important to note that in terms of stabilizing the variance of real wages, these multi-period contracts are not the best option to follow. This goal can be achieved more easily by agreements concluded for shorter periods. However, we have to bear in mind the considerable costs of recurrent wage negotiations and agreements, leading to longer-run wage contracts.

above wage contracts concluded in period $t-2$, the monetary policy in period t is based on pieces of information which were unknown two periods earlier, thus a monetary policy can react to developments which wage contracts, due to their rigidity, cannot adjust to. The parallels with the reasoning of Kydland and Prescott (1977) have to be underlined[61]: in this situation, monetary authority has the option to optimize its own policy after becoming familiar with the actions made by market participants (or some of them). The issue of time inconsistency will be revisited when tackling the problem of inflation targeting. If a monetary authority does not commit itself to the prohibition of the utilization of the trade-off between unexpected inflation and unemployment, its action will amount to little more than implementing a dynamically time-inconsistent policy; the fact that misleading is completely precluded for Kydland and Prescott is quite another story, i.e. by no means is a monetary authority in a position to exploit the rigidity of nominal wage contracts (KVA 2004; Szabó-Bakos 2005). Kydland and Prescott argue that wage contracts concluded in period $t-2$ would take additional inflation of period t into account. The underlying consideration is rather an interesting one. Being misled in situations like this is nothing but a quantitative issue. Although employees might be aware that due to the rigidity of wage contracts the monetary authority will not resist the temptation of inflating, through generating additional inflation, part of their real wages which were believed to be fixed—they will be unaware of the extent of this additional inflation, i.e. if the central bank is determined to carry on with its misleading policy, it can always exploit the trade-off between inflation and unemployment thanks to settled wage contracts. Let us assume that the monetary authority announces a 5 % inflation for the next period and does not commit itself credibly to achieve this target. Kydland and Prescott argue that in this case employees are aware that the monetary authority will generate additional inflation and, bearing this in mind, they expect, say, a 10 % inflation to avoid a decrease in their real wages. Kydland and Prescott are of the view that because of this a reduction in unemployment is unachievable for the central bank; moreover, social welfare will fall due to a higher rate of inflation (uncompensated by a decrease in unemployment). However, it should be noted that the monetary authority can always use the option of reducing real wages after employees set the nominal wages for the next period, since there is only one level of inflation that can be taken into account in wage contracts. That is, if this level is, for instance, 10 %, the monetary authority will be able to achieve the desired real effects by generating a 15 % inflation. This mechanism, however, is constrained by a social objective function underlying the decisions of the monetary authority (since familiarity with this objective function is a part of the common information set owned by the

[61] To draw the analogy between these two theories, what we need to do is to regard the inflation expected by the employees as announced by the monetary authority (and the implementation of which it adheres to or not).

central bank and market participants), because increasing inflation beyond all limits, even if it has positive effects on employment, will reduce its usefulness in the context of social welfare. Therefore, if the central bank does not commit itself to a specific inflation path, agents will know (on the basis of the social objective function and the actual level of unemployment) the extent of additional inflation still acceptable for the monetary authority in accordance with the maximization of social welfare and they will form their expectations accordingly. Thus the only result is an increase in inflation without any favourable employment effects. Therefore, we should stress the distinction between the *ability of misleading* and the *intention of misleading* with regard to the theory of dynamic time-inconsistency: the ability to mislead is always a possibility, i.e. the central bank is always capable of generating an additional inflation that causes unemployment to fall (for example, under its natural level). However, the intention to do this may be absent, since the necessary extent of this additional inflation would be so high that it would make the desire a complete nonsense—since such an action would deteriorate social welfare.

Of course, it also follows that monetary policy remains inefficient in case of contracts with *one period* maturity. Fischer, therefore, did not set rational expectations aside but completed the previous results only by introducing some additional assumptions. Even if the public during period $t-1$ does foresee the actions of period t monetary policy, and long-run wage agreements concluded in period $t-1$ make the monetary policy inefficient in the next period, the monetary authority can still affect real wages and output through contracts that mature only *after* the period t monetary policy step is taken. However, this is only a temporary effect, since economic structures will, of course, adapt to policy changes: if the employees realize that real wages never follow the desired and pre-set path due to long-run wage agreements, a completely new system of contracts will emerge to eliminate these adverse effects.

John Taylor (1975) investigated the situation in which agents make distorted predictions about the forthcoming policy actions for a while, due to a structural change in monetary policy—that is to say, there is a divergence between public beliefs about monetary policy and actual monetary policy steps. In cases like this, of course, a short-lived effectiveness emerges, in other words monetary policy can take some real variables under its control temporarily (for Taylor, this was unemployment), since this scrutinized situation, after all, is nothing else than an analogy of surprise policy actions. Taylor's purpose was not to define precisely the limits of rationality of expectations as Fischer did, since he stresses that monetary policy is not capable of affecting anything real if continuous market clearing and the REH hold (by doing so, his position is opposed to Fischer's, since Fischer, due to some temporary rigidities, did not preclude a systematic control over output, not even with rational expectations assumed). Instead, Taylor tried to create a situation in which *rational expectations are suspended for a while*, and in this way a monetary authority is offered the opportunity to exercise control over certain real variables. The situation Taylor investigated presents itself as a period of transition to rational expectations when the combination of former public beliefs[62] about a previous policy with information about a new policy (as distinct from the old one) can lead to distorted predictions, since a public analysis of the monetary policy steps taken by the new regime and, based on this, unbiased expectations about future actions are possible only after a longer period of observation[63] (the transition to rational expectations is explained by the biasedness of adaptive expectations realized by the public, plus the fact that the central bank actively tries to exploit this biasedness through the relationship between unemployment and expectation errors). Structural shifts in monetary policy are an excellent opportunity for the central bank to exploit these possibilities in practice. Naturally, though, these effects are only temporary.

In Taylor's model, new monetary policy rule is simple and can be easily characterized by the expected (mean) rate of inflation and the variance of the inflation path (as we have seen, these features are characteristic in the island models of Lucas). Public ignorance means that although people know the time-dependent variance (the variance path) but are unaware of the expected (mean or *target*) rate of inflation (by the end of this transition period the public will be aware of both data sets), and mean inflation is replaced by the expected value from the old rule[64] (this feature represents the combination of pieces of information pertaining to the old and the new monetary policy rule). There is only a limited control over real variables (not only in terms of the timeframe), and Taylor points out a few circumstances that set the limits of a monetary policy activism. He strongly warns against keeping the public in a complete ignorance via monetary policy pursued in the new regime (by announcing, for example, a specific variance path and following another, preventing the public from learning any of the parameters), since, in this case, effects of stimulating monetary policy actions would be uncertain and unpredictable. Taylor's clear message is that target inflation should be the only factor of uncertainty. A further limitation for the new variance path is that an excessive boost of this variance should be avoided. If not, a welfare loss may be experienced due to the unnecessary increase in expectation errors.[65] In this case, too, the instrument of delusion is generating an unexpected inflation (under

---

[62] These new beliefs and opinions ensure biasedness. Without them, expectations would be rational and predictions would be unbiased (so, there would be no transition period either), i.e. the public would know the rules followed by the new monetary policy regime from the very beginning.

[63] Bear in mind the lessons from Chap. 2 that helped us realizing that there is no way to find the relevant model in a stochastic environment. Since prediction errors may occur even if this relevant model is known, any modeller walks randomly among the different states and phases of his model, as he cannot judge whether these errors are due to contingencies or to the misspecification of the model. Thus Taylor's reasoning is dubious.

[64] A different case (leading to the same conclusions) is when the public tries to approximate mean inflation, chosen by the new policy, through estimations based on various factors (even on the personal character of new policy makers).

[65] This welfare loss is not caused by the increase in the variance of the error *per se*. Phillips curve is generally assumed to be convex, therefore it is true that although the expected value of the unemployment-dependent inflation is zero, a greater variance of prediction errors results in a

stochastic conditions, expected and actual inflation are not necessarily equal, *random* expectation errors may emerge, but Taylor argues that a fall in unemployment occurs only as a result of *systematic* prediction errors). To see the logic behind this a few structural relations should be considered:

$$f = \phi(u) + x; \tag{4.40}$$

$$\phi'(.) < 0; \tag{4.41}$$

$$\phi(u^*) = 0; \tag{4.42}$$

$$u > 0, \tag{4.43}$$

where f is actual rate of inflation and x is its expected level, $\phi(u)$ additional (positive or negative) inflation raised by unemployment and u^* designates the natural rate of unemployment where that additional inflation is zero (as it is illustrated, unemployment can only be reduced at the expense of a higher rate of inflation). If the actual inflation exceeds its expected rate, i.e. $f > x$, it is equivalent to a fall in unemployment. Under stochastic conditions, prediction errors will inevitably occur, even if the public is aware of the mean (target) rate of inflation as a characteristic of the inflation path to be followed. In this case, however, prediction errors (disappointments of expectations) do not correlate with the available information set (upon which expectations are based), therefore these are random errors. Consequently, it is also true that the conditional mean of unemployment-dependent inflation (where the information set upon which expectations are based constitute the condition) is zero.

Naturally, an option available to the monetary authority at all times is to influence expectations (in other words, it can always convince the public of its forthcoming strategies through announcements) then to move from this preannounced path, thereby surprising market participants. The positive and negative consequences of such a central bank policy are dubious and the balance of these costs and benefits is a further problem. Differently put, although a monetary policy based on deliberate deception can be successful in controlling real variables (naturally, the length of the time horizon is always a problem, since, after all, the monetary authority crosses the very expectations that it previously created—thus the question of political credibility is of considerable significance in this situation), it may prove destructive on macroeconomic structures in the long run. There may be serious negative consequences and, after all, monetary policy may lose its potential to influence expectations. Taylor comes up with the recommendation that a policy created in this manner should be avoided; but if the monetary authority refuses to give up its policy of deception, it should, at least, allow market participants to be disappointed in their self-created expectations. Similarly, Edmund Phelps (1972) also broke inflation down into two types, highlighting the case in

---

higher expected level of unemployment because of the Jensen inequality–and the welfare loss is its direct consequence.

which a kind of deal is assumed valid between market participants and the government, and, accordingly, in which the disappointment of expectations is no more than the breach of this agreement by the government. However, he also warned against government-induced disappointment (delusion) of rational expectations.

4.3.5 Summary: Monetary Policy in the New Classical Framework

Before moving on, it might be a good idea to summarize our conclusions so far. Talking about new classical macroeconomics, the axioms of (1) continuous market clearing and of (2) rational expectations were distinguished and, as it was realized, the inefficiency of systematic countercyclical monetary policy cannot be inferred from the latter on its own (the qualitative characteristics of business cycles are raised as a separate issue). However, it is also true that the incomplete flexibility of nominal quantities (that is, stickiness of prices and wages) is not the only friction that can lead to real fluctuations—this is the most important lesson drawn from the Lucasian island models. These frictions result in real shifts even if nominal quantities are completely flexible. We now know that, within the new classical world where there are only random (i.e. non-systematic) differences (errors) between actual and expected price dynamics, systematic countercyclical monetary policies *cannot be followed at all*, since monetary authorities do not have any information on the basis of which such actions can be planned and implemented. For new classical macroeconomics, real fluctuations cannot be resolved or eased through monetary policy interventions if expectations are rational and continuous market clearing is at play along with white noise errors.

However, it is worth reconsidering new classical thoughts on monetary policy from a different aspect. The traditional view on the long-run controlling capabilities of monetary policy over output was not altered considerably by the new classicals. Keynes himself had already argued for a limited, short-run efficiency of monetary policy. For him, the case of the liquidity trap prevented central banks from boosting output and decreasing unemployment in the long-run. This view was maintained by Milton Friedman (concentrating on the resulting destructive inflationary processes). Monetary policy was assumed to be able to deviate economy from the natural level of unemployment and output through surprises, but to maintain this as a long-term objective would only be possible at the expense of an ever-accelerating inflation, which appears to be a highly destructive kind of monetary policy intervention. If at all, monetary policy, even in orthodox monetarism, was capable of controlling real output (in a non-countercyclical, but rather cycle-generating manner) only in the short-run—if we disregard the fact that this potential was considerably eroded by the fact that its effects were to be felt with a time lag. Actually, we are far from countercyclical goals or purposes here, since as long as the real system remains at the natural level of unemployment and output, the need for a countercyclical (monetary) policy should not be considered as an option. Such a policy is needed when something dislocates the economy from its equilibrium state—that is, if business cycles emerge. The possibilities of countercyclical monetary policies are determined by the nature of these cycles. If these business cycles are white noises, a countercyclical intervention is needed which, however, is not possible. By contrast, exerting countercyclical influence through known (anticipated) actions is a prospect made plausible by long-lasting cycles. We have seen that, according to some results (models), the build-up phase of rational expectations, for example, or the maturity structure of nominal agreements could also cause monetary policy to have countercyclical effects.

New classical macroeconomics further elaborated and refined the barriers surrounding the potential of monetary policy control over output and unemployment. For new classicals, just like in the case of Friedmanian orthodox monetarism, monetary policy was allowed to make efforts to control output and unemployment in the long-run (or at least it was in a position to make an attempt to do so). Yet again, that aim can be achieved by generating additional inflation. Long-run impotency is a basic assumption for new classicals, too: systematic changes in the rate of inflation leaves real output unaffected if rational expectations and flexible prices can be assumed. Even if there are shocks presumed to be white noises, long-run shifts in the rate of inflation do not affect unemployment or output, only inflation becomes higher. The real problem is short-run efficiency. *If expectations are rational, prices are flexible and shocks are unforeseeable, not even in the short-run will monetary policy be able to exert a countercyclical effect on output and unemployment through changing the money supply*. However, we have seen that monetary policy regains its potential to trigger real adjustments through applying its available instruments *in systematic steps if* rationality of expectations is incomplete, *if* prices are sticky or *if* prolonged cycles are allowed. Nor should we turn a blind eye to the content of this countercyclical controlling. As it was mentioned earlier on, by no means will a monetary policy be able to stimulate long-run economic growth efficiently—such an effort could only result in an ever-worsening inflationary pressure (in Chap. 6, where the issue of supply side economics is tackled, we will review the instruments of long-run growth enhancing policy thought to be the most efficient by new classicals). In other words, no monetary policy should have the intention of fixing output above its potential (natural) level in the long-run. The most reasonable option to follow is to use the available countercyclical instruments of monetary policy with the aim of closing the output gap. It will become clear later on that it is this very intention which central banks operating within the context of inflation targeting are driven by. However, the purpose of closing the output gap is identical to the desire of directing the real economy towards the level of potential output, thus the well-known purpose of neoclassical-monetarist-new-classical economics in maintaining the equilibrium output can directly and unmistakably be identified here. The output target of monetary policy should be the equilibrium level, since it is the only level that can be sustained for longer periods without inflationary pressures.

In addition to the traditional questioning of the long-run output controlling potential of monetary policy, new classical macroeconomics refined the criteria for short-run efficiency. New classicals defined the circumstances under which monetary policy is ineffective and, simultaneously, highlighted the conditions under which short-run (countercyclical) achievements could be expected. The denial of the efficacy of systematic actions is justified only if some very specific and strict assumptions remain valid. In this way, new classical macroeconomics made an almost direct contribution to the evolution of the framework in which modern monetary policy exists.

References

Alexander SS (1940) Mr. Keynes and Mr. Marx. Rev Econ Stud 7(2):123–135

Arthur WB, Holland JH, LeBaron B, Palmer R, Taylor P (1996) Asset pricing under endogenous expectations in an artificial stock market. Working paper 96-12-093. Santa Fe Institute, Santa Fe

Barabás G, Hamecz I (1997) Tőkebeáramlás, sterilizáció és pénzmennyiség I–II. Közgazdasági Szemle 44(7–8):653–672 and 44(9):757–781

Barro RJ (1977) Unanticipated money growth and unemployment in the United States. Am Econ Rev 67(2):101–115

Barro RJ, Rush M (1980) Unanticipated money and economic activity. In: Fischer S (ed) Rational expectations and economic policy. University of Chicago Press, Chicago, pp 23–48

Begg DKH (1982) The rational expectations revolution in macroeconomics–theories and evidence. Philip Allan, Oxford

Benczes I, Kutasi G (2010) Költségvetési pénzügyek–hiány, államadósság és fenntarthatóság. Akadémiai, Budapest

Blinder AS, Fischer S (1979) Inventories, rational expectations, and the business cycle. NBER working paper No. 381. National Bureau of Economic Research, Cambridge, MA

Bowen WG, Berry RA (1963) Unemployment conditions and movements of the money wage level. Rev Econ Stat 45(2):163–172

Brandis R (1985) Marx and Keynes? Marx or Keynes? J Econ Issues 19(3):643–659

Brimmer AF, Sinai A (1981) Rational expectations and the conduct of monetary policy. Am Econ Rev 71(2):259–267

Carns F, Lombra R (1983) Rational expectations and short-run neutrality–a reexamination of the role of anticipated money growth. Rev Econ Stat 65(4):639–643

Chadwick O (1993) The reformation. Penguin Books, Harmondsworth, Hungarian edition: Chadwick O (1998) A reformáció (trans: Szabó I). Osiris, Budapest

Chan CK, Steiglitz K (2008) An agent-based model of a minimal economy. Princeton University, Department of Computer Science, Princeton

Dosi G, Fagiolo G, Roventini A (2008) The microfoundations of business cycles–an evolutionary, multi-agent model. J Evol Econ 18(3):413–432

Ellison M (2003) Lectures in quantitative monetary economics. University of Warwick, Warwick

Erdős T (1998) Infláció. Akadémiai, Budapest

Erdős T (2003) Fenntartható gazdasági növekedés, különös tekintettel a rendszerváltást követő magyar gazdaságra. Akadémiai, Budapest

Fischer S (1977) Long-term contracts, rational expectations, and the optimal money supply rule. J Polit Econ 85(1):191–205

Fisher I (1928) The money illusion. Adelphi, New York

Franses PH (1998) Time series models for business and economic forecasting. Cambridge University Press, Cambridge
Friedman M (1953) The methodology of positive economics. In: Friedman M (ed) Essays in positive economics. University of Chicago Press, Chicago, pp 3–43
Friedman M (1968) The role of monetary policy. Am Econ Rev 58(1):1–17
Friedman M (1970) The optimum quantity of money and other essays. Aldine Publishing Company, Chicago
Friedman M (1977) Nobel lecture: Inflation and unemployment. J Polit Econ 85(3):451–472
Gaffeo E, Gallegati M, Gostoli U (2012) An agent-based "proof of principle" for Walrasian macroeconomic theory. Cognitive and Experimental Economics Laboratory, Trento
Gordon R (1979) New evidence that fully anticipated monetary changes influence real output after all. NBER working paper No. 0361. National Bureau of Economic Research, Cambridge, MA
Grey GB (2002) Federal Reserve System–background, analyses and bibliography. Nova Science Publishers, New York
Gruca TS, Klemz BR (2003) Optimal new product positioning–a genetic algorithm approach. Eur J Oper Res 146(3):621–633
Hahn FH (1980) Monetarism and economic theory. Economica 47(185):1–17
Hansen AH (1953) A guide to Keynes. McGraw-Hill, New York, Hungarian edition: Hansen AH (1965) Útmutató Keyneshez (trans: Andorka R). KJK, Budapest
Heckman JJ (2008) The many contributions of Edmund Phelps–American Economic Association luncheon speech honoring the 2006 Nobel laureate in economics. Capitalism Soc 3(3):1–11
Hirabayashi A, Aranha C, Iba H (2009) Optimization of the trading rule in foreign exchange using genetic algorithm. Paper presented at GECCO '09, Montréal, 8–12 July 2009
Howitt P (2007) Macroeconomics with intelligent autonomous agents. In: Farmer REA (ed) (2008) Macroeconomics in the small and the large–essays on microfoundations, marcoeconomic applications and economic history in honor of Axel Leijonhufvud. Edward Elgar Publishing, Cheltenham/Northampton, pp 157–177
Hume D (1752) Of money. In: Hume D (ed) Political essays. Cambridge University Press, Cambridge, pp 115–125
Kahneman D, Tversky A (1979) Prospect theory–an analysis of decision under risk. Econometrica 47(2):263–291
Kaldor N (1970) The new monetarism. In: Kaldor N (1978) Further essays in applied economics. Duckworth, London, pp 1–21
Kalecki M (1935) The mechanism of business upswing. In: Kalecki M (1969) Studies in the theory of business cycles 1933–1939. Basil Blackwell, Oxford, pp 26–33. Hungarian edition: Kalecki M (1980) A tőkés gazdaság működéséről–válogatott tanulmányok 1933–1970 (trans: Balogh A). KJK, Budapest, pp 64–73
Karni E (1980) A note on Lucas's equilibrium model of the business cycle. J Polit Econ 88 (6):1231–1236
Keynes JM (1936) The general theory of employment, interest and money. Palgrave Macmillan, London. Hungarian edition: Keynes JM (1965) A foglalkoztatás, a kamat és a pénz általános elmélete (trans: Erdős P). KJK, Budapest
Király J (1998) A makroökonómia vége, avagy egy megkésett Nobel-díj? Közgazdasági Szemle 45(12):1082–1095
Király J (2000) Robinson és a közgazdászok. Café Bábel 10(2):107–113
Király J (2005) Robert E. Lucas. In: Bekker Z (ed) Közgazdasági Nobel-díjasok. KJK–Kerszöv, Budapest, pp 631–649
KVA (1995) The scientific contributions of Robert E. Lucas, Jr. (Laudation). The Royal Swedish Academy of. Sciences, Stockholm
KVA (2004) Finn Kydland and Edward Prescott's contribution to dynamic macroeconomics: the time consistency of economic policy and the driving forces behind business cycles (Laudation). The Royal Swedish Academy of Sciences, Stockholm

KVA (2006) Edmund Phelps's contributions to macroeconomics (Laudation). The Royal Swedish Academy of Sciences, Stockholm
Kydland FE, Prescott EC (1977) Rules rather than discretion–the inconsistency of optimal plans. J Polit Econ 85(3):473–491
Lin L, Cao L, Wang J, Zhang C (2007) The applications of genetic algorithms in stock market data mining optimisation. University of Technology, Faculty of Information Technology, Sydney
Lipsey RG (1960) The relation between unemployment and the rate of change of money wage rates in the United Kingdom, 1862–195. A Further Analysis. Economica 27(2):1–31
Lucas RE (1972) Expectations and the neutrality of money. J Econ Theory 4(2):103–124
Lucas RE (1973a) Some international evidence on output-inflation tradeoffs. Am Econ Rev 63 (3):326–334
Lucas RE (1973b) Econometric policy evaluation–a critique. Carnegie-Rochester Conference Series on Public Policy 1(1):19–46
Lucas RE (1975) An equilibrium model of the business cycle. J Polit Econ 83(6):1113–1144
Lucas RE (1980) Rules, discretion, and the role of the economic advisor. In: Fischer S (ed) Rational expectations and economic policy. Chicago University Press, Chicago
Lucas RE (1995) Nobel lecture: Monetary neutrality. J Polit Econ 104(4):661–682
Mátyás A (1963) A polgári közgazdaságtan története. KJK, Budapest
Mátyás A (1984) A polgári közgazdaságtan napjainkban. Kossuth, Budapest
Mátyás A (2003) A modern közgazdaságtan története. Aula, Budapest
Mérő L (2004) Az élő pénz – a gazdasági vállalkozások eredete és az evolúció logikája. Tericum, Budapest
Mihalik I, Szigeti E (1984) Fejezetek a marxista-leninista politikai gazdaságtan történetéből. Kossuth, Budapest
Morton AL (1969) The English utopia. Lawrence and Wishart, London, Hungarian edition: Morton AL (1974) Angol utópia (trans: Békés Á). Kossuth, Budapest
Németh G (1995) Az aranykortól Utópiáig. Café Bábel 5(1–2):105–119
Nicolaisen J, Petrov V, Tesfatsion L (2001) Market power and efficiency in a computational electricity market with discriminatory double-auction market. Iowa State University, Ames
Nigg W (1979) Thomas More ou la conscience d'un saint. Centurion, Paris
Oeffner M (2008) Agent-based Keynesian macroeconomics–an evolutionary model embedded in an agent-based computer simulation. Inaugural dissertation, University of Würzburg
Patinkin D (1989) Money, interest, and prices. MIT, Cambridge, MA
Phelps ES (1967a) Phillips curves, expectations of inflation and optimal unemployment over time. Economica 34(8):254–281
Phelps ES (1967b) Money-wage dynamics and labor-market equilibrium. J Polit Econ 76 (4):678–711
Phelps ES (1969) The new microeconomics in inflation and employment theory. Am Econ Rev 59 (2):147–160
Phelps ES (1972) Inflation policy and unemployment theory. Macmillan, London
Phelps ES (2006) Nobel lecture: macroeconomics for a modern economy. Am Econ Rev 97 (3):543–561
Phelps ES, Archibald GC, Alchian AA (1969) Microeconomic foundations of employment and inflation theory. Norton, New York
Phillips AW (1958) The relation between unemployment and the rate of change of money wage rates in the United Kingdom, 1861–1957. Economica 25(11):283–299
Plato (380 BCE) De republica. English edition: Plato (2006) The republic (trans: Allen RE). Yale University Press, New Haven
Samuelson PA (1958) An exact consumption-loan model of interest with or without the contrivance of money. J Polit Econ 66(6):467–482
Samuelson PA, Nordhaus WD (1998) Economics, 16th edn. McGraw-Hill, New York
Sardoni C (1986) Marx and Keynes on effective demand and unemployment. Hist Polit Econ 18 (3):419–441

Sargent TJ, Wallace N (1974) Rational expectations and the theory of economic policy. University of Minnesota, Minnesota

Sargent TJ, Wallace N (1975) "Rational" expectations, the optimal monetary instrument, and the optimal money supply rule. J Polit Econ 83(2):241–254

Shaw GK (1984) Rational expectations–an elementary exposition. Harvester Press, Brighton

Szabó-Bakos E (2005) Finn E. Kydland - Edward C. Prescott. In: Bekker Z (ed) Közgazdasági Nobel-díjasok. KJK–Kerszöv, Budapest, pp 893–913

Szalkai I (1995) A monetáris irányítás. Bankárképző, Budapest

Taylor JB (1975) Monetary policy during a transition to rational expectations. J Polit Econ 83 (5):1009–1021

Taylor JB (1979) Staggered wage setting in a macro model. Am Econ Rev 69(2):108–113

Tesfatsion L (2000) Structure, behavior, and market power in an evolutionary labor market with adaptive search. Iowa State University, Ames

Tversky A, Kahneman D (1974) Judgement under uncertainty–heuristics and biases. Science 185 (4157):1124–1131

Tversky A, Kahneman D (1986) Rational choice and the framing of decisions. J Bus 59(4):S251–S278

Tversky A, Kahneman D (1992) Advances in prospect theory–cumulative representation of uncertainty. J Risk Uncertain 5(4):297–323

Ungvári T (1956) Daniel Defoe. In: Defoe D (1956): Moll Flanders örömei és viszontagságai (trans: Vas I). Új Magyar Könyvkiadó, Budapest, p i–xxiv.

Varian HR (1999) Intermediate microeconomics. W.W. Norton, New York/London, Hungarian edition: Varian HR (2001) Mikroökonómia középfokon–egy modern megközelítés (trans: Bara Z, Temesi J). KJK–KERSZÖV, Budapest

Vigvári A (2004) Pénzügy(rendszer)tan, 1st edn. KJK–KERSZÖV, Budapest

Vigvári A (2008) Pénzügy(rendszer)tan, 2nd edn. Akadémiai, Budapest

Vincze J, Madarász K (2005) George A. Akerlof – a közgazdasági Nobel-díj 2001. évi díjazottja. In: Bekker Z (ed) Közgazdasági Nobel-díjasok. KJK–KERSZÖV, Budapest, pp 775–787

Weber M (1904) Die Objektivität sozialwissenschaftlicher und sozialpolitischer Erkenntnis. Archiv für Sozialwissenschaft und Sozialpolitik 19(1):22–87, English edition: Weber M (1949) Objectivity of social science and social policy (trans: Shils EA, Finch HA). In: Shils EA, Finch HA (eds) The methodology of the social sciences. Free Press, New York, pp 49–112

Weber M (1934) Die protestantische Ethik und der Geist des Kapitalismus. J.C.B, Mohr, Tübingen, Hungarian edition: Weber M (1982) A protestáns etika és a kapitalizmus szelleme (trans: Gelléri A, Józsa P, Somlai P, Tatár G). Gondolat, Budapest

Weeks J (1989) A critique of neoclassical macroeconomics. Macmillan, London, Hungarian edition: Weeks J (1998) A neoklasszikus közgazdaságtan kritikája (trans: Szalai Z). Aula, Budapest

Weintraub R (1980) Comment on Barro–Rush. In: Fischer S (ed) Rational expectations and economic policy. University of Chicago Press, Chicago, pp 63–70

Westerhoff F (2009) A simple agent-based financial market model–direct interactions and comparisons of trading profits. Bamberg Economic Research Group, Bamberg

Chapter 5
Fiscal Policy and New Classical Macroeconomics

The trouble with the world is that the stupid are cocksure and the intelligent are full of doubt.

(Bertrand Russel)

5.1 Introduction

Scrutinizing monetary policy above, we stressed that for new classical macroeconomics, countercyclical stimulation of aggregate demand through monetary policy instruments is neither possible nor beneficial *if the assumptions of the theory hold*. *If* expectations are rational and *if* markets are characterized by completely flexible nominal quantities and *if* shocks are unforeseeable white noises, then macroeconomic systems can deviate from the equilibrium level only under contingencies (i.e. random shocks). However, no systematic countercyclical monetary policy can be built on these conditions, since even monetary policy makers cannot foresee these shocks hitting economies, so no planned response is possible. According to the common and traditional judgement, new classical macroeconomics brought the inefficiency of economic policy into the limelight. Moreover, these statements are always undermined by the fact that new classical assumptions are too far from life-world conditions to plausibly underlie the theorems. So, it has to be realized that *the precise design of the assumptions underlying the inefficiency-theorem makes the most influential (though highly ignored and misunderstood) scientific development of new classical macroeconomics*. New classicals did not assert simply that activist economic policy (in a narrow sense: monetary policy) is ineffective. Lucas and his group drew the attention to *the conditions under which this inefficiency probably emerges*.

It has to be noted: for new classicals, the inefficiency of systematic monetary and (general) economic policy is not a *shortcoming* but a *merit*. If rational expectations work well, for instance, disinflation can be performed without real economic sacrifices (cf. Erdős 1998). So, there is no need to try to exploit the Phillips curve or to figure out if there are negatively sloped short- and long-run Phillips curves at all–and any actions taken by economic policy have no effects on anything real. This characteristic of a macro-system can be regarded as a merit since otherwise (that is, in a system without rational expectations) inflation can be broken down only in the

P. Galbács, *The Theory of New Classical Macroeconomics*, Contributions to Economics, DOI 10.1007/978-3-319-17578-2_5

"traditional" way, at the expense of a considerable employment loss (cf. Sargent 1986). *The faith in the efficiency of economies let loose is clearly unveiled.*

In the beginning, we will follow the previous line of reasoning and scrutinize whether fiscal policy, for new classicals, has a higher potential to control aggregate demand under the usual and known conditions (rational expectations, completely flexible quantities). After this, we turn our attention to the supply side to investigate the possibilities of supply side stimulation of aggregate income (output).

Monetarism shows an unhidden doubt for traditional instruments of fiscal policy. For example, Friedman (1968) detailed the limited reaction capabilities of fiscal policy, stressing that the general denial of the efficiency of fiscal policy does not stem from its alleged impotency to control aggregate demand. *So, we face the argument again that any intent of countercyclical demand management is completely useless.* Monetarism (both its orthodox wing and new classical macroeconomics) tried to support its findings by a well-elaborated system of arguments. Either the permanent income hypothesis or the Ricardian equivalence theorem is considered, the weak potential of (countercyclical) demand management follows from both.

The theory on the possibilities of demand side fiscal stimuli has done a very special trajectory during the last seven decades. It is generally known that Keynes built his theory to highlight the efficacy of fiscal instruments and to recommend fiscal policy. Compared to it, as far as demand side management is considered, monetarism can be labelled by "fiscal policy nihilism" at best. This scepticism, however, was not uniform: reasoning emerged even years after orthodox monetarism and new classical macroeconomics had begun that tried to phrase wisely and precisely the (in)efficiency of fiscal policy. Investigations called our attention to the fact that the final effect of a budget deficit evolves under the joint influence of multiple factors; and though it could remain true that fiscal policy actions cannot affect nominal GDP either in the short- or the long-run, we have a completely different setting if we divide nominal GDP into real and price processes and scrutinize these components separately. Budget deficit, emerging either from a drop in revenues or an increase in expenditures, leads to a boost in aggregate demand that enhances national real income, but these positive effects get partly offset by the resulting rise in prices. A further negative consequence of increased aggregate demand may be a rise in real interest rates, caused by the monetary conditions not accommodated to demand side developments (the effect of a constant or non-sufficiently growing money supply exerted through real balances) that (viz. a rise in real interest rates) may unfavourably affect the interest-rate-dependent elements of aggregate demand (crowding-out). Digging deeper, the expansionary effect of a budget deficit gets dampened, as a result. The new equilibrium is short-living, since the higher price level and real output trigger further reactions in factor costs and investment, for example. In terms of long-run outcomes, it is of fundamental importance whether favourable supply side effects emerge or not–for instance, states may enhance employment through tax cuts or, through expenditures, finance activities that have improved productivity compared to the case in which the public would be entitled to spend fiscal resources. These positive supply side effects can moderate the extent of the demand-side-driven increase in prices–

however, without them, the increase in the price level will be higher. This rise in prices, through expectations adjusted upwards, can drive macroeconomic systems into inflationary spirals, and, as a result, an accordant rise in real interest rates may also occur, depending on the monetary policy steps taken. The final outcome may be even a convergence of real output towards the level that had prevailed before the deficit increased (cf. e.g. McElhattan 1982).

Crowding-out can be interpreted as well as in the context of a changing financial structure underlying government expenditures. If a government covers its revenues fallen short due to a tax cut through issuing bonds (*debt-for-tax-swap*–this financial mechanism will get special importance later, when studying Ricardian equivalence), according to the traditional interpretation, a crowding-out effect emerges, since tax-cut-increased income boosts consumption as well, and, hence, national savings get exceeded by investment demand, leading to a rise in the rates of interest eventually. Moreover, these higher interest rates result in a smaller stock of productive capital available in the future (Barro 1989). In this interpretation, the public debt makes an intergenerational burden because of its negative consequences on the stock of capital, i.e. generations, living after the public borrowing has been completed, carry the burdens manifested in a smaller stock of capital, while the advantages are realized at the time of borrowing (Modigliani 1961). Martin Feldstein (1974) also stresses the negative effects on savings referring to the case of some social security programs. The idea was supported both by theoretical and empirical investigations according to which anticipated social security payments boost consumption and exert adverse effects on savings–so, here again, interest rate has to rise to restore equilibrium, and crowding-out occurs. We will see that if the Ricardian equivalence works, then crowding-out fails, since consumption and savings do not respond to these fiscal actions under those conditions as the equivalence-theorem has specified them.

We have seen that the messages of new classical macroeconomics (taking the risk of oversimplification) can be confined to the theorem that only unexpected monetary shocks can exert influence on real output. Similar arguments are lined up in terms of fiscal policy: according to them, fiscal policy actions can be affective only if they affect the quantity of money–so, the same arguments can be referred to both fiscal sphere and monetary policy. Accordingly, authors sometimes claim that completely anticipated fiscal actions are as ineffective as anticipated monetary policy interventions. Of course, we can advance an argument against this extreme view, saying that relative prices can be modified through tax changes which are capable of triggering real effects on their own. Similarly, boosts in the intensity of some activities can be achieved through certain kinds of tax cuts–and the public ignorance on part of the affected agents over the policy steps taken is not a prerequisite for these effects at all (Shaw 1984). However, these two systems of reasoning do not override each other. Both orthodox monetarism and new classical macroeconomics were sceptical about the success and effectiveness of fiscal policy in stimulating real output through demand management–though, in different ways. In spite of this, new classical macroeconomics still gives some leeway for systematic economic policy actions (either as demand-manipulating fiscal steps or growth-enhancing measures suitable to supply side principles), avoiding the dangers of restoring the rough Keynesian fiscal prescriptions, of course. New classical macroeconomics responded to limited effectiveness by focusing on supply side considerations. So, we are going to identify the arguments supporting the inefficiency of demand side fiscal policy actions (as we have already reviewed monetary policy of similar kind) and to realize the key factors of success of supply side policies. Even if economic policy is not able to eliminate cyclical effects, it is still possible to stimulate real output through modifying the natural rate of unemployment (and natural level of output). Our primary problem here is to pinpoint the circumstances (assumptions) that lead to the inefficacy of fiscal policy.

5.2 Permanent Income Hypothesis: The Basics

Permanent income hypothesis (PIH) is the theory of consumption eventually (cf. Friedman 1957). In his theory, Keynes supported economic policy makers by his showy argument emphasizing their capability of macroeconomic fine-tuning. The only problem was that *actual consumption time series were much less volatile than the predictions derived from the theory of Keynes*. For Keynes, consumption expenditures are linked to disposable income (measured in terms of wage-units) that characterizes a level of employment with some limitations[1] by a parameter called marginal propensity to consume. Marginal propensity to consume in itself is also a function (of income), as Keynes mentioned–though it shows great extent of stability, so it should be regarded as determined by factors other than income.[2] However, since marginal propensity to consume itself is a function of income, it is also true that additional increases in disposable income lead to diminishing increases in consumption expenditures (in other words, marginal propensity to consume is in a reverse relation with real income).[3] This is not the right time to give a fundamental and comprehensive revision of the consumption theory of Keynes, so only the elements relevant for the scrutinised monetarist doctrines will be highlighted. Accordingly, it must be stressed that the relation characterized by a substantial stability (or, at least, by predictable dynamics) links current consumption expenditures to current disposable income–and, on these grounds, a considerable leeway is provided for aggregate demand stimulation, since a change of income (exerted primarily by government expenditures) immediately results in a multiplied shift in aggregate demand (this is the essence of the Keynesian case of

[1] The different distributions of the same level of employment across sectors may lead to different levels of real national income, so a direct functional relation between employment and real national income should be accepted only as a first and rough approximation (Keynes 1936).

[2] Keynes (1936) divided these explanatory variables into *objective* and *subjective* factors, assuming the latter to be unchanging (i.e. given).

[3] As marginal propensity to consume is a function of income, it follows eventually that average and marginal propensities do not coincide. We expect it to be the general case that average propensity exceeds marginal propensity to consume (cf. Houthakker 1958).

the multiplier effect). The same is true of tax cut policies, of course. According to the basic theory of Keynes, governments are always capable of countercyclical fine-tuning of macroeconomic systems through demand management as a part of wise and reasoned fiscal policies.[4]

Permanent income hypothesis questions this ability of governments. In one of his works, even Friedman (1957) stresses that one of the most serious consequences of his theory is a multiplier much smaller than Keynes assumed. However, it is also true that permanent income theory is concentrated mainly on long-run dynamics and relations, while Keynes focused primarily on short-run considerations (it follows that PIH was not simply a critique of the consumption theory of Keynes, but it tried to create a fundamentally new conceptual framework). The emergence of the PIH raised serious debates, and the authors tried either to verify or to falsify the theory of Friedman–in the latter case, arguments were directed mainly towards stressing that the relation between consumption and disposable income still follows (more or less) the mechanism supposed by Keynes. According to some hints dropped in the literature, PIH has the advantage (among others) that it can help us resolve the (alleged) inconsistency between occasionally arising large-scale fluctuations of disposable income and the considerable stability of consumption expenditures (Shaw 1984). An illuminating review phrases in the way that Friedman relates measured (that is, actual) consumption to permanent and not to measured (that is, not to actual or disposable) income. In other words, Friedman's PIH implies that a household plans consumption in advance over quite a long period on the basis of its income expected for that period–and that consumption will be stuck to this planned path in spite of fluctuations in disposable income (Houthakker 1958).

Friedman starts elaborating his theory under the assumption of *complete certainty*. Under these conditions, a consumer unit[5] precisely knows each definite sum it will receive in each of a finite number of periods and knows in advance the consumer prices plus the deposit and the borrowing rates of interest that will prevail in each period. Under such circumstances, for Friedman (1957), there are only two motives for a consumer unit to spend more or less on consumption than its income: The one is to smooth (or to "straighten out", as Friedman says) its consumption expenditures through an appropriate timing of borrowing and lending; and the second is either to realize interest earnings on deposits if the relevant rate of interest is positive, or to benefit from borrowing if the interest rate is negative. The concrete behaviour of a consumer unit under the joint influence of these factors depends on its tastes and preferences. So, these explanatory variables exert their influence on

[4] The scepticism shown by Keynes regarding the efficiency of monetary policy instruments is not independent of this statement.

[5] It is not a mistake to identify the concept of the "consumer unit" mentioned by Friedman with "household", the extent of which ranges from households consisting of only one member to multinuclear households. The definition of household perfectly suits our purposes since it means a social group sharing accommodation and cost of living, *consuming together* (cf. Andorka 1997; Giddens 1993).

consumer units not uniformly, i.e. behavioural patterns may show discrepancies (in other words, by these mechanisms are meant to highlight tendencies rather than give a precise description of behaviour).

If complete certainty about future developments (or, in other words, perfect foresight) was set aside, we would have a fundamentally different setting. Under complete certainty, first period consumption uniquely and concretely determines second period consumption–however, under uncertainty, only the probability distribution of second period consumption will be known for the household on the basis of its first period consumption. In this situation, the uncertainty about consumption goes back to a twofold root: on the one hand, there is a direct effect from the lack of perfect foresight into future consumer prices and incomes; and, on the other hand, an indirect effect from the lack of complete certainty of future possibilities of lending or borrowing (cf. Friedman 1957).

According to PIH, the distribution of consumption across consecutive periods is the result of an optimizing method by which each consumer tries to maximize his utility. At the same time, whatever ratio of income one devotes to consumption in each period, all these consumption expenditures are allocated in the course of an optimization process–that is, consumer units try to optimize not only across periods but within each period [→ **ecological economics**]. This optimization technique characterized by indifference curves and a budget line is of secondary importance with regard to our present purposes, so it will not be scrutinized here (moreover, Friedman elaborated it relying strongly on works of other authors[6]). However, the element that describes the explanatory variables underlying the consumption function is of crucial importance. If R_1 and R_2 denote the incomes expected for the first and second period, respectively, and i is the relevant interest rate, then first period wealth of the consumer unit (W_1) can be described as

$$W_1 = R_1 + \frac{R_2}{1+i}. \tag{5.1}$$

Those changes in R_1 and R_2 that do not affect W_1 do not affect consumption either. Since, for Friedman (1957), consumption function can be summarized as

$$c_1 = f(W_1; i), \tag{5.2}$$

any change in current income has effects on current consumption expenditures only if income expected for the next period (or each subsequent periods) does not change direction and extent so that this change could (taking as well as i into consideration)

[6] At this point, Friedman cited a book by Irving Fisher and another by Kenneth E. Boulding. Fisher laid the foundations of the theory specified for two periods about half a century before Friedman the essence of which is that incomes available for the different periods can be modified through borrowing and lending (Fisher 1982). By the way, Boulding, in his explication, also refers to Fisher (Boulding 1966). This analysis entered microeconomics as “intertemporal utility maximization” (or “intertemporal choice”) (Varian 1999; Kopányi 2003).

offset that shift in current income. So, this circumstance designates the most important difference between the consumption theory of Keynes and Friedman, since, as we have seen, for Keynes, *any* change in current income induces *a* shift in consumption, dependent on (marginal) propensity to consume. In the theory of Friedman, such a shift requires an appropriate change in a multi-period income stream (as outlined above).

Eventually, these are the foundations on which PIH was built. If y denotes the measured income of a consumer unit for a period (e.g. a year), then, for Friedman, it is true that

$$y = y_p + y_t, \tag{5.3}$$

where y_p is the permanent while y_t is transitory[7] component of measured income. Permanent income lies and changes under the influence of factors that determine the wealth of a consumer unit (here, Friedman lists the owned non-human wealth, certain personal attributes and the attributes of the economic activity of the earners in the unit).[8] The transitory component reflects the effect of all other factors. This component is determined by contingent though still predictable factors (e.g. a business cycle) that can be specific explanatory variables of the income of a particular consumer unit (for instance, a sudden sickness) or may affect even all or larger groups of the units (for example, bad weather or shifts in the demand for a product). It is natural that a favourable factor induces a positive transitory income. On these grounds, it is also true that

$$c = c_p + c_t, \tag{5.4}$$

that is, consumption also consists of a permanent and a transitory component (Friedman 1957).

After all, it follows from this theory of consumption outlined above that *intertemporally optimized choices by households mellow the relationship between current incomes and current consumption expenditures.* Of course, budget constraints of each household have to be taken into account in PIH as well, since the present value of planned consumption expenditures has to be equal to the present value of the expected income stream plus the initial wealth owned by the household (since the latter also can be utilized for covering consumption expenditures). However, the exact nature of the fundamental motive that induces consumer units to smooth their consumption expenditures in this way is still problematic. Friedman's answer focuses on the diminishing marginal utility of consumption. Accordingly, if consumption responded flexibly to the changes in disposable

[7] When studying the mechanisms and effects of automatic fiscal stabilizers (Sect. 5.7), we will see that transitory income does not need to be factor income, it may also come from government transfers.

[8] In lack of complete certainty, permanent income should be interpreted as the expected value of a probability distribution, of course.

income, then the utility gains from additional consumption could not compensate for the utility losses of periods with drop in consumption triggered by lower income. Under these circumstances, the only way of utility maximization is to avoid fluctuations in consumption expenditures by adjusting these expenditures to the level that appertains to long-run average, that is, permanent income. The optimal extent of this smoothing, of course, depends on the costs and benefits of lending and borrowing plus the relation between the rate of time preference of a household and the relevant real rate of interest (cf. Begg 1982)[9] If, for instance, real rate of interest exceeds the rate of time preference, then expected future utility (and its present value) can be maximized by planning a rising consumption path, since, under such circumstances, the returns on present savings accumulated in order to cover future consumption exceed the costs (that is, the negative utility consequences of consumption are delayed for later). Of course, it is also true that if there is only a negligible difference between the rate of time preference and the real interest rate, then maximal utility can be achieved by a constant stream of consumption expenditures [→ **ecological economics**].

The assumption of perfect capital markets is a crucial element in the permanent income hypothesis: the smoothed consumption path as PIH postulates can be realized only if frictions in borrowing and lending do not hinder the optimization process. As it follows from the theory of Friedman, when disposable income drops sharply, consumption expenditures may even exceed this level of income–of course, this outcome requires households to own marketable assets or to be capable of borrowing against future earnings. If both of these options are out of reach, there is nothing to do but cut consumption expenditures back (Flemming 1973). We get the same situation when current income of a household falls short of its expected future income–it follows from PIH that such a household spends more than its current income, bridging over the difference between income and expenditures by borrowing. Although Friedman admits that households may lend and borrow at different rates of interest, but, for the sake of simplicity, he assumes these complications away (Friedman 1957). It is exactly the point on which Flemming exerts his critique. He argues that if the difference between permanent and current (disposable) income is caused by an increase in unemployment and by the resulting drop in household income, and if capital market imperfections prevent the unemployed from borrowing, the shape of the consumption function and, hence, the multiplier-effect will be much closer to the case Keynes depicted. Under such circumstances, consumption cannot be adjusted to the permanent income (for further details see Houthakker 1958). According to Flemming's considerations, unemployment, apart from periods of serious economic depressions, usually affects those workers the worst who are less likely to own marketable assets. So, for them, borrowing against future incomes may turn out to be difficult, thus the interest rates at which they can borrow may be extremely high. Accordingly, the present discounted value of their

[9] The rate of time preference determines the rate of interest a consumer unit applies to calculate the present value of future utilities.

future incomes is lower, since, because of the higher rates of interest for borrowing, their future earnings (through unfavourable credit conditions) allow a lower level of current consumption (cf. Begg 1982). These consumer units usually possess as many liquid resources as can cover the previous level of consumptions only for a few weeks at most. Moreover, uncertainty aggravates the situation further:[10] since the duration of unemployment is not known in advance, households are expected to reduce outgoings even though the previous level of consumption expenditures could be actually covered by the marketable assets and the liquid resources they own (Flemming 1973). So, on these grounds, we should infer that *actually* (i.e. if the circumstances as Friedman defined them are not expected to hold) consumption responds to shifts in income more sharply than Friedman assumed–so, as the multiplier is higher as a consequence, fiscal policy may turn out to be capable of exerting an efficient control over real output.

Therefore, it is hard to overstate the limitations of PIH. Uncertainty about future and the lack of perfect capital markets do not make households able to keep consumption on an optimized path (cf. Heller and Starr 1979). For the former, they cannot predict the dynamics of their permanent incomes–and for the latter, not every consumer unit and not under all conditions can borrow against future incomes.

5.3 Permanent Income and Adaptive Expectations: An Algebraic Explication

The assumed mechanism of expectation formation is also crucial for PIH. Friedman originally hypostatized an adaptive scheme, but the consequences can be reconsidered under REH as well. As any new piece of information that can affect future income stream and, hence, permanent income becomes known to the consumer unit, household is forced to adjust the permanent income immediately.

Now, our problem is the information set that is supposed to affect the predictions on permanent income–or, in other words, what pieces of information are utilized in the formation of expectations about period *t* permanent income? In case of adaptive expectations, predictions are based on past income data; accordingly, actually non-observable stream of permanent income is calculated on the basis of past data of actual (disposable) income in a way that the more distant an income in time is, the lower the weight it gets is. It follows exactly from this weighing process that an increase in current income has fundamentally lower impact on consumption expenditures, since these outgoings are calculated on the basis of incomes from a longer (past) period of time. So, though current disposable income may increase, this shift affects consumption only in an extent to which permanent income rises as

[10] Bird and Bodkin (1965) also stress that the conclusions Friedman drew require the assumption of complete certainty.

well. So, Friedman does not surprise us by his evaluation according to which *current consumption is mainly dominated by past incomes* (Friedman 1957).

First, let us consider PIH under adaptive expectations. Assume that the following finite distributed lag model describes the way a prediction on permanent income is calculated:[11]

$$y_{t+1}^{p} = \beta_0 y_t + \beta_1 y_{t-1} + \ldots + \beta_k y_{t-k}, \tag{5.5}$$

where y_{t+1}^{p} denotes the prediction of permanent income for period $t+1$ while y_{t-i} is the actual income measured in each relevant period. As it could be seen, expectations reflect the joint effects of data from multiple periods, where each weight β_i is assumed to be a decreasing function of i. If it is assumed that weights form an infinite geometric series, then we get

$$y_{t+1}^{p} = \sum_{i=0}^{\infty} \beta_i y_{t-i} = \sum_{i=0}^{\infty} (1-\lambda)\lambda^i y_{t-i}, \tag{5.6}$$

utilizing the characteristic of infinite geometric series that $\beta_i = \beta_0 \lambda^i$, where $0 < \lambda < 1$ and the sum of the weights are assumed to be

$$\frac{\beta_0}{1-\lambda} = 1. \tag{5.7}$$

This structure is referred to as Koyck (or geometric) distributed lag model in the literature. Figure 5.1 displays the weights for three possible pairs of β_0 and λ.

We get the following equation from the one-period lag of permanent income of period $t+1$ multiplied by λ:

$$\lambda y_t^{p} = \lambda \sum_{i=0}^{\infty} (1-\lambda)\lambda^i y_{t-i-1} = \sum_{i=0}^{\infty} (1-\lambda)\lambda^{i+1} y_{t-i-1}, \tag{5.8}$$

that can be boiled down to

$$\lambda y_t^{p} = \sum_{j=1}^{\infty} (1-\lambda)\lambda^j y_{t-j} \tag{5.9}$$

by introducing $j = i + 1$. Subtracting Eq. (5.9) from the pre-lagged state we get

[11] In other words, actually non-observable permanent income is described on the basis of lagged, actually observed incomes (cf. Bilson 1980).

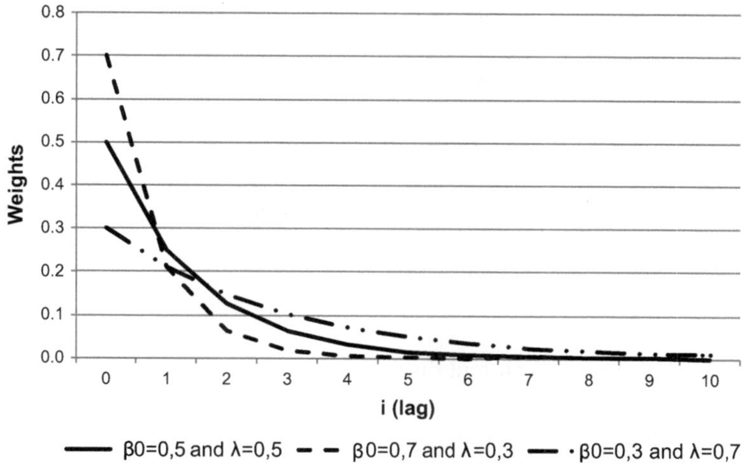

Fig. 5.1 A Koyck distributed lag structure with different parameters

$$y_{t+1}^{p} - y_{t}^{p} = (1 - \lambda)\left(y_{t} - y_{t}^{p}\right) \tag{5.10}$$

implying that agents forming adaptive expectations correct their period t expectations for the next period (i.e. $t + 1$) proportionally to current (period t) prediction error. The extent in which errors are taken into consideration is characterized by $1 - \lambda$ (cf. Maddala 2001).

For simplicity, let us take $1 - \lambda = \phi$ as follows. According to PIH, current consumption (c_t) does not depend on actual (disposable) but on permanent income. In formal terms, the consumption function can be written as

$$c_t = \alpha + \beta y_t^{p} + u_t, \tag{5.11}$$

where u_t is a white noise, α is autonomous consumption and β is the marginal propensity to consume out of permanent income. Friedman details the factors determining marginal propensity to consume and, hence, the relation between permanent income and consumption. Accordingly, β is independent of the level of permanent income; while, on the contrary, the rate of interest, the relative importance of property and non-property incomes[12] plus tastes and preferences of

[12] At this point, Gallaway and Smith (1961) raise the possibility of introducing the real balance effect into PIH. As the real balance effect implies, the money stock held by agents is evaluated (in real terms) by the price level (this is real balance itself) and it is true that aggregate consumption is not independent of the price level (a drop in the price level enhances consumption through increasing real balances). Through elaborating the theoretical conception, Arthur Cecil Pigou tried to respond to Keynes' suggestions according to which a macroeconomic system may get into equilibrium state before full employment is reached, saying that the mechanism of real balances can (and should) be regarded as an automatism directing the system towards full employment. Gallaway and Smith argue that if the value of certain financial assets increases due

the consumer unit about consumption and treasuring do affect the relation designated by β (Friedman 1957).

Permanent income expected for period t can be rewritten as

$$y_t^p = (1 - \phi)y_{t-1}^p + \phi y_{t-1}. \tag{5.12}$$

Let y_t^t denote period t transitory income. Now, it is clear that transitory income can be written as

$$y_t^t = y_t - y_t^p. \tag{5.13}$$

It can be seen that neither current disposable nor transitory income have effect on consumption expenditures.

If the consumption function is solved for y_t^p, that is,

$$y_t^p = \frac{c_t - \alpha - u_t}{\beta}, \tag{5.14}$$

then we can write down the following on the grounds of the last two equations:

$$\frac{c_t - \alpha - u_t}{\beta} = (1 - \phi)\frac{c_{t-1} - \alpha - u_{t-1}}{\beta} + \phi y_{t-1}. \tag{5.15}$$

After some reductions, we get a form of the consumption function that can easily be treated also in econometric terms, since it contains only directly observable and measurable variables:

$$c_t = \alpha\phi + (1 - \phi)c_{t-1} + \beta\phi y_{t-1} + u_t - (1 - \phi)u_{t-1}, \tag{5.16}$$

the final setting of which (after reductions) is:

$$c_t = \beta_1 + \beta_2 c_{t-1} + \beta_3 y_{t-1} + v_t, \tag{5.17}$$

where $v_t = u_t - (1 - \phi)u_{t-1}$ is a moving average error term (cf. Maddala 2001). This functional form bears a strong resemblance to the basic equation of *habit persistence hypothesis* given by Brown that stresses slow adjustment capacity of consumer habits (cf. Brown 1952). For Brown, using a lagged dependent variable as explanatory variable is not a result of the algebraic deduction but a deliberate effort to express and highlight the slack reaction of consumption to shifts in income. Additionally, it has to be mentioned that this is not the only consumption function in

to a drop in the general price level, this development will affect the ratio between non-human wealth and income, and, eventually, real balance effect can exert influence on permanent income, and hence on consumption. If this change proves to be long-lasting (and, on this ground, agents following the adaptive scheme expect it to survive in the long run), then (planned) expenditures also increase. However, empirical results showed this effect to be only of moderated importance.

literature that emerges as a combination of PIH and Koyck lag structure. For instance, Singh and Ullah (1976) specified an equation without a constant term:

$$c_t = \alpha_1 y_t + \alpha_2 c_{t-1} + w_t, \tag{5.18}$$

where c_t and y_t is period t consumption and income, respectively, while w_t is transitory consumption in period t (that is, *permanent* consumption is determined by the first two terms on the right). However, this function is debatable in theoretical terms, since it links consumption with current income after all.

So, we could realize on the grounds of this simple mathematical model how shifts in current income affect expectations about permanent income (it is always the current income of the preceding period that has the largest weight and hence the strongest influence), we could learn the essence of error correction procedure and that marginal propensity to consume should be interpreted in relation to permanent income.

It is worth considering in short what dynamics are followed by permanent and transitory incomes in case of a generated income stream.[13] Disposable income (DI) was allowed to fluctuate between 1.000 and 1.100 during a 40-year-long period (hence, consisting of 480 months).[14] We set the Koyck structure arbitrarily in a way that $\beta_0 = 0.3$ (and, accordingly) $\lambda = 0.7$.[15] For the sake of simplicity, a 26 period long lag structure was taken into account in the course of the presented simulation process (between β_0 and β_{25}), so, the sum of weights is

$$\sum_{i=0}^{25} \beta_i = 0.9999061252 \approx 1.0. \tag{5.19}$$

The length of the applied lag structure seems to be good enough for our data not containing a considerable bias compared to the infinite structure assumed in theoretical terms (the effects is completely ignorable since $\beta_{25} = 0.0000402321$ yet). On these grounds, permanent income (PI) becomes computable, and transitory income (TI) is also given as the difference between permanent and disposable income. Dynamics of each time series is depicted in Fig. 5.2.

[13] It has to be mentioned here that (though the initiative was raised by REH) the literature of PIH consists mainly of empirical tests through which authors try to verify or falsify this theory in statistical terms. Among them, the efforts of Chulsoo Kim are very remarkable, putting his emphasis on theoretical-economic significance of PIH rather than on statistical significance. His analysis was aimed at the question of how and to what extent U.S. consumption data deviate from the path predicted by PIH. Although deviations seem to be significant enough for the statistical tests to reject the theory (since a statistical rejection could equally stem from measurement errors or from some simplifying presumptions), Kim infers that PIH still offers the best approximations of data among the alternative theories (cf. Kim 1996).

[14] Random numbers are used.

[15] Since an infinite lag structure is assumed in which the sum of weights is unity.

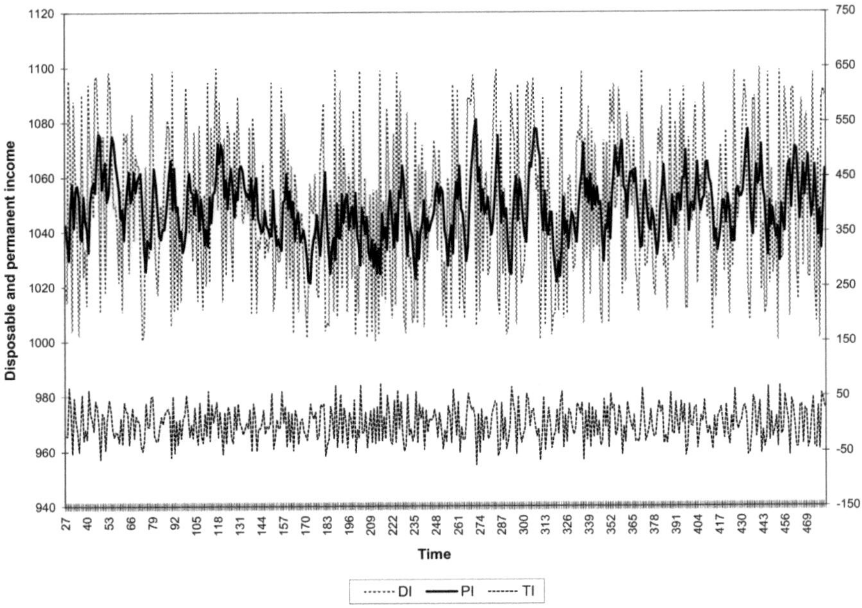

Fig. 5.2 Dynamics of each income category computed on the basis of a randomly generated time series of disposable income, assuming adaptive expectations

It can be seen that, due to the assumed mechanism of expectation formation, permanent income can significantly level out the fluctuations in disposable income. While $E(DI) = 1049.0098$ and $E(PI) = 1048.7043$ (that is, mean disposable and permanent income is quite the same), then standard deviations of income types are $\sigma_{DI} = 29.2760$ and $\sigma_{PI} = 11.6697$, respectively. It is obvious that the distribution characteristics of consumptions calculated from DI and PI data would be similar. It is far from surprising that expected values of disposable and permanent income fall close to each other. Since, if (using the former notations) we want to find period t permanent income in case of an income series free from expansionary or contractionary trends, we get the equation

$$PI_t = \sum_{i=0}^{\infty} \beta_i DI_{t-1-i} \tag{5.20}$$

It is true of mean permanent income that

$$E(PI_t) = \sum_{i=0}^{\infty} \beta_i E(DI_{t-1-i}) = \sum_{i=0}^{\infty} \beta_i E(DI), \tag{5.21}$$

and, as $\sum_{i=0}^{\infty} \beta_i = 1$, we get

$$E(PI_t) = E(PI) = E(DI) \tag{5.22}$$

for the expected value of permanent income.

It is worth looking over some of the most important distribution characteristics of our three income categories (Fig. 5.3).

Most of the curiosities can be found in the first two sections. As it was expected, randomly generated disposable income series indicates no definite tendency; skewness index shows the distribution to be roughly symmetric. On the contrary, the second part indicates that permanent income distributes more closely around its mean (as it was expected), and hence we get a normal distribution for the time series generated through our Koyck structure. Skewness is still close to zero, indicating a slight right tail, and kurtosis index also approximates satisfactorily to the reference value of normal distribution. Normality of permanent income is acceptable on the basis of the Jarque-Bera test statistic calculated from skewness and kurtosis indices. On these grounds, we can infer intuitively that permanent income is a variable with more steady dynamics compared to disposable income (it is free from large-scale fluctuations) and the whole income stream is more concentrated around the mean.

Getting back to the series of disposable and permanent income, we can draw the conclusion (as it has already been done in the theoretical discussion above in advance) that permanent income responds to shifts in disposable income with less flexibility. The mechanism can be well displayed by studying the expectations about permanent income in the case of a disposable income series free from volatility–apart from a single level shift (Fig. 5.4).

We used the same weight structure as in the previous case. It can easily be realised (even in formal mathematical terms) that disposable and permanent income coincide in longer periods (that is, if multiple periods are considered) in case of a constant (unchanging) disposable income, but permanent income responds only slowly to the level shift in disposable income (of course, the exact length of the reaction time directly depends on the weight structure: the more close β_0 falls to 1.0, the shorter this reaction time is).

After these simple numerical examples, it is time to turn back to the logical explication of PIH. We have seen that Friedman (1957) divided both income and consumption into permanent and transitory components. Moreover, he stipulates important caveats in the definition-conceptual phase. It is true in his theory that

$$\rho_{y_t y_p} = \rho_{c_t c_p} = \rho_{y_t c_t} = 0, \tag{5.23}$$

permanent and transitory components of income and consumption are uncorrelated,[16] and, similarly, transitory income and transitory consumption are

[16] It is a common practice in the empirical literature to apply the Koyck distributed lag scheme to estimate permanent income, though Friedman (1957) suggested the method of Philip Cagan. In algebraic terms, the Koyck-model is a discrete period analogue of the Cagan-lag (Singh and Drost 1971). However, literature sometimes calls our attention to the fact that applying the Koyck-model

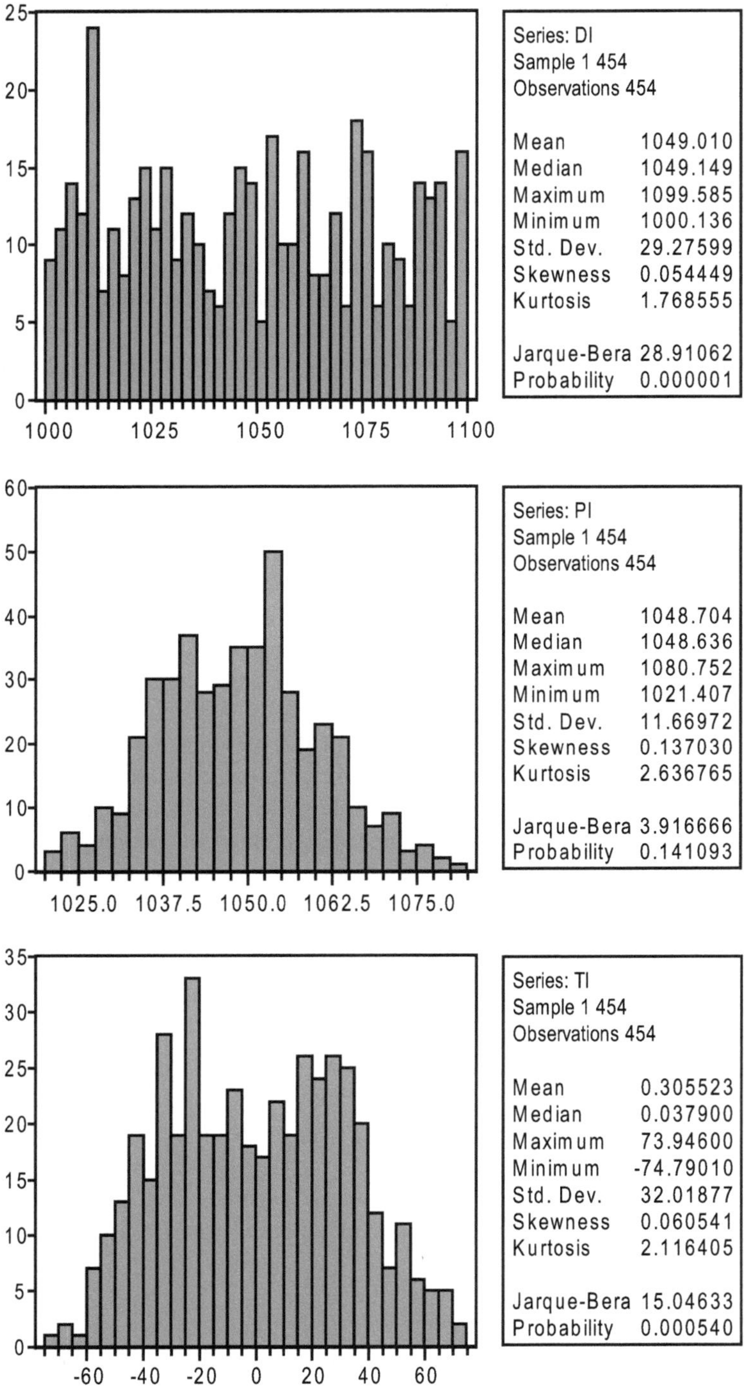

Fig. 5.3 Major characteristics of the distribution of our generated income series

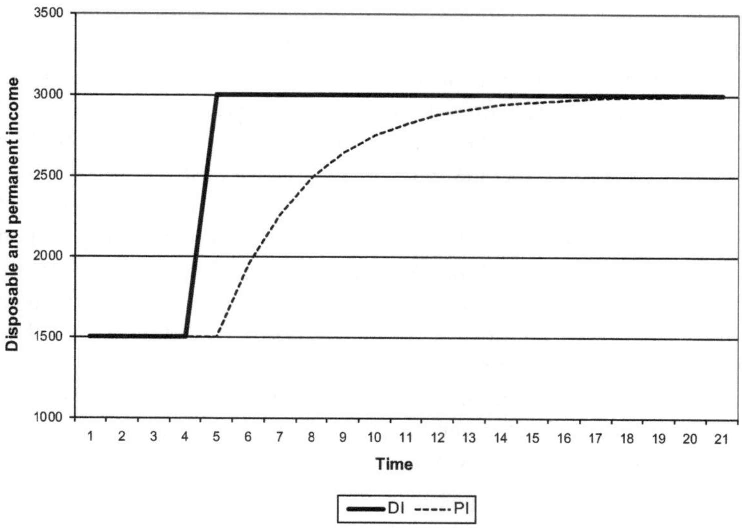

Fig. 5.4 Changes in permanent income (PI) in the case of a level shift in disposable income (DI), with adaptive expectations

also independent of each other.[17] This latter is the most serious assumption, leading to the consequence that marginal propensity to consume out of transitory income is zero (cf. Laumas 1969). Through these developments, it can be realised as well why only a marginal propensity to consume out of permanent income was taken into consideration in our mathematical model above.

The stipulation that transitory income and transitory consumption are uncorrelated can be interpreted in practical terms meaning that a regression model on consumption data of a household in which permanent income is the explanatory variable contains also an error term that can be identifiable with transitory consumption, and this error term is independent of transitory income (cf. Holbrook and Stafford 1971). It was the third among the three correlation coefficients above that stood in the crosshairs of most of the relating empirical investigations. Here, it has to be noted that a loose definition of PIH was also formulated on empirical grounds not constraining the marginal propensity to consume out of transitory income to zero, though admitting that it is smaller than its counterpart out of permanent income.[18] This phrasing implies the realisation

to PIH violates some of its assumptions of crucial importance. As it was stressed above, the nature of the relation (correlation) between permanent and transitory components of income and consumption is not ignorable in conceptual terms, however, these required independencies cannot be held when using a Koyck-lag. It has importance with regard to the acceptability of PIH, since the statistical tests built on the Koyck-lag scheme are likely to refuse PIH (cf. Singh and Ullah 1976; Holmes 1970, 1971).

[17] On the methodological difficulties related to these stipulations see Bowden 1973.

[18] The theory of Holbrook and Stafford (1971) has the same roots: they hypothesized that different marginal propensities to consume out of incomes from various sources can be found.

(assumption) again that consumption responds to short-run (non-permanent) changes in income as well. According to the arguments of Bird and Bodkin, for example, there are situations in which *it is exactly the receiving of a transitory income that allows a household to adjust its consumption to the level appertaining to the permanent income*; for the authors, particularly when the household expects an increase in disposable income, but the accordant boost in consumption, in the lack of any kinds of borrowing, can be carried out only by spending out of an unexpected windfall gain, so we should expect a positive marginal propensity to consume out of transitory income as well (Bird and Bodkin 1965). However, it should be noted that this reasoning of Bird and Bodkin can be held only under REH, since, with adaptive expectations, a permanent positive shift in disposable income triggers a *gradual* convergence of permanent income, and, during this (accumulation-treasuring) period, permanent income systematically remains below the level of disposable income (a period of accumulation). However, it means that consumption, during this adjustment period, is not prevented from following the path allowed by permanent income (a negative shift in disposable income can be accompanied by a consumption appertaining to permanent income only in case of flexible borrowing conditions). As it will be seen in the following section, only rationally expecting agents can modify their predictions about permanent income on the basis of *ex ante* information (that is, before any actual shift in disposable income). The amendment given by Bird and Bodkin refers rather to terminological problems (we can hardly talk about REH taken into account a couple of years before the emergence of new classical macroeconomics), according to which *unexpected(ly realized)* and *transitory* incomes are synonyms. By the way, this terminological confusion ("short-run" vs. "transitory" income) emerges at other places in literature as well (cf. e.g. Holbrook and Stafford 1971) in spite of the fact that they are synonyms–though it is undoubted (as it will be seen below) that Friedman circumscribed some income categories that do not affect predictions about permanent income. However, let us assume a disposable and a permanent income series following the paths depicted in Fig. 5.5.

Spending transitory income in relation to permanent consumption purposes is a problem of borrowing and lending again (liquidation of deposits accumulated from previous transitory incomes in order to cover permanent consumption), if short-run components are also included in disposable income.

Rather, the right question is whether Friedman's assumption according to which transitory income and transitory consumption are uncorrelated can be held–since if *not* (i.e. if transitory income still correlates with transitory consumption), then measured consumption (containing transitory consumption as well) can be significantly explained not only by permanent but also by transitory income. As far as our final results are considered, it is not really an *ex ante* speculative inventory of the relevant situations that matters but rather the empirical tests justifying the assumption of a significantly positive marginal propensity to consume out of transitory income (cf. Laumas and Mohabbat 1972). The story described by Bird and Bodkin corresponds pretty well to a later example by John Shea: if a household expects its disposable income to change, then it tries to smooth the consumption path through

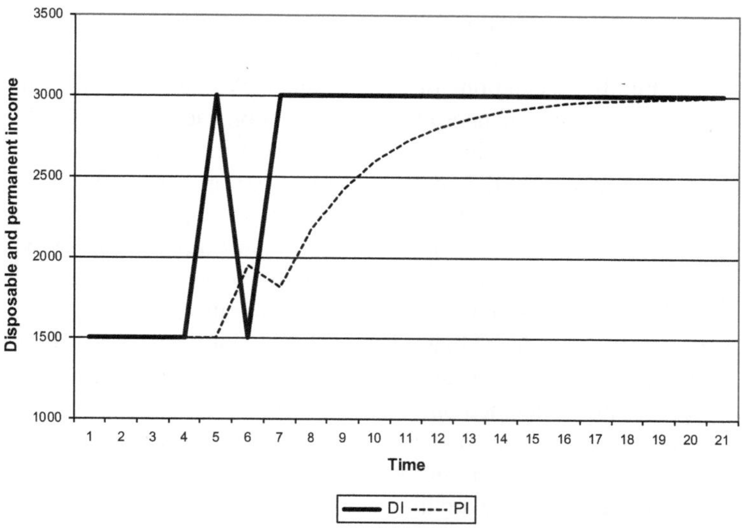

Fig. 5.5 Changes in permanent income (PI) following from short-run unexpected fluctuation and level shift in disposable income (DI), with adaptive expectations

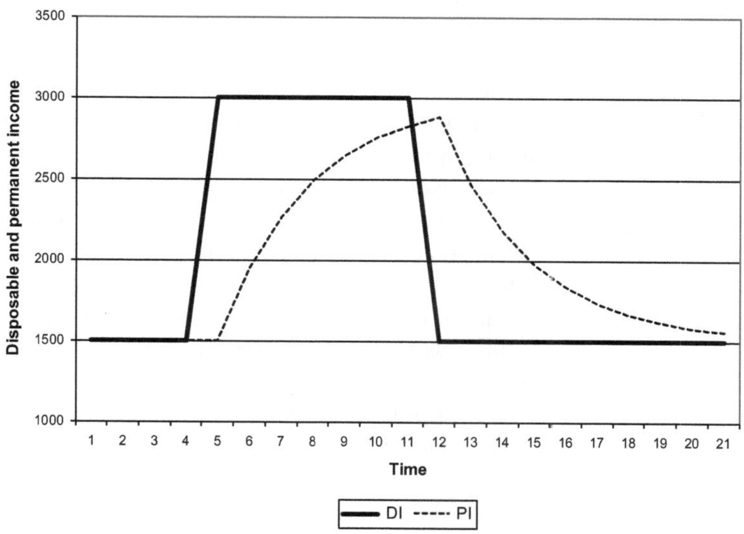

Fig. 5.6 Changes in permanent income (PI) following from a temporary rise in disposable income (DI), with adaptive expectations

borrowing and lending (Shea 1995). However, Shea stands on the ground of REH, following Hall. By the way, assuming the adaptive scheme raises some problems that require some theoretical consideration. Take a look at Fig. 5.6.

If the disposable income of a consumer unit is higher for a while, this development changes permanent income as well because of the assumption of an adaptive expectation operator. In this case, the consumer unit cannot distinguish the usual (stabile) and extraordinary (short-run) components of disposable income, and, eventually, it builds expectations of both of them. We could say that an income available only temporarily cannot change expectations about permanent income–however, information on the durability of additional income is not utilized when forming expectations under the adaptive scheme, since, here, only the trend-like attributes of past factual data affect predictions. *To keep permanent income at an unchanging level or, at least, to avoid large-scale modifications requires possessing and processing information sets about the future*–however, these data are (and can be) possessed and processed only by agents forming rational expectations.

We have reached the limitations of the Friedmanian approach to PIH through these questions by now. The source of the terminological confusion mentioned above is that Friedman included only such elements in disposable income that can be planned well in advance, i.e. that can easily be predicted even on the basis of past trends (cf. Friedman 1957; Holbrook and Stafford 1971). However, on these grounds, we can say that there are incomes being unable to exert any influence on the predictions about permanent income, but should be considered to be the components of transitory income in their total amounts. In such a case, of course, such a situation may occur in which (e.g. in a calendar month) only a transitory ("short-run") but no "long-run" income is available, and *it is exactly this extraordinary income that will be spent to cover permanent consumption*. But even if these feasible cases are disregarded, it remains a question for what purposes the transitory incomes become utilized–and, as we have seen, a non-zero marginal propensity to consume out of transitory income cannot be precluded on empirical grounds. According to some interpretations, spending transitory incomes obeys a radically different system of purposes: as Friedman (1957) ruled out durable goods from his category of permanent consumption, it was a reasoned scientific effort to outline the relation between the purchasing of this kind of goods and the realization of transitory incomes. Empirical results showed that shocks in transitory incomes mainly bring about the purchase of durable goods (cf. Thurston 1977).

The distinction between the categories of income (i.e. long-run and short-run incomes) requires us to assume some kind of rationality (meaning: forward-looking behaviour).[19] *Agents forming expectations only on the grounds of past data are incapable of such evaluations. Friedman, though implicitly, advanced and prepared the assumption of REH.*

Either REH or the adaptive scheme is assumed, permanent income is nothing more than the ordinary annuity equivalent of the income stream the consumer unit expects over a long horizon of time. However, the information set underlying these expectations is not of secondary importance. The effects of the adaptive scheme assumed by Friedman originally were reviewed above: we could see that *any deviations of disposable income from its previous trend is capable of triggering modifications in predictions about permanent income*. However, it is time to turn to REH.

[19] The literature sometimes explicitly stresses that PIH implicitly assumes consumer units to have such an ability (cf. e.g. Pistaferri 2001).

5.4 Permanent Income and REH

We have a fundamentally different framework if expectations are rational. Under these circumstances, not only some past but also all information about the future available at the moment is utilized in forming expectations about permanent income. To revise the level of consumption expenditures it is not enough to realize the changes in current income, since if this shift could be foreseen, rationally expecting agents built this development into their expectations in advance. As we have seen it when studying monetary policy lessons of REH, it is true again that *only surprising changes in expected incomes can modify predictions about permanent income*. Expectations contain all (known) relevant information, so only unexpected events and pieces of information can trigger modifications. Of course, it has to be stressed that it is of no secondary importance in terms of this revision what agents think about these shocks on the grounds of their information set. If the effects are judged to be long-lasting, agents transform their expectations about permanent income–by the way, this modification can be quite minute or completely fails. So, we have to distinguish between the case when a fiscal policy action (e.g. a tax change) *affects* expectations about permanent income and the case when this action *changes* these expectations. It is highly true that all fiscal policy steps affect predictions, since rationally expecting agents definitely take information about these interventions into consideration–but a decision (or a turn) modifies these expectations only if the information has been unknown to agents so far.

It has to be noted here that Robert Hall (1978), taking a stand on these theoretical grounds, demonstrates that consumption follows a random walk path under REH (for Begg 1982, so does permanent income, though it is of little avail to us in terms of empirical investigations). According to this, the best performing prediction about consumption in the next period is current consumption, with a white noise difference between successive data, being a simple prediction error with a zero mean, though, of course, the trend of economic growth may be also taken into account[20] (Hall's theory is also referred to as "martingale hypothesis" in the literature; cf. Christiano et al. 1991). And, moreover, we have a further troublesome consequence that lagged values of income, (either lagged or current) unemployment and any other "traditional" explanatory variables do not contribute to predictions about the future dynamics in consumption at all (Begg 1982). By the way, empirical results quite definitely verified Hall's approach.

[20] The most bothering problem with PIH is that it refers to a macroeconomic variable that cannot be directly observed, as it has been already mentioned in Chap. 1. Consumption and income (or some categories of income) are evidently assumed to be correlated. It follows that it is only the path of consumption that can be compared with the dynamics predicted by the theory, and if a considerable (or, at least, a rough) consistency is observed, we infer the validity of the theory. So, because of the fact that permanent income as a variable is unobservable, researchers are (and were) forced to subject some of the implications drawn deductively to empirical tests. Such an implication is the random walk of consumption.

Consumption expenditures respond to (unexpected) changes in current income to different extents under adaptive and rational expectations. With the adaptive scheme, preceding (pre-current) disposable income is only one element of the relevant income stream (though having the highest weight), that is, its instant effect on consumption is small. In such a case, consumption expenditures drop significantly only if the decrease in income proves to be long-lasting on the grounds of subsequent incomes being realized. With rational expectations, if the (unexpected) drop in income promises to be permanent, agents *immediately* adjust expectations about permanent income downwards–so, under REH, changes in the consumption are much more drastic, by which (however strange it may sound) *the original and traditional conclusions of the theory of Keynes are restored*, since consumption expenditures are supposed to respond considerably to shifts in current income (cf. Shaw 1984; Bilson 1980).

If expectations are rational, permanent income is nothing more than the ordinary annuity (that is, the payments of which arise in equal amounts at equal intervals) the present discounted value of which is equal to the initial wealth possessed by the consumer unit plus the present value of its expected incomes. The circumstance that permanent income is interpreted as an ordinary annuity does not arise an inconsistency with the element of the theory concerning revision of expectations. The annuity constituting permanent income should be defined always on the grounds of a particular set of information. If pieces of information do not make a revision necessary, permanent income is actually a stream of constant payments. These payments change only under the influence of new information that makes it necessary to modify predictions–in such a case, a new ordinary annuity emerges remaining unchanged as long as a further modification turns out to be unavoidable. Taken all this into account, permanent income under REH can be easily interpreted as a series of subsequent (and, partly, overlapping and overwriting) ordinary annuities (of course, only if revisions of predictions are made necessary by newly emerged pieces of information). We can write it in formal terms as

$$W_t + Y_t + \sum_{i=1}^{\infty} \frac{{}_tY^e_{t+i}}{(1+\delta)^i} = Y^p_t \times \left[1 + \frac{1}{(1+\delta)} + \frac{1}{(1+\delta)^2} + \frac{1}{(1+\delta)^3} + \ldots\right], \quad (5.24)$$

where δ is the constant rate of interest on deposits and credits, Y^p_t is permanent income expected for period t, W_t is real wealth possessed at the same time, Y_t is current (period t) real income, while ${}_tY^e_{t+i}$ is a prediction of expected future income conditional on currently available information. Of course, it still holds that only those changes in expected income matter that are capable of triggering shifts in permanent income (Begg 1982), that is, changes in payments levelling out each other are irrelevant.

Depressing conclusions for the demand management capabilities of fiscal policy are drawn on these grounds. Stimulation of consumption (e.g. through tax cuts) can be carried out only if consumer units expect their income to rise for a long period. So, if the government, in course of a fiscal intervention, issues bonds to compensate the tax-cut-driven drops in revenues, agents know in advance that the debt service is going to impose heavy burdens on the state budget that will have to be financed probably by future tax increases. Of course, it means that currently higher actual (disposable) income is levelled out by an expected future drop, and this development, eventually, keeps consumption expenditures and predictions about permanent income intact. In other words, *any change in the tax system affects consumption expenditures only if this change exerts influences on permanent income*. So, a fiscal policy action affects (or can affect) current and permanent income at the same time–and the shorter the length of the period is to which the tax-cut-driven shifts in income are constrained, the more true it is that effects on permanent income are ignorable (cf. Shaw 1984). These considerations do not preclude the efficiency of fiscal policy of course, but definitely define the circumstances under which fiscal policy can turn out to be effective.

This limitation of fiscal policy is far from being as insurmountable as it seems to be–since our previous theorem can be rephrased in such a way that stimulation of consumption (and of aggregate demand) can be definitely carried out if consumer units expect their income to be higher for a long period. So, if an *unexpected-unforeseen* (in our terminology: surprising) fiscal policy step results in a permanent rise in disposable income, its consequence will be a rise in permanent income and hence consumption. Now, our question is whether fiscal policy has the potential to exert a long-lasting income effect through a surprising intervention.

Through these fiscal policy considerations, we get to a highly important semantic difference. It has to be noted that the meanings of *surprising* and *random* actions/effects are not *the same* (either for monetary or fiscal policy), though distinctions can be easily blurred. Monetary policy also could have been able to operate through *surprising* steps, if macro-economic systems had not contained a *random* component that was not predictable (not even for monetary policy) and, therefore, could have been answered by monetary policy only through another random action. So, a systematic monetary policy should have been built on contingent bases which is hopeless. That's why surprising and random effects were identified in semantic terms (using them as synonyms) in case of monetary policy. In other words: we have seen when scrutinizing monetarist considerations on monetary policy that *a surprising effect could have been only random*. However, as soon as prolonged business cycles are not precluded, senses of "surprising" and "random" immediately become different. It means that monetary policy, under such circumstances, becomes capable of following growth-enhancing (or growth-discouraging[21]) countercyclical purposes. Initially surprising actions still exert influences even after they have become known–and since there are long-lasting business cycles built in the system, they are non-random shocks that have to be answered by

[21] Countercyclical economic policies have to be definitely symmetrical. It means that the cushioning of recessions and the cooling of overheating boosts are equally important goals. It is extraordinarily damaging if an economic policy precludes only contractions, intensifying growth only (for further details see Csaba 2008).

using monetary policy instruments. Something surprising does not have to be something random–in other words: the criteria of efficiency do not contain randomness anymore. And all this is true of fiscal policy as well. If there are long-lasting cycles, that is, when actual real output is below or above the potential level for longer periods, *the real system proves to be controllable through non-random fiscal surprises* [→ **output gap**]. Therefore, also in the case of fiscal policy, we are talking about definitely *non-random*, but (only) surprising, that is, unforeseen actions–that can be still planned deliberately. The fiscal authority is not forced to construct a systematic policy out of random effects either. *So, in this case, a surprising effect can be a part of a systematic policy*. We will scrutinize these surprising actions of fiscal policy later when analyzing the Ricardian equivalence.

Output gap A commonly used index to describe briefly the conjunctural stages of national economies is the output gap (OG). Eventually, output gap is a ratio that measures the difference between period t potential and actual real output to period t potential output in the following form:[22]

$$OG = \frac{Q_t - Q_t^*}{Q_t^*}$$

The advantages are that its sign directly informs us about the direction of the difference, and the percentage form makes the comparison of various macro-systems possible, while disregarding measurement differences.[23]

The most difficult methodological element in the calculation of the output gap is the estimation of potential GDP. For a long time, the path of potential GDP was identified with the long-term trend of GDP, so HP-filter or other de-trending and smoothing methods were good enough to get the necessary time series. GDP-smoothing was offered by EU and OECD methodological recommendations for long (Kiss Gábor 1998). However, there is no doubt that these techniques are completely unconscious, so higher-order information about a given national economy plays no role in the analyses. In this case, economists can consider those conditions at most that do not concern either the quality of the utilized data or the structural characteristics of a given national economy. For instance, HP-filter requires us to choose arbitrarily a parameter that determines how strong the smoothing will be. Although the authentic methodological paper announces definite recommendations (Hodrick and Prescott 1997), the literature often denies them, so researchers are divided about which analytical principles are right to follow. This parameter can be chosen in a way that, for example, the cyclical fluctuations shown by the smoothed series are consistent with the expert opinions formed previously on the (actual) cycles of the given macro-system. So, the bottom-line of the problem is that researchers are forced to make arbitrary decisions on the smoothness of potential GDP dynamics (for further details see OECD 2001). Moreover, in case of the HP-filter, it also has to be taken into account that the starting point and the end of the period to be analyzed should be contained by similar cyclical phases, since,

[22] In order to eliminate the effects of price changes, data measured in real terms (on a base year) are needed both in the case of actual and potential GDP.

[23] Over and above this notion, output gap can be interpreted in a non-ratio form as well, as a simple difference between actual and potential real GDP–though its applicability is limited in this way. The sign is still informative, though makes no comparison possible.

otherwise, the resulting trend will be distorted. A possible solution is to use expert predictions on future GDP dynamics, extending the sample period.

Calculating and analyzing the output gap is a routine component of the research carried out by the Central Bank of Hungary (demonstrated by the series of its inflation reports–see e.g. MNB 2010), while also playing a key role in the practice of international organizations. The methodology that has been preferred by the OECD and the European Commission from the mid-1990s computes potential output on the basis of a Cobb-Douglas production function,[24] utilizing much more data and details than a simple de-trending method and giving far more leeway for expert predictions and judgements. *The advantage of the production-function-based approach is not that it delivers more reliable results than the de-trending techniques*–it stems from the broader set of information that is actually applied. In this approach, potential real output is the level of GDP that is produced at the equilibrium utilization of productive capacities and labour force. Results are more established in any way–but more dubious as well at the same time, so alternative estimations made on the grounds of different assumptions have to be involved in economic policy and methodological discussions [→ **structural balance**].

Whatever the shortcomings of PIH are[25] and no matter how important the empirical difficulties it raised are, the basics of this theory have already loomed behind the critical remarks by now. If our concerns are set aside, we can see that PIH can appropriately demonstrate that the efforts of fiscal policy to control aggregate demand are predestined to fail. Although PIH is originally linked to the orthodox monetarism, the theory can easily be applied to underpin the new classical doctrines as well. However, new classicals focused on another theoretical construct, namely the Ricardian equivalence when arguing for the ineffectiveness of fiscal policy (or for the limitations of its effectiveness). Yet we have to see that permanent income theory and Ricardian equivalence theorem are very close in terms of the theoretical message. So, it is time we turned to the Ricardian equivalence.

---

[24] This conceptual change is not independent of the concept chosen for potential output. If potential output is a trend around which actual production fluctuates (forming business cycles), potential GDP can be described well through various de-trending techniques. If, however, potential output is linked rather to the supply side, then potential GDP will be that level of production that emerges with optimal capacity utilization (labour force, productive capital, human resource). In this case, numerical estimates require a production function (for further details, see Benk et al. 2005). The difference between these two aspects illustrates well also the bottom-line of the difference between the methods of estimating the potential GDP.

[25] For a paramount summary see Carroll (2001). Carroll places Friedman's seminal innovation (Friedman 1957) with a good sense in the context of the history of economic thought reaching even back to the 1950s. The author's methodological-empirical efforts deserve our attention because he has a considerable turn for scrutinizing consumption theory with liquidity constraints through simulation techniques. Carroll approaches a traditional research area by using a new method that promises to deliver a great bulk of new insights–similarly to agent-based models.

5.5 The Ricardian Equivalence: Basic Considerations

The Ricardian equivalence, similarly to PIH, models individuals' consumption and savings decisions–moreover, some authors stress that the modern re-formulization of the Ricardian equivalence may be regarded as an influential generalization of PIH (cf. Seater 1993). We will see that the REH-based modification of PIH can be interpreted in a way that actually echoes the Ricardian equivalence. However, the significance of this theory should not be understated, since the Ricardian equivalence originates in the soil of rational expectations, while PIH was formulized within the adaptive expectations framework in the beginning with REH held back only for later considerations.

Although the initial considerations of Ricardo were first discovered by Patinkin (1956) and then Barro (1974) for modern analyses and economic theory, the concept "Ricardian equivalence" became generally known not by them but by Buchanan (1976). Albeit this present investigation is not aimed at taking us on a wearisome and roundabout journey in the history of economic thought,[26] an extreme view definitely deserves attention: according to it, the Ricardian equivalence (as far as the technical term is concerned) is the result of a huge theoretical blunder, since Ricardo himself did not regard those two competing financial settings as equivalent (cf. O'Driscoll 1977). This view could never gain wide acceptance. To understand the mechanism embedded in the Ricardian equivalence theorem, it is very useful to take a look at Fig. 5.7.

Assume that government wants to stick government purchases of goods and services to an *unchanging path*, but commits itself to cut taxes at the same time. The assumption of keeping government purchases intact means that government expenditures do not change, accompanied by an also unchanged budget revenue demand (so, theoretical consequences are easy to generalize). According to the parable, the unreduced revenue requirement forces the government to issue bonds in a value equal to the sum of per capita tax-cuts (by the way, the Ricardian equivalence could be demonstrated as well by assuming transfer payments equal to the cuts in lump-sum taxes). Bonds are assumed to have a coupon rate that is equal to the market rate of interest r (i.e. the public and government can borrow at the same interest rate, plus the rates of interest on deposits and credits are also assumed to be equal). In the simplest examples, the term to maturity is assumed to be 1-year long or infinite sometimes. The debt service of bonds (interest payments and principal repayment) is covered by lump-sum taxes levied later. The Ricardian equivalence tries to answer the question how this new financial-fiscal setting (particularly the tax-cut) affects consumption demand.

[26] Since, for us, all the reviewed theories are descriptions of living mechanisms working under actual or assumed conditions–hence being completely relevant for theoretical analyses or statistic and econometric researches, and not to be taken as dead relics in the museum of the history of economic thought.

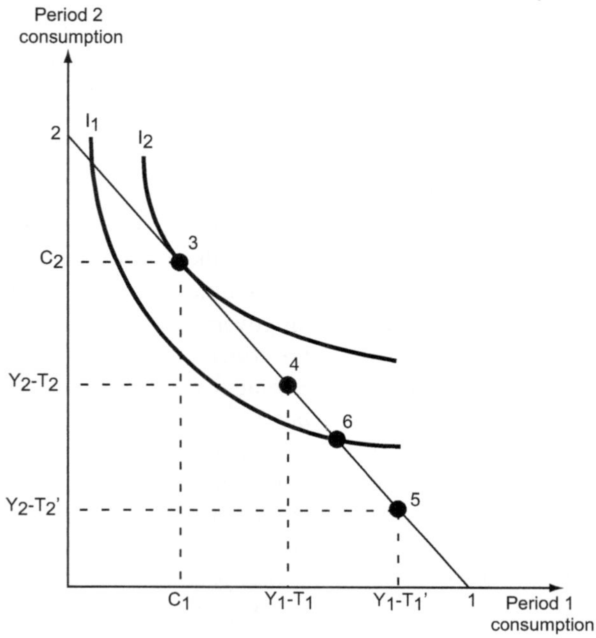

Fig. 5.7 The basic mechanism of the Ricardian equivalence (Cullis and Jones 1998)

Disposable incomes in period 1 and 2 are $Y_1 - T_1$ and $Y_2 - T_2$, respectively. If the representative individual does not want to spend all his income on consumption in period 1, then he can save, of course. This decision of his is symbolized by budget line $1 - 2$, the incline of which is $1 + r$, where r is the rate of interest–every dollar of income saved in period 1 pays this interest. This line, after all, summarizes the consumption options available for the representative individual, since if he decides not to consume in period 1, then the amount spendable on these purposes in the next period will be $[Y_2 - T_2 + (Y_1 - T_1)(1 + r)]$. As for the opposite extremity, he may opt for taking period 2 income forward to the preceding term in toto by borrowing, increasing his income in period 1 available to cover consumption expenditures, which is now $[Y_1 - T_1 + (Y_2 - T_2)/(1 + r)]$. Disposable incomes and the rate of interest determine the position of the budget line together, so the point symbolizing the optimal income allocation, similarly to PIH, can be found by means of indifference curves (I_k). These indifference curves show the extent to which the representative agent is ready to cut back his consumption in any of the periods in return for realizing additional resources in the other term. Utility can be maximized by a distribution of consumption across periods in a way that the privileged point of the budget line coincides with a point of the highest reachable indifference curve (in other words: budget line is tangential to the highest possible indifference curve); in our graphic example, this is indifference curve I_2.

The Ricardian equivalence aims at highlighting *the incapability of fiscal policy in stimulation of individual consumption.* If the government wants to finance its purchases in period 1 by issuing bonds, then this decision (according to the parable) is accompanied by a simultaneous cut of taxes–so, period 1 incomes grow to $Y_1 - T_1{'}$. However, if the individual forms rational expectations and is aware of the fact that the accordant debt service is going to burden the budget considerably in the next period, then he expects taxes to be increased in the forthcoming term and hence disposable income to drop to $Y_2 - T_2{'}$ (holding that the tax increase $T_2{'} - T_2$ is exactly the future value of tax cut $T_1 - T_1{'}$). Nothing changes but the starting point of the optimization process relocates from 4 (appertaining to the pre-change setting of fiscal policy) to 5, while the optimum stays in 3.[27] The circumstance that the subsequent tax increase is the future value of the tax-cut after all results from the way the budget constraint of government is defined. This, in each period, requires total government expenditures (including interests to pay on state debt) to be made equal with total revenues coming from taxation and other sources plus net government borrowing. By the way, one of its consequences is that the *present value of taxes can change only if government modifies the present value of total expenditures*. So, there is no free lunch: the cover of government expenditures has to be definitely paid either now or later, since the present value of government revenues is fixed by the present value of government expenditures (Barro 1989). *That is exactly why the assumption stating that public expenditures are to be kept intact by the government in the future is required to explain the Ricardian equivalence–a circumstance that has to be stressed to throw light on the fact: emergence of real effects following from the shifts in government expenditures is not strange to the Ricardian equivalence.*[28] It was not debated either by orthodox monetarism–of course, this instrument was not recommended to be used in countercyclical stabilisation, since any intention to stabilize strengthens rather the instability of macro-systems (because of the lagged effects). Even if government changes the level or the composition of public expenditures, it can carry it out only by following the changes in social preferences (cf. Friedman 1948; Tarafás 2001).

It is of crucial importance, as far as the mechanism embedded in the Ricardian equivalence is concerned, whether fiscal authority consistently abstains from mon-

---

[27] Heller and Starr (1979) calls our attention to it because of the liquidity constraints (that is, the limitations of borrowing against future incomes), a tax-cut can trigger a boost in current consumption even if future tax obligations related to bond-financing are fully anticipated. After all, these limitations imply current consumption to be determined by currently available resources (cf. Hubbard and Judd 1986). So, we have to draw a distinction between *wealth* and *liquidity*: fiscal policy actions do not affect the former, while the latter definitely responds to tax-cuts. If households are not able to keep their consumption through borrowing on the optimal path, then their additional liquidity resulting from the tax-cut will be devoted to these purposes. Accordingly, current consumption still responds to tax-cuts, that is, the potential of fiscal policy to control aggregate demand is restored.

[28] Moreover, actually, it was the relationship between the changes in the policy the fiscal authority follows and the reactions of the real system that form the basis of supply side economics.

etizing its debt (or, at least, what the public thinks about the future behaviour of government). The Ricardian equivalence is valid only if market participants expect the fiscal authority not to finance its tax-cut-raised obligations by issuing fiat money–but if it still opts for this action, the initial tax-cut will have *temporary* real effects. Aggregate demand rises for a while–supply side developments depend on what we think about the flexibility of prices and wages. Moreover, the rate of interest bid up by the increased government demand for credits will affect (public) demand for money and velocity in order that the lack of liquidity as distracted by government bonds could be compensated and money in circulation could transfer the increased purchasing power. This remains the situation when the government postpones the offsetting tax increase for long. Of course, it is a little bit complicated, because if the government still opts for a tax increase after an extended period of prolonging, it may push the economy into a serious recession, since the cover of the increased taxes has already been exhausted by the previously increased consumption. At a place, James Tobin (1981) stresses that budget deficit can be financed through either borrowing or money-issuance. It is obviously true as well that there are options also in terms of repayment, so governments can choose between tax increases and monetizing (even if the freedom of choice may be constrained by certain conditions). And there is a further important aspect: the Ricardian equivalence is not conditional on rational expectations, that is, real effects do not fail just because the public anticipates later fiscal policy actions. In other words: if the government is prepared in the very beginning to monetize the debt effect of the initial tax-cut, real effects will emerge (that is, real effects are not caused by the lack of anticipation). Delusion can help us only with the timing of real effects indeed: if the public expects the fiscal authority to finance the debt through tax increases, but, later, the government opts for monetizing after all, real effects will emerge *at this time*. In modern central bank routine, the monetization of government debt is forbidden, manifested in a prohibition of buying government bonds by the monetary authority in the new issue market. We will return to this topic in the following chapter when scrutinizing the problems of credibility of economic policy. Soon, several limitations will be reviewed that undermine this theory. It has to be mentioned that the way Thomas J. Sargent (1986) scrutinizes the Ricardian equivalence in several chapters of his summarizing booklet while he gets stuck at the inflationary effects of budget deficit shows how bound new classical reasoning is–the possibility of real effects does not occur to him at all.

The most important consequence of the Ricardian equivalence is *that individual intertemporal consumption decisions cannot be changed by fiscal policy instruments unless government expenditures shift*. Savings in period 1 increase exactly by the amount of the tax-cut in order that these higher savings could cover the tax increase while keeping consumption intact in period 2.[29] Taxation and government

---

[29] Here, the relation between permanent income theory and the Ricardian equivalence becomes quite clear. In both cases, agents try to keep consumption at the optimal level (this path is not affected by the chosen fiscal policy mix), and the shifts in disposable income lead to fluctuations in savings. It does not seem to be an oversimplification to sum up the essence of both theories in this single sentence.

borrowing are equivalent, borrowing does not trigger further burdens. Choosing between taxation and bond-issuance does not affect agents' life-time budget constraints and hence consumption decisions–and it is exactly the essence of the Ricardian aspect. In other words: replacing of taxes by bond-issuance has no influence on aggregate savings, since there is an offsetting effect from private savings against the drop in budget savings (that is, the deterioration of budget position) (Barro 1989). All this holds only if taxation is carried out exclusively by lump-sum taxes. As a benefit, governments can avoid tax distortions by applying lump-sum taxes, since individuals cannot do anything to reduce their tax obligations. Income-proportional taxes and excises always have distorting effects–governments can always get higher tax proceeds through setting them aside, that is, through replacing them by lump-sum taxes (cf. Stiglitz 1988). By the way, the repressing of distorting effects as much as possible is a fundamental principle of tax policy (cf. Vigvári 2002). If the government tries to finance the debt service of bonds by levying distorting taxes (income taxes or excises), the equivalence fails (Cullis and Jones 1998).

In the previous example, debt service of bonds was assumed to raise taxes only for one period–but, on the contrary, we can consider bonds the maturity of which tends to infinity as well (cf. e.g. Seater 1993). Whatever assumption is chosen, equivalence requires, among others, market participants, who get higher incomes in period 1 thanks to the tax-cut, to live through the whole period of higher taxes necessitated by the repayment phase (we will return to this assumption below).

Pondering our results over, we can realize that there is nothing extraordinary in the findings the theory suggests. Tax burden drops in period 1, but individuals devote additional income to savings in order to create cover for taxes to be levied later–so, in this case, consumption remains intact. Although there are more taxes to pay in the next period, revenues from interest payments and principal repayments help agents keep consumption on an unchanged path.[30] This intuitive analysis confirms the results drawn from a complex mathematical apparatus.

5.6 Assessment and Conclusions

The literature offers an abundance of arguments *for* and *against* the Ricardian equivalence. It is typical that while permanent income theory had generated mainly methodological-technical literature, then authors tried to reinforce or refute the

[30] This is the point where the literature calls our attention to the fact that analyses built on the Ricardian equivalence break with the conclusions of simulations carried out within the traditional IS-LM framework, since it is assumed in the story that the willingness to save and the supply of bonds move hand in hand–however, it follows that no crowding-out occurs, since the rate of interest (if, abandoning the genuine intention of Keynes, regarded as an equilibrium category standing under the influence of loan demand and supply of credit) remains unchanged even after the financial scheme is rearranged. So, if the Ricardian equivalence holds, we do not need to reckon with crowding-out (cf. Seater 1993).

Table 5.1 The Ricardian equivalence in the case of a two-period long time horizon

Period	Tax-related position	Bond-related principal revenues and expenditures	Interest earnings
1.	+B	–B	0
2.	–(B + Br) = –B–Br	+B	+Br

Table 5.2 The Ricardian equivalence in the case of an *H*-period long time horizon ($H \rightarrow \infty$)

Period	Tax-related position	Bond-related principal revenues and expenditures	Interest earnings
1.	+B	–B	0
2.	–Br	0	+Br
3.	–Br	0	+Br
...	...	...	...
H–2	–Br	0	+Br
H–1	–Br	0	+Br
H	–(B + Br) = –B–Br	+B	+Br

Ricardian equivalence on theoretical grounds as well. According to a thorough summary, the Ricardian equivalence is built on a number of excessive and too rigorous assumptions (regarding consumer behaviour and the economic environment), so it cannot emerge under life-world conditions in the purity the theory implies–however, as a more or less relevant approximation to reality, it should be taken into consideration. It is truer since the empirical tests aimed at delivering arguments against the theory are fraught with methodological problems (Seater 1993).

Serious concerns can be phrased when the maturity of bonds is considered. In the simplest case, bonds issued in period 1 were assumed to repay the principal in the next period, parallel with the interests. If per capita tax-cut in period 1 is B, that gives the per capita face value of bonds as well and r is the market rate of interest, then the according cash flow of a representative taxpayer is the following (Table 5.1).

Our representative agent gets a quasi-tax-proceed in the amount of B in period 1 due to the drop in his tax obligations, devoting it to buying bonds. In the next period, personal tax obligations will be raised in order that the budget could meet the debt service of bonds, that is, tax-related additional liabilities emerge. However, they are compensated, first, by releasing the principal tied up in bonds and, second, by the interests paid. To put it simple, the budget constraint does not move.

We get similar results if an H-period long term is assumed, that is, if $H \rightarrow \infty$. Now, Table 5.1 can be rewritten as Table 5.2.

Either a 1-year long or a finite term is considered, it is similarly true that the present value of the debt service of bonds is equal to the initial tax-cut. In case of an infinite term, a reduction can be applied, since

$$\lim_{H\to\infty}\frac{B}{(1+r)^H}=0, \tag{5.25}$$

that is, the present value of principal repayment due at the end of the term is zero. As $H \to \infty$, interest earnings from bonds can be taken as perpetual annuity, and it is true of its present value that

$$PV=\frac{Br}{r}=B, \tag{5.26}$$

i.e. the present discounted value of takings is equal to the initial tax-cut again.

Although our results do not change if a finite term is assumed, the infinite term can remarkably help us consider the outcomes if the lifetime of agents is (expected to be) shorter than the term of bonds (since the situation transforms here to the opposition of infinite-term bonds and finite-living individuals). If the life-time is shorter than the term, the initial tax-cut exceeds the present value of the later tax increases, hence the real wealth of the involved agents shifts and, as a further result, the individual consumption paths also change.[31] According to the story, the tax-cut is devoted to paying for bonds, even if this tax-cut exceeds the present value of the expected additional taxes. This is the situation in which agents' net wealth increases due to the tax-cut. Using the terminology of Barro (1974), *government bonds are net wealth if their value exceeds the present value of the later (debt-service-related) tax increases*. After all, a rise in net wealth triggers (may trigger) an increase in consumption expenditures–this is the essence of *wealth effect*. If it occurs, it implies indeed that the stock of bonds owned in period 1 is not needed to finance the taxes increased by the principal repayment of bonds with a term of H years–that is, from the time of bond issuance, consumer units finance consumer demand increased due to the wealth effect through a gradual and continuous dissaving of amounts accumulated in period 1. It has to be mentioned now that the debate over the empirical validity of the Ricardian equivalence can be identified with the theoretical controversy that was held on whether these bonds make *inside* or *outside wealth* (Weeks 1989). If market participants discount their future debt-service-related tax liabilities and if the resulting present value is equal to the tax-cut (i.e. the bond-buy), then this stock of bonds is inside wealth and there is no wealth effect, so the Ricardian equivalence holds. Bonds make outside wealth (triggering, accordingly, a wealth effect) to an extent or in a ratio with which the present value of later tax liabilities is exceeded by the value of bonds bought in the present.

---

[31] Seater (1993) demonstrates this effect with an example that is built on the two-period model of Samuelson (1958). According to it, if taxes that burden producers' generation of period 1 are reduced, then the later tax increase required by the debt service of bonds compensating lost tax proceeds is going to burden another group of people, that is, the young of the next period. So, workers of period 1 can enhance consumption and they are not forced to provide a cover for later tax increases in advance.

So, the Ricardian equivalence does not hold in these cases. However, Robert J. Barro could demonstrate that it is not necessary to abandon the equivalence-theorem even if lifetime is finite. Starting from the framework of Samuelson (1958), Barro studied whether an increase in the stock of government bonds triggers a rise in wealth as it is *perceived* by the finite-living households. On model grounds, he concluded that, under these conditions (that is, with finite lifetime), it remains true that households act as if they operated on an infinite time-horizon when overlapping generations are assumed to be interconnected by intergenerational transfers (e.g. by bequests). So, government bonds do not affect net wealth, that is, consumption path does not change either (Barro 1974), since, after all, households take their children (descendants) as their own extensions. Supporting and arguing for the Ricardian equivalence require all of these only if the term of bonds is longer (that is, the debt service is more extended) than the (expected) lifetime of the currently alive–as it is also true that only those people will feel wealthier after new bonds are issued who expect the debt service not to burden them entirely (cf. Barro 1989). To put it in other words, it is not possible to carry out an intergenerational redistribution by applying these fiscal policy instruments, since individual actions always cross it. Barro could attain this assumed effect by making the perceived utility of generation t dependent on the perceived utility of generation $t+1$. In algebraic terms:

$$u_t = v_t(c_t; u_{t+1}). \tag{5.27}$$

The mechanism is clear: in this case, finitely living households choose saving in order to create cover for later generations against increasing future tax liabilities (Seater 1993). *After all, altruism extends the framework of finite-time-horizon models to infinity.*

It has to be stressed that Barro asserted not only the equivalence of the cases of *bond-* vs. *tax-finance*, but also expanded his arguments to some social security programs. The mechanism remains the same here as well: in the framework of the Samuelsonian overlapping generations, an increase in the social security payments to the old does not affect their consumption path, only the amount of assets they hold rises–and these increased savings are intended to compensate the higher expected social-security-related tax liabilities of the young (assuming, of course, that social security payments to the old generation are always financed through lump-sum taxes levied on the currently young). Generalizing these results further, Barro (1974) talked about the inefficiency of all fiscal programs that appear, eventually, as imposed intergenerational transfer schemes. For example, in case of public support of education, the old are forced to finance (or at least to contribute to) the education costs of the young–however, this income redistribution becomes neutralized by transfers paid in the opposite direction. No matter how serious these accusations are, it has to be realized that they do not question the existence of the welfare state *in a direct way. Barro and new classical macroeconomics never asserted governments to be unable to carry out income transfers between social groups and generations.* If a government wants to devote higher amounts to public education, it can do it without further ado–even if this action can be financed through an increased tax burden. Similarly, pension increases are not missions impossible per se. New classical reasoning was focused on the circumstance that these government actions do not affect the structure of consumption and, in the individual cases, the optimal consumption path–that is, fiscal stimulation of consumption is fraught with difficulties. So, preferences are still crucial.

Of course, all this is not independent of the traditional arguments for the welfare state, though the relationship is very complicated. New classical macroeconomics did *not* talk about the *impossibility*, but the *futility* of government fiscal actions (in certain special cases). However, all this on its own is capable of undermining the conception of the welfare state, since recent arguments for the welfare state are strongly linked to the concept and requirement of efficiency (cf. Barr 2004), implying that intentions and actions that have limited effects and efficiency can hardly be subsidized. However, we can recall the things said above, according to which the Ricardian equivalence does not preclude real effects coming from changes in the level of government expenditures (cf. Barro 1989). The only thing that matters is the question of the intention of fiscal policy actions: countercyclical initiations remain inefficient *under specified conditions*, since agents forming rational expectations refuse to alter behaviour *in the lack* of long-lasting shifts in the structure of fiscal policy. According to it, with an unchanging expenditure background, a currently decided transitory tax-cut turns out to be a useless government attempt. If fiscal policy makes its attempts through long-lasting (i.e. non-temporary) modifications in its expenditure and revenue structure, theory predicts these attempts to exert influences mainly through changes in the growth potential (that is, in the natural rate of unemployment and the natural level of output) rather than in a countercyclical way.

The literature scrutinizes in detail how uncertain lifetime affects planned consumption path. Since it is true that living a little longer has a non-zero probability at any age (even if this probability tends to zero as time is passing by), so building our theories on an infinite time-horizon does not seem to be excessive–especially as this infinite time-horizon can be perfectly adjusted to the theoretical requirements of finite expected lifetime (since a death at any age also has a non-zero probability). Sometimes, it is stressed that the effect of uncertainty about lifetime on consumption stands under the joint influence of two opposing factors: following from the need to provide for the expenditures that emerge during a possibly longer lifetime, this uncertainty reduces current consumption, while increases it at the same time for the need of a certain current consumption (Levhari and Mirman 1977). If individual lifetime is expected to be finite, this condition affects consumption path. Yaari (1965) examined the case in which the representative individual faces a finite lifetime that is expected to terminate at a time previously unknown (so, he knows only a probability distribution). Since the utility realized through consumption is not independent of the lifetime itself, and since lifetime is a stochastic variable, thus consumers, according to Yaari, maximize expected utility, i.e. plan a consumption path that is expected to maximize utility. In his far-reaching analysis, Yaari also took the case in which individual's (expected) utility is conditional not only on his own consumption but also on the value of the bequest inherited to his descendants into account. Olivier Blanchard (1985), using a little bit cumbrous mathematical apparatus, could demonstrate that if agents have finite lifetimes, a present tax-cut can boost current consumption–also taking the circumstance into consideration that anticipated future tax increases, for finite horizons, are expected to burden the next generations only. Moreover, Blanchard also highlights that, in case of finite lifetimes, fiscal composition (i.e. spending, taxes and debt) exerts real effects, since government debt is a net wealth now in an extent to which this debt, exactly for finite lifetimes again, is not offset completely by tax liabilities expected for later periods. Obviously, these liabilities are going to rise, if the path of government purchases of goods and services are not modified. *If altruism is not assumed, the Ricardian equivalence fails.*

We could see how important altruism and the assumption of intergenerational transfers are with regard to whether the Ricardian equivalence holds or not. Some models highlighted the fact that these transfers are not necessarily due to altruistic motives: intergenerational transfers (that can be even reciprocal, as it has to be stressed) are not necessarily intended to ease future tax liabilities. Bernheim and his co-authors assumed and modelled a bequest motive according to which a testator (the parent) tries to influence the behaviour of his beneficiaries (his children) through his bequest.[32] As it is assumed, though the testator is altruist, his utility is not independent of the behaviour of his chosen beneficiaries–under these circumstances, the consequence is not surprising that parents use expected and promised inheritance to affect their children's behaviour. This strategic bequest motive probably violates the Ricardian equivalence (cf. Bernheim et al. 1985), since future tax liabilities are not offset (not completely at least). However, Barro (1989) calls attention to the fact that if the threat of exclusion is not empty, so the parent is able to control the children's behaviour in the desired way, the equivalence is not violated. A representative section of researches considers not only the parents', but also the children's strategic attitude.[33] So, the opposition of an altruist parent and an altruist child also raises gripping problems: in this situation, welfare of the other is a concern for both parties, so they probably cannot agree about the amount of net transfer (i.e. the balance of the opposing transfers) defined in their relation–hence the balance required by the Ricardian equivalence does not necessarily emerge. Moreover, the children's strategic behaviour may be rooted in an intention to encourage parents to be more generous, by rejecting transfers from them, and hence reducing their own and, through it, parents' utility (Kotlikoff et al. 1990).

[32] The authors claim that the distinction between the cases of only one or more beneficiaries is of high importance. In the former case, the threat may sound incredible (Bernheim et al. 1985), though the parent can run through his savings completely even under this circumstance, realizing the exclusion in this way.

[33] It has to be noted that Barro (1974, 1989) also investigated the case of children's altruism but transfers in his models were always assumed to be *non-reciprocal*. Accordingly, offsetting future tax liabilities in case of an altruistic child is carried out through appropriate reductions in transfers the child devoted to his parents–however, two-way flows are missing.

Finally, we should not forget that important result either that puts some light on the fact that intergenerational transfers are not necessarily due to altruistic or other strategic considerations.[34] If an individual forms untrue expectations concerning his lifetime, the precautionary savings he has accumulated will result in an unintended bequest (cf. Abel 1985).

The comprehensive assumption of altruism was confuted by Bernheim and Bagwell (1988) in a high quality analysis. As a start, they questioned the dynastic family model of Barro in the centre of which stood the intention to model the chain of finite-living individuals, i.e. the *dynastic family* itself as only one, infinite-living agent. For Bernheim and Bagwell, the simplification that tries to approach macro-economic mechanisms through parables on the only representative agent is untenable. We can easily realize that such representative agents are, first, the individual who is considered to make up a whole generation and, second, the dynastic family itself. So, according to Bernheim and Bagwell, such models are fundamentally and inherently inappropriate to analyze the effect of macro-policies. Instead, they suggest a model in which generations consist of large numbers of individuals (and not only one, as opposed to the models of Barro), being interconnected by family linkages (e.g. marriage). *So, altruism connects not only members within a family but also a multitude of families.* Accordingly, society can rather be described as a complex network than by a model of co-existing linear families–and, in these complex networks, individuals can belong to even several dynastic (hence altruistic) families simultaneously. It is interesting that Bernheim and Bagwell, after all, deduce the inefficiency of various government actions (e.g. income redistribution), though highlight that the way towards these results went through the assumption of intergenerational altruism.[35] However, it is not of marginal importance that the *consequences* drawn on the basis of their applied framework do not prevail with a considerable consistency. Well, since the neutrality achieved as a result within the model cannot be observed under life-world conditions, it may follow that altruism does not play as a significant role in explaining behaviour as Barro assumed, and, eventually, family structures, as far as the case of altruism is concerned, could be regarded only as a general framework at most. Whatever we think about altruistic behaviour, whether its generalized case is taken as an acceptable and tenable assumption or not, it should not be forgotten that only families with children can

[34] Feldstein (1988) suggested that an uncertainty about future incomes may discredit the Ricardian equivalence even if both altruism and an accordant bequest are assumed–consequently, present tax-cuts affect current consumption.

[35] It is worth stressing the distinction Seater (1993) draws between *altruism* and the *intention of actual bequeathing*. Accordingly, altruism maintains the Ricardian equivalence only if altruism is accompanied by the assignment and transference of bequest. However, this slight difference does not invalidate the foregoing comments on altruism–it is only needed (as done implicitly by the authors) to regard altruism as the synonym for the intention of actual bequeathing (the case of an operative bequest motive). However, Barro (1989) notes that altruism can manifest itself not only in bequests transferring at death; *inter vivos* transfers (e.g. education-related expenditures) are all capable of guaranteeing the Ricardian equivalence.

behave altruistically towards *some* members of the next generation (i.e. towards their own children). Staying within the Samuelsonian framework, it would be a mistake to assume a family with children in the older generation to worry about the whole next generation. *Families without children can hardly be assumed to follow an altruistic behavioural pattern.* However, it does not invalidate the Ricardian equivalence, since if a world is assumed in which families with and without children simultaneously exist, then families with children (i.e. those with more than the average number of children) will experience new government bonds as a *drop* in their wealth, taking later tax liabilities into account. And though adults without children respond by boosting consumption, families with children decide on saving more in order to cover descendants' extra-proportional tax liabilities to the highest degree possible (Barro 1989).

Concerns about the quality of capital markets are left to the end. First, we should mention the difficulties coming from the circumstance that government and agents have access to credits at different rates of interest. Let us assume it is true of the interest rates on credit relevant for the government and the public (r_G and r_P, respectively) that $r_G < r_P$. Additionally, for the sake of simplicity, let us assume the term to be $H \to \infty$ (a finite term would not alter our findings). If the initial tax-cut is B again, then the later increase in tax liabilities induced by the debt service is Br_G (there are only interest payments to consider, since the term is assumed to be infinite; so Br_G is the amount of interest to pay in each period). It is true of the present value of additional tax liabilities that

$$\frac{Br_G}{r_G} = B > \frac{Br_G}{r_P}, \tag{5.28}$$

because of the difference between the rates of interest, i.e. the initial tax cut exceeds the present value of later additional tax liabilities. Thus, the Ricardian equivalence is violated, since the tax-cut triggers a positive wealth effect, so it becomes possible to stimulate consumption through the tax-cut.

Similarly, the Ricardian equivalence does not hold again when rates of interest on deposits and credits relevant for the public are different, that is, when the relation $r_{P_{deposit}} = r_G < r_{P_{credit}}$ can be assumed. It is worth depicting the latter case in Fig. 5.8.

Consumer preferences are characterized by indifference curves I_i. If there was no difference between the rates of interest (in other words, budget line was still 1–2), optimal consumption allocation would be in 5, independently of the fiscal policy actions. However, under present conditions, this state is unattainable, since possible options are contained by broken line 2–3–6, because of the different interest rates. Equilibrium is in 3, since this point is tangential to the highest possible indifference curve. Tax-cut in period 1 relocates optimum into 4, since the representative consumer can realize a higher utility this way (after the tax-cut and bond issuance, his new budget constraint would have been 2–4–7). Under these circumstances, the chosen fiscal mix exerts influences on consumption, i.e. period 1 tax-cut boosts current consumption expenditures, since consumption is driven by disposable income (cf. Cullis and Jones 1998).

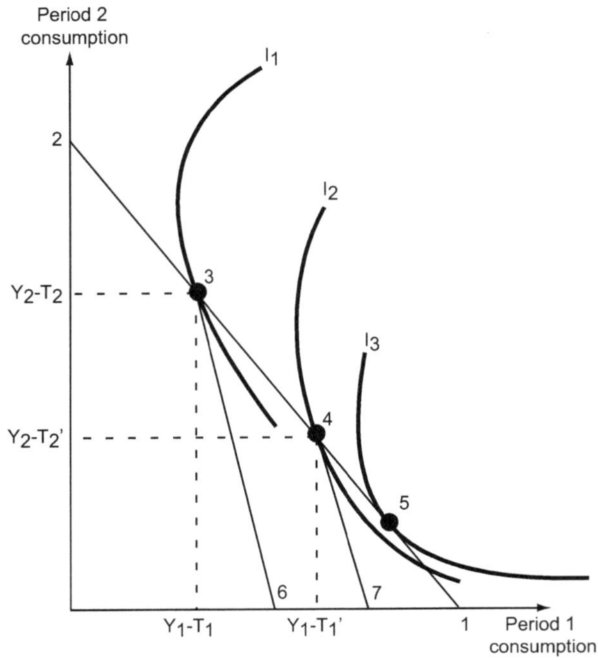

Fig. 5.8 Violation of the Ricardian equivalence in the case of different rates of interest on deposits and credits (Cullis and Jones 1998)

5.7 From the Equivalence-Theorem to Non-Keynesian Effects

Important economic policy consequences can be outlined on the grounds of the Ricardian equivalence and its limitations. *If a tax-cut can be carried out in a way that it is not needed to be offset by new bond debt, so, as a result, a later tax increase is not necessary either, then the initial tax-cut can be perceived and regarded as a permanent increase in income by the society, responding through a boost in consumption*. We have seen above that, under REH, shifts in permanent income are sudden, resulting in considerable changes in consumption, as Keynes postulated. So, in this case, it is possible to enhance aggregate demand through the tax-cut. Moreover, all this does not weaken the Ricardian equivalence, since this equivalence required the initial tax-cut *not* to be accompanied by any kind of *public finance rearrangement processes*.[36] So, if the tax-cut is accompanied by a fiscal reform that, *by* enhancing efficiency, makes a permanent reduction in government expenditures possible, then there is a free lunch for customers. However, it is important that a fiscal reform should not destroy the consumption-generating

[36] For now, this comprehensive category designates both short-run *fiscal consolidation* (which is a simple correction actually) and *structural reforms* also taking long-term concerns into account. We will see, how important this distinction is, as far as real effects are concerned.

effects of the tax-cut while cutting back public expenditures. In other words: such measures should not hold back payments from segments of the society the members of which are forced to reduce consumption because of, for instance, a drop in government transfers. It is not a good answer either if a tax-cut somewhere in the society is compensated by increases in taxes to be paid by other segments if these additional increases reduce their consumption.

The exact nature of this *rearrangement process* is a significant circumstance, since it is always a question whether a short-term fiscal consolidation is capable of influencing expectations to end up in a boost in aggregate consumption. *The equivalence theorem guarantees a large extent of freedom for fiscal policy.* In the simplest example, government chooses a tax-cut in period t in order to stimulate aggregate demand and finances lost revenues through bond issuance so that the new bonds will burden the expenditure side only in period $t+1$. Period t tax-cut is (expected to be) followed by tax increases in the next period, and agents forming rational expectations, anticipating this development, do not respond through any boosts in demand at all. A further possible scenario can be found according to which, again, government replaces tax by bonds in the same period of time, but these papers with one-period long term are not going to be settled at maturity, but policy makers, undertaking new bond debts, postpone the repayment. So, the tax increase evoked by the repayment requirement occurs only in period $t+2$(Table 5.3).

Up to period 2, our story proceeds in the way already known. Government in period 1 issues new bonds in the amount of B' maturing in period 2 with an interest to pay according to the rate of interest r; bond debt is $B'(1+r)$ now. Debt service (that would imply an appropriate tax increase in period 2) is settled by issuing new bonds in the amount of B'' so that $B''=B'(1+r)$. So, the public escapes tax increase; though people formally get the principal and interests of bonds issued in period 1, they buy bonds B'' issued in period 2. In period 3, therefore, bond debt is $B''(1+r)$, and now, government decides on repaying, so it commits itself to an appropriate tax increase. Equivalence holds, since $B''(1+r)=B'(1+r)^2$, that is, the initial tax cut is the present value of the later tax increase again. It is a remarkable moment of this story that the prolongation of repayment of the initial bond debt (actually, the rollover of debt) is possible only with an increasing stock of debt. To put it in other words, the price of prolongation of debt clearing is a perpetual rise in debt. Of course, this rolling routine cannot last till eternity, though our parable shows as if government voluntarily opted for period 3 repayment and as if further accumulation of debt was not constrained.

Table 5.3 The Ricardian equivalence in the case of a two-period long time horizon and prolonged bond debt service

Period	Tax-related position	Bond-related principal and interest revenues and expenditures
1.	$+B'$	$-B'$
2.	0	$+B'+B'r-B''=0$
3.	$-B''(1+r)$	$+B''(1+r)$

These issues make the bottom-line of Ponzi gamble played by governments. Ponzi gamble means that a government pays transfers to a generation (or cut back lump-sum taxes) replacing lost revenues by new debts and postpones the offsetting tax increases and undertakes further debts (rollover) as long as the stock of debts reaches a critical (threshold) value that is proportional to national income. This threshold and the prohibition of crossing it make the exit criterion of Ponzi gamble. Ponzi gamble is successful if (or *as long as*) the rollover succeeds–while the gamble fails if this strategy cannot be maintained anymore. How long the government can follow this strategy built on debt accumulation depends on the interest rate on public debt and the growth rate of real economy. If the growth rate is higher than the relevant rate of interest, government is free to finance its prevailing debt (both principal and interests) further by incurring new bond debt. If the real economy grows at a higher rate than the interest rate on public bond debt, then the debt-income ratio continuously declines in time, i.e. real economic growth gives a solution to the problem without taking any particular measures. Under these circumstances, it actually seems to be unnecessary to repay the public debt out of tax increases. In this case, public debt is moving away from the threshold underlying the exit criterion. However, the deficit gamble cannot be played further if growth rate falls below the relevant rate of interest–in such a case, debt repayment necessitates a tax increase that cannot be replaced by new debts already. So, a successful Ponzi gamble offers only a temporary escape from the necessity of settling the debt raised by an initial bond-replaced tax-cut through tax increases. The probability of failure is increased by, for example, the circumstance that the rate of interest may be dependent on the volume of accumulated public debt. As it has already been mentioned in the previous paragraph, overcoming the pressures towards tax increases, i.e. undertaking more and more debt, may be accompanied by boosts in the stock of debt (even to a radical degree). If it happens, this development may affect the rate of interest relevant for public debt: higher levels of debt imply higher and higher degrees of risk that may force up interest rate on government bonds. As the success of Ponzi gambles depends on the relation between economic growth and the rate of interest, this rise could be equivalent to the failure of the game and the inevitability of tax increase (Ball et al. 1995). However, it has to be mentioned as well that, in our simple model, where economic growth is assumed away, rolling over and accumulating bond debt cannot occur, or occurs during only very limited periods of time at most–so, a successful Ponzi gamble is not possible.

But what happens if a government gives an answer to a drop in tax revenues with a short-run consolidation? The direct result of the tax-cut is a *resource gap*, as a response to which government may withdraw resources from some budgetary subsystems without easing their financial needs. It has to be realized that it is a possible approach of deficit reduction: a tax-cut *would* result in a boost in budget deficit *ceteris paribus*. In order to avoid it, fiscal policy tries to resolve the problem by cutting back government expenditures. However, simultaneously, neither operation was made more efficient nor lost revenues were offset out of other resources, so the operative processes of the aggrieved subsystems become endangered.

Exactly this short-run aspect is the essence of fiscal consolidation programs (cf. Benczes and Kutasi 2010a). So, resource gap permanently abides that does not allow revenue needs to be absorbed. So, even if government opts for a countercyclical tax-cut in period t, and even if government is able to defer to compensate the lost tax revenues by bond issuance through a fiscal consolidation (through an appropriate reduction in expenditures), the story passes off without doing violence to the equivalence. *Revenue needs are to be met sooner or later*. For example, fiscal policy in period $t+1$ may decide on withdrawing the tax-cut of the previous period and getting additional tax resources, which, however, would be equivalent to the case in the basic story. If the amount of the initial (period t) tax-cut is B, and if fiscal policy makers want to refund the resource gap in the next period, preserving the real value (relevant rate of interest is still r), then the tax increase in period $t+1$ is exactly $B(1+r)$. The equivalence holds: the initial tax-cut is the present value of the later tax increase. Again, agents forming rational expectations cannot respond to the initial tax-cut through boosting consumption expenditure.

A mixture of the previous two strategies is also feasible: a tax-cut of B in period t with a simultaneous fiscal correction (resulting in a resource gap), which is going to be refunded by the government in $t+1$ through issuing new bonds, maintaining the real value. Thus, new bond debt in period $t+1$ is $B(1+r)$, from which a repayment requirement of $B(1+r)^2$ occurs for period $t+2$, financed out of tax increases by the government. Tax-cut is the present discounted value of the tax increase again. Agents with rational expectations cannot respond through enhancing consumption. Our story obviously trickles towards the end game already known. It should be noted that government may even try to get out of the trap of the equivalence by avoiding bond issuance, but as long as the initial tax-cut raises a resource gap and hence revenue needs, the freedom of the government consists in deferring the threshold at most until which refunding can be postponed (financing budget deficit). Of course, finite lifetimes, the presence or absence of altruism and the complications already learned with regard to the rate of interest affect the Ricardian equivalence again–however, the bottom-line is that the tax-cut leading to revenue needs results in financial difficulties that can be settled, after all, only out of tax revenues. Government may even try to follow a scenario that is different from the combination of tax-cut and bond issuance ending up in equivalence–however, as long as the reduction of the deficit is possible only with keeping the resource gap unchanged, consumption-enhancing tax-cut remains ineffective.

So, as long as basic operations of public sector are kept intact, as long as market participants do not experience the easing of expenditure needs (or, at least, cannot expect them), they identify the tax-cut (even if it results in an increase in disposable income) with additional resource gaps only to which agents link a later *necessary* (unavoidable) fiscal action intended to provide refund. It is completely indifferent whether this refunding action (tax increase) is needed to fill the resource gap or to finance the debt service of bonds undertaken to narrow the resource gap. *In this case, timing is the only question for market participants*: from the fact of a resource gap (veiled by only a fiscal consolidation) raised by the tax-cut, agents infer that filling this gap is inevitable, which can be carried out only through burdening their incomes (i.e. only through relieving the earlier tax-cut). *In case of corrections that keep fundamental operations of the public sector intact, equivalence holds in the basic model* (taking known limitations into account as well). However, interesting results emerge if rationality is incomplete and if, for example, from the postponement of meeting the financial needs, agents gather that these needs are going to disappear. In this case, aggregate demand can be stimulated through the initial tax-cut–however, it is another problem that the later tax increase (either it is necessitated by the direct filling the resource gap or by the debt service of bonds filling this resource gap), generating a fluctuation in the opposite direction, takes a cyclicality of dubious value in real output dynamics. In Chap. 6, the case of Reaganomics excellently demonstrates how the postponing-deferring strategy of budget policy can convince market participants that the tax-cut has not resulted in a resource gap and how public can respond to it through enhancing aggregate demand.

Now, the question is what consequences emerge with regard to the Ricardian equivalence from those fiscal rearrangement processes through which the resulting resource gaps can be avoided. In our previous stories, this resource gap was caused by the circumstance that the government did not offset its lost revenues by issuing bonds in order to get out of the trap of the equivalence theorem. Agents forming rational expectations infer a later increase in tax burden from the mixture of an initial tax-cut and a compensating bond issuance, and, as a result, consumption-enhancing effects of the tax-cut completely fail. In our basic model, fiscal policy can cure the deficit that comes from a tax-cut aimed at enhancing consumption through not only a simple correction but through linking it to a comprehensive reform covering the fundamental operations and characteristics of government sector. Abandoning this short-run aspect of fiscal corrections is exactly the essence of structural reforms, focusing on *qualitative* changes over and above simple *quantitative* corrections. The distinction we apply here is also grounded on the separation of (1) mere quantitative changes from (2) the joint execution of qualitative and quantitative modifications. Adverse effects of quantitative changes (that lead to resource gaps) can be eliminated exactly through these qualitative modifications (cf. László 2001). The literature on public finance reforms is vast–however, it is obvious that the fundamental purpose of such a reform is to enhance efficiency (and performance) of government sector operations through considering the special features of a given national economy plus the country- and cultural-specific circumstances and problems of a given public sector (cf. Vigvári 2006a). The hierarchy and system of goals of a public sector reform may be very diversified and complex. These purposes may embrace a transformation (or establishment) of a national economy planning system; stressing the need of strategic attitude; a competitiveness-improving reform of tax and contribution systems (cf. e.g. Báger et al. 2010); enhancing financial discipline (cf. e.g. Báger and Vigvári 2007); the rationalization of public services; and even the development of a public finance management information system providing well arranged and clear-cut data on public sector operations (e.g. Vigvári 2006b; Demény 2008) together with starting the necessary and accordant regulation processes. This list is far from being complete. So, we should not think that the purpose of a structural reform is *only* a rearrangement of public sector operations that facilitates a tax-cut without creating a resource gap–but it is obvious that *this outcome* could be still regarded as a possible and desired result. Similarly, it is also evident that, within our simple framework to look into the Ricardian equivalence, such a change (i.e. a tax-cut not resulting in a resource gap) can be really qualified as a structural reform. So, as the improvement of operative efficiency and some qualitative features makes it possible to implement the period *t* tax-cut without a resulting resource gap, i.e. not to force agents to link this tax-cut with a necessary subsequent (and, at best, only deferrable) tax increase, it seems that on the grounds of the equivalence theorem *a key element of demand stimulation through fiscal instruments is made exactly by these reforms.*

However, we have further problems if we try to realize that fiscal policy, with rational expectations, can execute the desired stimulation of aggregate demand through a public sector reform. If agents anticipate that government is going to execute a tax-cut in an unfavourable conjunctural situation and to accompany this measure by a reform process that can make a later offsetting tax increase avoidable, the public takes this fiscal policy action into account in advance when calculating expected permanent income. The postponement of the compensating tax increase should be unexpected (surprising) in order that permanent income and hence consumption could change under REH. So, the question is whether this surprise could be triggered through a fundamental reform.

However strange it may sound, this surprise could be carried out, since *reforms tend not to occur*. As, under REH, market participants expect the outcome with the highest probability, *agents forming rational expectations predict comprehensive fiscal reforms not to happen*. It follows (no matter how absurd it is) that fiscal government can surprise the agents if it carries the reform through. So, if we stand on the grounds of PIH specified under REH, a tax-cut and a related reform (that makes the subsequent offsetting tax increase unnecessary) works as a surprise, hence it becomes possible to exercise influence on consumption.

On the basis of the vast literature cited, Benczes and Kutasi (2010a) give a marvellous summary of the possible causes for the failures of reforms. It is not necessary to echo their results, it is enough to drop some hints. We can argue for the failure of fundamental structural reforms along different lines of reasoning. Executing such reforms is automatically impeded by the contradiction of interests of social (and political) groups affected by the reform. The concrete and exact form taken by a reform program itself eventually is dependent exactly on the interest enforcement capacities of the conflicting groups. The games the coalition partners play against each other while trying to represent the interest of heterogeneous voters predestine the same outcome. The result is not only the procrastination of a reform but also a gradual leaking of its content: even if, formally, the reform is executed, it is likely to become more meagre in its content than its planned version was. Moreover, reforms are *public goods*: nobody can be excluded from the effects (economically); such an exclusion would be unnecessary because of the lack of rivalry (for summaries of the theory of public and other goods see Vigvári 2002; Stiglitz 1988). It is exactly the lack of excludability that leads to the free rider problem: market participants are actually somehow interested in backing out of the contributions to the reform (e.g. by defeating government efforts to increase tax revenues through broadening the tax base), since they can still benefit from it. So, the advancement of a reform is always fraught with dangers. It is another interesting remark that if a government (or the chain of successive governments) is successful in postponing a structural reform for a long time, the problems becoming more and more urgent, possibly accompanied by an unfavourable fluctuation in world economy, will force policy makers to make a short-run and immediate intervention. The result is only a fiscal consolidation, only a correction promising quick and showy effects. *The postponement of a reform seems to reduce the chances of a reform further*.

Although consequences do not concern directly the likeliness of deferring reforms, the existence of political business cycles can be still interpreted as an argument for higher probabilities of failure. Interesting results emerged e.g. from the considerations of the deficit effects of party composition of governments or of the bias of governments towards expansions and the breaking with fiscal discipline or of the postponement of long-term structural reforms in order to be re-elected (for more details see Kutasi 2007). By doing so, governments utilize the incomplete situation awareness of agents who tend to overestimate the present gains of popular measures and underestimate the future costs of the elimination of fiscal consequences (e.g. Csillag 2007; for further details see Bokros 2009). Related researches have been carried out since the 1970s, and the system of hypotheses getting more and more elaborated is leading to more and more convincing results. By now, only the underlying mechanism to be grasped theoretically has remained dubious. For sure, it is increasingly clear that political elections leave marks on public sector operations, typically enhancing a bias towards deficit (cf. Benczes and Ürögdi 2008). However, if fiscal policy displays cycles subordinated to political processes and mechanisms and if the desire to be re-elected leads to the erosion of fiscal discipline, we get a further effectual argument for the delay of public sector reforms concerning the fundamental operations. Accordingly, experiences sometimes report that the deficits accumulated at election times are not compensated later due to the lack of the required intentions to turn developments back. Although governments tend to loosen their fiscal policy in order to be re-elected, a complete elimination of the negative effects may not happen after all (cf. Kraan 2007), so fiscal policy can easily find itself permanently stuck on the wrong track. While a comprehensive public sector reform is more and more urgent, paradoxically, fiscal policy does not take any steps towards it.

Additionally, we should not forget the negative effects of governmental bureaucracy as regards reforms, even if this insight is claimed mainly by sociology and political science and not by economics. Public sector operations cannot even be imagined in the lack of a governmental bureaucracy, but its portrays rarely depict civil servants to be flexible and adaptable (cf. Weber 1923). According to some interpretations, the essence and the immanent attribute of bureaucracy is *rigidity*, so a process of changing and adapting to changes in environment is a hard row for them to hoe. The very reason for the existence of bureaucracy is its ability to handle stable and routine tasks (see Perrow 1986). But how all these affect the execution process of public sector reforms? It also follows from the above circumstances that bureaucracies are typically conservative organizations striving for conformity and stability. A resistance against any reforming efforts follows automatically (such reforms would influence their own operative activities), accompanied even by an intention to protect their own institutional interests. It is very interesting how Almond and Powell (1992) writes about the operation of governmental bureaucracies. These state bureaucracies, as a consequence of their operative activities, could successfully monopolize the task of execution and enforcement of laws and legal rules, and their influence covers political decision making as well.

We should not forget about the ability of selecting information, by the utilization of which it is possible to qualify and distort decision alternatives and, after all, to influence resolutions. Decisions are made following the bureaucratic interpretation lines and even may be far from the goals that were set in political judgements in advance. It seems as if public sector reforms were susceptible to fail, flatten out and lose their real and substantial content in spite of explicit political wills and efforts. In this picture, bureaucracy hardly gives way to these intentions if any element of the planned reforms interferes with its organizational-institutional interests. The overall success is dependent on whether the political power can find and utilize the ways by which bureaucracy can be *forced* to execute the process (for further considerations see Almond and Powell 1992). On second thoughts, the picture depicted above of bureaucracy is not considerably far from the interpretation given by János Kornai (1984). For example, András Hegedüs (1970), relying on vast literature when analyzing the activity of bureaucracies, talks about a *professional incompetence* and the almost evident conservatism of bureaucracy that mostly manifest themselves as efforts to preserve the *status quo*. In this context, the phenomenon that administrations extend their skills (and even procedures) necessary for efficient performance in certain areas to scopes where these competences become rather obstacles to accomplishments and efficiency is also remarkable.

We may have sufficient arguments by now for the high probability of failures of public finance reforms. PIH specified under REH helps us interpret these results. We have seen that a surprising-unexpected pressure that forces agents to reassess permanent income triggers an instant adjustment of consumption (Sect. 5.3). We could see it on the grounds of the nature of public finance reforms that a non-occurrence of a compensating tax increase can surprise even agents forming rational expectations. *Public sector reforms are always more likely not to occur.* Assuming complete rationality of expectations, a fiscal reform produces effects through its *unexpected* nature. We have seen that a fundamental public finance reform can surprise market participants since the necessity may be known even to everyone, but governments can still roll the problem over by deferring the solution (cf. Vigvári 2006a). Even if the program is launched in the end, this outcome will be taken as a surprise by market participants not expecting it. So, if the tax-cut is accompanied by a fiscal reform as well, i.e. agents are not forced to be afraid of a later tax increase to compensate lost revenues, then a reassessment of permanent income is also done and government succeeds in stimulating aggregate demand through fiscal policy instruments.

According to the equivalence theorem, even less is enough to launch a fiscal stimulus. Since the theorem sets, as a fundamental assumption, that public sector operations are not going to change in the relevant time-horizon (interpreting public sector as an unopened black box), if these operations are changed after all, i.e. it is not needed to offset the tax-cut later, the public interprets this tax-cut as a rise in income and a boost in demand starts. *In this case, no surprise is needed: according to the equivalence theorem, a public finance reform automatically helps fiscal policy to achieve its countercyclical purposes through the stimulation of aggregate demand.* We should realize that *new classical macroeconomics, laying great emphasis on the Ricardian equivalence, tried to specify and define the limitations of the efficiency of economic policy again.* While, as for the theory of Keynes, a fiscal policy applying a tax-cut automatically contributes to the elimination of unfavourable cyclical effects, this mechanism is still not precluded from the new classical theory, only the criteria of success has become more rigorous. So, it is essential to realize: *a fiscal expansion on its own is not sufficient to stimulate aggregate demand anymore.* REH, as we have seen, does not preclude the potential success of a fiscal action appeared as a tax-cut if it is accompanied by an appropriate reform process that enables agents to perceive the tax-cut as a permanent rise in income. As it was stressed when analyzing PIH specified under REH, *the result will be the consumption response already learned from the theory of Keynes.* At this point, new classical macroeconomics is extraordinarily close to the theory of Keynes. *While monetary policy showed only limited effectiveness in short-run manipulation of economic growth (*i.e. *in countercyclical control), fiscal policy, even under REH, can remain effective in the stimulation of aggregate demand in the case of efforts made by a fiscal government towards system modifications.*

However, we may go farther. Either PIH specified under REH or the equivalence theorem is considered, changes in demand were due *not* to the tax-cut but to the public finance reform after all. As it was phrased above: a fiscal expansion on its own is not sufficient. In the end, following these lines, we get to the mechanisms referred to as non-Keynesian effects in the literature.

In the story analyzed hitherto, consumption-enhancing action carried through as a tax-cut was accompanied by two alternative approaches to cope with the budget deficit. Of course, there was a third option: to replace the lost tax revenues by bond debt. However, this simple case resulted in the Ricardian equivalence within the basic framework, so we can hardly talk about it as a successful attempt to stimulate aggregate demand. Short-run correction was aimed only at a simple numerical reduction in budget deficit, resulting in financial pressures (a resource gap) and mainly proved to be unable to convince agents of a permanent increase in income. However, it was argued that a public finance reform that makes it possible to enhance aggregate demand through a tax-cut without raising a resource gap could be successful.

Non-Keynesian effects are special fiscal policy mechanisms that lead to boosts in macroeconomic activity (typically in aggregate demand) through *reductions* in budget deficits. However, qualitative features of cuts in deficit seem to be of crucial importance. Although the literature links the launch of non-Keynesian effects only with a *fiscal contraction*, *fiscal correction* or *fiscal consolidation* (Rzońca and Ciżkowicz 2005), it is also obvious that the *actual emergence of positive effects definitely requires credibility*.[37] According to the demand-side interpretation[38] of the non-Keynesian effects, (rational) market participants decide on enhancing their consumption expenditures because fiscal policy is able to increase their streams of current income in a way that it simultaneously triggers shifts in expectations about permanent income.[39] To put it in other words: agents infer a higher level of permanent income from an increase in their current income, since they take the fiscal correction *persistent* and *significant* (cf. Giavazzi and Pagano 1995; Giavazzi et al. 1999). It is obviously not possible through a fiscal correction aimed at mere *quantitative* adjustments–therefore we are justified to think that linking the launch of non-Keynesian effects with *credible* and *persistent* deficit-reducing actions is only the result of an inconsistent and unspecified terminology. As we have seen, without qualitative changes, simple quantitative corrections can hardly be expected to be successful, and it is exactly the foreseen lack of success that determines these drastic, but still superficial fiscal actions to fail to influence public expectations about permanent income. If the public does not anticipate a fiscal action to have long-lasting effects, it forms beliefs in expecting a rearrangement that can be deferred at most but not irrevocably annulled. Of course, it has to be underlined that the actual (effective) degree of rationality of expectations influences the outcomes of real economic policy situations. In Chap. 6, we will have some examples according to which fiscal policy, with incomplete rationality, may be able to enhance aggregate demand even in the lack of credibility. However, it is not a consideration of marginal importance that the (temporary) success of demand-enhancement is of dubious value in this case, since the later tax increase turns back the evolution of demand, so, as a result of government action, real economic performance starts large-scale fluctuations (Fig. 5.9).

---

[37] The importance of quality considerations is also the part of the big picture in a sense that the deficit reduction strategy is far from being irrelevant in terms of efficiency: cutting expenditures back is far more promising than the enhancing of revenues.

[38] Later, in Chap. 6, we will return to supply-side explanations of these non-Keynesian effects with regard to the results of supply-side economics.

[39] On the ground of the vast literature cited, Afonso (2001) gives a great review of possible interpretations and explanations. For instance, he mentions that, over and above the dimension (size) and permanence of a fiscal correction, its composition is a further significant determinant of starting a non-Keynesian effect and that deficit-reducing fiscal actions focusing on expenditure-cuts promise to be more successful than tax increases. A drop in interest rates also follows from the lower levels of deficit, that, through the wealth effect (in addition to a transmission through the expectations channel), also leads to a rise in consumption expenditures (cf. Giavazzi et al. 1999).

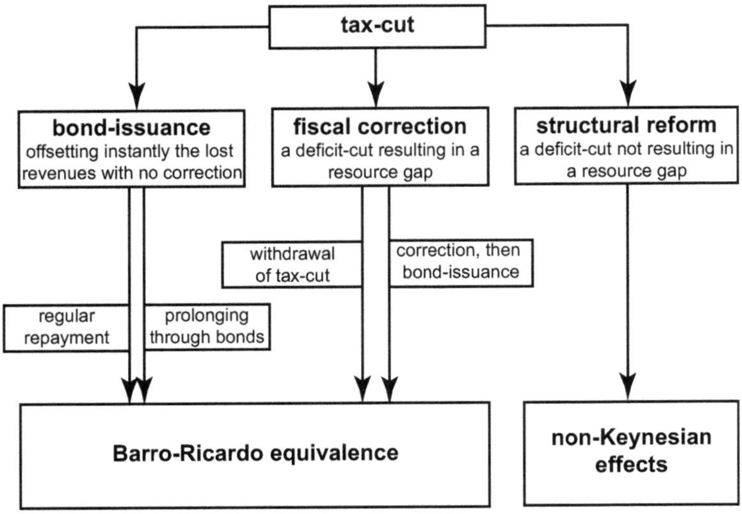

Fig. 5.9 Possible fiscal strategies accompanying a tax-cut and consequences within the theoretical framework

However, it has to be stressed that our present argument was not intended to portray comprehensive public finance reforms as instruments of countercyclical demand management–as our line of reasoning may be interpreted, in spite of our actual purposes. Rather, it was argued that *structural reforms contribute to the success of some short-run efforts over and above the favourable long-run effects.* We have seen that, within our basic model, structural reforms become the prerequisite of a successful demand management when the public, after a planned tax-cut implemented in accordance with enhancement purposes, does not perceive an increase in permanent income–so, the desired consumption reactions fail to occur. If the playing-field of fiscal policy is narrowed, then the demand-side effects of a stimulative tax-cut do not emerge. In this case, a structural reform (e.g. an improvement in competitiveness) promising medium- and long-run effects as well, i.e. the modification of some qualitative parameters of fiscal operations is needed to achieve success. The ultimate purpose of a reform is not to conduct short-run demand management, it only *contributes* to it. It is aimed at relieving budgetary-fiscal pressures that deteriorate and erode the potential of fiscal policy and that undermine, for example, the effectiveness and the transmission channels of demand management. However, it is undoubted that we are justified by the Ricardian equivalence and PIH specified under REH to argue for the favourable short-run effects of comprehensive structural reforms aimed at shifts in fundamental qualitative features on aggregate demand. Following this line of reasoning, *new classical macroeconomics can hardly restore the potential and possibility of countercyclical policies built on activist demand management.* Success is possible, though fiscal policy may be forced onto a very narrow playing-field. Structural reforms do not need to be adjusted to the patterns of macroeconomic cycles in order that a countercyclical control of aggregate demand could be attained. A fiscal intervention with stabilization purposes is not impossible, just the playing-field could be so narrow that fundamental operational changes may be required to regain efficiency. An outcome of structural reforms could be the restoration of the potential of short-run demand management. Macro-economic systems may favourably respond to reform purposes of fiscal policy even in the short-run.[40] However, it is obvious that new classical macroeconomics phrased a well-grounded critique against the theory of activist fiscal policy outlined by Keynes. New classicals did not argue for the predestined failure of activist economic policies, but it is also true that the criterion of efficiency had become more rigorous in their hands. Only a sound, healthy and long-run sustainable fiscal policy (and budget) is capable of exerting a countercyclical control over aggregate demand, since, otherwise, market participants strip all interventions of the possibility of the desired outcomes through their concerns and doubts. Although we can create a situation in which fiscal policy can trigger consumption responses claimed by the theory of Keynes, the success of these actions is not guaranteed evidently but requires some necessary conditions to hold good.

Non-Keynesian effects offer a great opportunity to overcome some of the limitations of the demand-enhancing initiatives of fiscal policy, summed up under the label of *crowding out*. Crowding out refers to the circumstance when demand-enhancing efforts of expansionary budgetary policy crowd out private investments by increasing the rates of interest. With a given cost structure and a level of expected sales revenues, an increase in the rates of interest on credits naturally makes some investment programs unprofitable–in other words, the behaviour of budget makes them fail to come into existence, that is, they get crowded out of the components of aggregate demand owing to fiscal policy. In practical terms, crowding out is said to be related to budget deficits and budget borrowing needs[41] (P. Kiss et al. 2004). Beyond adverse effects on private investments, we also have to consider the consequence according to which public sector (as a whole), through its expansion, appears as a producer of some goods and services, supplanting private sector. Crowding out effect is invariant to the strategy of fiscal expansion: either taxes are cut (which leads to a drop in savings through an increase

---

[40] It should be added that aggregate demand may also respond to a structural reform along different mechanisms. Up to now, we have scrutinized the case in which the public perceives an initial tax-cut as a rise in permanent income due to a reform process ending up in qualitative changes (as well). However, it is also a possible scenario according to which demand may increase *even without* an initial tax-cut, if the structural reform triggers improvements in efficiency that turn market participants to expect the tax burden to drop permanently in the near future. In this case, structural reform itself becomes a stimulant of aggregate demand–though it is more urgent to stress that structural reforms are not necessitated by demand management purposes. These favourable demand-side consequences should be regarded as only beneficial by-products.

[41] There were conceptualizing efforts to extend the crowding out mechanism to cover not only the effects of budgetary expansions deteriorating private investments but the rivalry of government bonds with different risks in competing for resources, when low-risk papers crowd out high-risk counterparts (Kiss and Mák 2009).

in consumption, pushing up rates of interest) or government investments and purchases increase (which boosts the interest rates due to a decrease in budgetary savings), both of them force rates of interest to rise. At a place, Tibor Erdős (2000) argues that the increasing government expenditures launch the crowding out effect even when state spending does not raise the rates of interest, since higher government purchases necessarily hold back private investment in the case of full capacity utilization (i.e. when a short-run expansion of aggregate supply cannot be expected). Simply saying, budget deprives other activities (e.g. private investments) of some goods. It seems to be a powerful argument against government demand management if prices in a macro-economic system are flexible and there is full capacity utilization (see e.g. Benczes and Kutasi 2010b). Of course, if these circumstances cannot be assumed, the purpose of demand stimulation proves to be rational (for further details see Benczes and Kutasi 2010a). So, the basic mechanism is very simple: demand-enhancing purposes of expansionary fiscal policy are offset (at least partly[42]) by the drop in private investment, so the desired positive output outcomes may not occur (real output does not rise as much as it would follow from the characteristics of a given fiscal stimulus, or does not rise at all). As a result, the transmission mechanism of multipliers stressed by Keynes gets modified by the crowding out effect, so the evaluation of multipliers requires us to take the crowding out effect into account as well (cf. Benczes 2009). However, non-Keynesian effects show that if the traditional and well-known ways of fiscal demand management are impassable[43], then what other lines and mechanisms can be found along which demand-side control over real output can be carried through. So, one channel of non-Keynesian effects works exactly through the exploitation of the crowding out mechanism (i.e. of the relationship between spending bias of budget and the rate of interest): one of the consequences of a credible cut in the resource gap is a reduction in budget borrowing needs, resulting in a drop in the rate of interest, which, after all, may enhance private investments. This mechanism is referred to as *crowding in effect* (P. Kiss 1998).

Do we need a more cogent evidence to argue that the lessons of new classical macroeconomics are still relevant? Are we justified to think after all that Lucas and his followers did not suggest guidance useful in everyday economic policy routines? The day-to-day debate on the planning and the effects of fiscal policy actions and on the need for public finance reforms has become enriched by important details and

[42] It means that though private investment programs may fall victims to a demand-enhancing intervention, but this negative effect may be overcompensated by the boost in real output driven by the fiscal action (cf. e.g. Kőrösi 2009). Of course, all these findings are dependent on the model framework or the structural parameters of a macro-economic system.

[43] Since, for instance, the fiscal policy playing-field is narrow and a successful fiscal expansion cannot be even imagined, or since the economy, due to the fiscal policy action, is likely to respond to higher rates of interest through a cut-back in private investments. As far as the final balance of effects is considered, the interest rate elasticity of private investments is of high importance, i.e. (staying within model frameworks) the shape of the investment function built on the rate of interest as an independent (explanatory) variable.

arguments. At the same time, the question why it is useful (e.g. in political terms) to apply a macroeconomic model constructed as a pure theory has also been answered, since the results drawn from the equivalence-theorem can be rephrased in such a way that a *fiscal stimulus intended as a tax-cut is inefficient only under the tight conditions specified by the theorem itself*. As we have seen, if we loosen one of these conditions (the tax-cut gets accompanied by a fiscal reform), the tax-cut proves to be efficient. Theorems of new classical macroeconomics have been really proclaiming the inefficiency of activist economic policies–but, doing so, the theory has defined and specified the conditions required for this inefficiency to hold. *If these conditions* (assumptions) *do not hold, economic policy cannot be ineffective. This* is the most important, though mainly unrecognized message of new classical macroeconomics.

5.8 An Outlook: Automatic Fiscal Stabilizers

Up to now, mainly the role of economic policy surprises and the criteria of their efficiency have been scrutinized. These are the measures that are labelled by the literature as *discretional economic policy instruments*. Discretional steps refer to the case when economic policy decides on an unusual action in a given macroeconomic situation. However, we should not regard economic policy surprises (unexpected actions) and discretional measures as synonyms. Every economic policy *decision* is qualified as a discretional measure in the course of which an attempt is made depending on the macroeconomic situation to influence a given macroeconomic system (e.g. national economy). *The key feature is the decisionary character of the intervention*: though economic policy is situation-dependent, i.e. different actions are needed in a recession or in an overheated boom, the interventions actually applied are not automatic. Fiscal and monetary policy makers make up and execute a set of measures considered the most appropriate with regard to an actual macroeconomic situation. These actions even may surprise market participants, but it is not necessary. There is always a system of economic policy rules that advises economic policy makers to take given measures under given conditions. Of course, these rules are *recommendations* in a literal sense: the concrete content of measures, or their intensity and persistency are always subject to consideration (decision), and economic policy, on the grounds of such sets of advice, draws up the therapy on an occasional basis. It is exactly the extent to which a system of economic policy recommendations is known to market participants that makes a crucial border-line between the theories studied here. Within the adaptive scheme, almost each measure results in a surprise deviating macro-economic systems from their previous trends. However, we have a much broader set of anticipated economic policy actions if expectations are rational. In this case, economic policy is in a far more difficult position if it tries to take unexpected-surprising-unforeseen measures in a given situation that is adequately perceived and interpreted by the public as well (as we have seen above in this chapter, it is not necessarily needed). But, anyway, it is sure that surprising measures are necessarily discretional, though not every discretional measure ends up in a surprise.

Over and above discretional measures, economic policy can intervene through automatisms as well. Taylor-rules (discussed in Chap. 6) are such automatisms in monetary policy. In fiscal policy, this role is played by *automatic fiscal stabilizers*. Actually, these stabilizers are revenue and expenditure rules that contribute to cyclical smoothing in accordance with the fluctuations in macroeconomic performance, mainly through the stimulation of disposable income and household consumption. In a recessionary environment, certain government expenditures increase and vice versa: in good times, certain revenues flow into the budget in higher amounts. Therefore, cyclical fluctuations leave definite marks on budgetary position [→ **structural balance**]. For instance, a progressive tax system is a revenue-side automatic stabilizer since it guarantees tax revenues from increasing incomes to outgrow incomes. So, simultaneously with the boost in macroeconomic performance, tax revenues increase accordingly, exerting countercyclical effects (there is a rise in the average tax rate). Along these lines, a progressive tax system can also produce countercyclical effects during the downward phase of business cycles, since tax revenues from contracting incomes slacken at a higher rate than at which incomes fall (the average tax rate, that is, tax burden drops). Typical examples regarding the expenditure-side are the various kinds of unemployment compensations. Such paying increases under unfavourable macroeconomic conditions to cushion the drop in aggregate demand. During upswings, these benefits gradually fade away (cf. P. Kiss et al. 2004).

Structural balance The relation between macroeconomic performance and the budgetary position has been a hot methodological topic for decades. The idea is very simple: the same budget may end up in a surplus or a deficit, depending on fluctuations in national income (Mackenzie 1993). An improvement in the fiscal position, a drop in budget deficit may follow from favourable macroeconomic developments. If real economy is in the middle of an upswing, thanks to it, government budget balance is likely to amend automatically–however, these fiscal processes should be expected to turn back under deteriorating macroeconomic conditions. In other words: the same budgetary act may lead to highly various balances depending on the conjunctural position of real economy. Of course, under these conditions, it cannot be argued that fiscal policy is on a right track, even if the indices depicting fiscal position report improvements, since these improvements follow only from the favourable conjunctural stance (if budget-balancing fiscal measures have not been taken). So, a rise in budget deficit can be traced back to various roots: it may equally follow from a recession or from unsound budgetary management. According to an illuminating comment, it is actually not possible to set government revenues and expenditures during budgetary planning. In this phase, only a set of acts can be established so that revenues (and expenditures as well) will develop as a function of national income (Solomon 1964). Thus fiscal policy is only partially determining. According to this line of reasoning, a favourable conjunctural stance may present a (purely cyclically fed) improvement of the budgetary position as a performance of fiscal policy–while these outcomes may switch over to the opposite direction if real economy gets on a downward track.

These insights also affect the set of actions required of fiscal policy. So, since a lot of government revenues and expenditures are dependent on macroeconomic performance, we may get to the oversimplifying view that the necessary extent of fiscal correction can be calculated on the grounds of budget deficit (Tanzi 1993). This fiscal correction may be much minor if the deficit rise can be attributed (mainly) to some unbeneficial conjunctural developments. So, if we want to judge the extent of a deficit, the indicator we study is required to reflect *only* the budgetary position and all those factors should be precluded from the big picture that affect the fiscal position but are out of the control of fiscal policy. So, the result of a cyclical "weed-out", the *cyclically-adjusted balance* (as a special indicator of the fiscal position) supports acumen. If the budgetary balance is cleared from cyclical effects,[44] all the fiscal rearrangements and adjustments that should be attributed to discretional economic policy actions will be apparent[45] (cf. P. Kiss and Vadas 2005). *Structural balance* takes a further step, disregarding extraordinary items as well, *in addition to* cyclical effects. Acumen naturally gets disturbed by the extraordinary items (cf. P. Kiss 1998), since the cyclically adjusted balance may contain items that are not raised (or cancelled) by cyclical effects but cannot be regarded as structural yet, as they are the extraordinary items. The ultimate purpose of structural balance (saying again, that excludes the effects of both cycle-related and extraordinary items) is to give a clear picture of the regular and normal operations of the public sector. It helps researchers and policy makers identify the fundamental budget operations and tendencies and makes required measures much easier to be planned since it is not possible anymore to veil the necessity of corrective actions through cyclical or extraordinary items.[46] [→ **output gap**]

First of all, automatic fiscal stabilizers got special attention because their negative balance effects emerging in a recessionary environment can be expected to fade away (moreover, to turn into the opposite direction) even without any special and related intervention when macroeconomic performance improves. Expansionary fiscal policy and the sustainability of fiscal finances sometimes are contradicting (Spilimbergo et al. 2008), but automatic fiscal stabilizers can smooth conjunctural fluctuations in a way that, simultaneously, the structural balance of a healthy (non-deficit-biased) budget stays on a sound path without any special measures–even if an unfavourable cyclical position may force the deficit to exceed the desired medium- and long-term target in the *short-run* (Capet 2004). Findings highlight, for instance, that stabilizers, in contrast to discretional measures, supported the recov-

[44] Because of this purport, estimations of the structural balance bear a close relationship to measuring potential GDP and output gap. So, we always have a twofold problem: as a first step, we need to judge the cyclical position of the given macro-economic system then, as a second step, structural deficit is calculated on the grounds of data gained in the previous stage. These two phases are connected by those pieces of information that describe the dependency of budget revenues and expenditures on real economic performance.

[45] It is interesting that literature sometimes demands too much from the index of cyclically adjusted budget balance. For instance, Muller and Price (1984) argue that fiscal sustainability can also be properly measured on the grounds of the structural balance position, since cyclical correction can separate conjunctural effects from the consequences of discretionary fiscal policy measures. According to them, structural balance is a good indicator of aggregate demand management of fiscal policy as well. Over and above finding the right methodology, it is similarly important to phrase the adequate questions and to find the proper scope for the structural balance (for further considerations, see Blanchard 1990).

[46] Because of the close conceptual relationship to core inflation, literature occasionally refers to structural balance as "core deficit" (cf. e.g. Ize 1983).

ery from the 2007 to 2008 world economic crisis more efficiently, resulting in a higher level of additional growth (Tóth 2010). Of course, these consequences are perhaps due to the more sizeable stimuli that might have come from the stabilizers. Measuring the size (and hence the impact) of demand-side impulses induced by the automatic stabilizers is really a difficult and marshy methodological problem far from being settled (Noord 2000).

So, automatic fiscal stabilizers mainly do not raise the problem of *permanent deficit impact*, since their impacts, as we have seen, fade away as soon as the conjunctural position improves. When economic growth starts, the prior additional deficit turns into the opposite direction. We have a more dubious setting to judge in the case of discretional demand-enhancing measures, since planning the extinction or neutralization of their impacts is always a special and separate task. *Provisional discretional measures* have temporary effects on deficit, but exert permanent influence on the level of state debt (i.e. triggering level shifts), while *fiscal actions remaining in operation for long* (e.g. a tax-cut not withdrawn after the hard times) have permanent deficit impacts, affecting the level of debt cumulatively, so it is continuously forced to rise. The third type of discretional fiscal measures has only temporary effects on both the deficit and the debt level–a typical example is the moving of capital investment expenditures over years (OECD 2009). Moreover, on the grounds of their overall effects, some *packages of measures* can also be listed here, for example, if the deficit and debt impacts of a tax-cut are offset by revenue-increasing measures later. The negative effects of the ideal fiscal stimulus on deficit are only of temporary nature. The purpose is to strengthen (i.e. to complete) aggregate demand which leads, of course, to a rise in budget deficit–however, the unfavourable budgetary developments should be restricted only to the length of recession, and deficit-reduction should be launched immediately after the hard times in order to prevent the debt level from changing in the long-run. The extent and the content of countercyclical discretional measures are always constrained by the current fiscal position. However, automatic fiscal stabilizers do not deteriorate the medium- and long-term sustainability of fiscal policy, not even without any interventions (cf. IMF 2009).

It is not surprising that the unconstrained and undisturbed operation of appropriately installed automatic fiscal stabilizers recently has been claimed in literature as a fundamental economic policy recommendation, since, along these lines, a stable, transparent and sound, an actually rule-based fiscal policy can be outlined and carried out without occasional (discretional) interventions (Aarle et al. 2003). Moreover, to establish a balanced budget through discretional interventions (i.e. through recurrent changes in revenue and expenditure rules) with the aim of replacing automatic fiscal stabilizers (OECD 1999) seems to be much more difficult, since we shoot on a moving target in such cases: shifts in the rules affect real economy in an unforeseeable way, so it becomes uncertain whether revenues and expenditures required for the improvements in fiscal position can actually be realized the way they have been planned in advance. In other words: occasional-discretional interventions can contribute to the confusion of economic policy makers as far as macroeconomic mechanisms and environment are concerned. However, stabilizers do not prove to be efficient in all situations: they work with less efficiency in the case of supply-side shocks and may lead to considerable inflationary pressures (Brunila et al. 2002). By the way, automatic stabilizers can exert influence with no (information and implementation) lags, since actuation needs no decision-making process and, as experiences testify, real economic effects also emerge with far minor impact lags, in contrast to discretional interventions (cf. Buti and Noord 2004). However, it is clear that (demand-side) stimuli coming from automatic stabilizers may not be sufficient to boost a macro-economic system–and in such a case, of course, demand-enhancing discretional measures may also be needed. This option should not be necessarily avoided or regarded as damaging (cf. Báger 2006), assuming that fiscal policy could be disciplined enough to recover the fiscal position (that was deteriorated discretionally in the interest of the upturn) through (further) discretional measures (i.e. strengthening and contributing to the positive deficit impacts of automatic fiscal stabilizers). The size of discretional demand-enhancing measures can be limited only by the fiscal position at most–however, according to the new classical theories studied here no inefficiency-related obstacles urge us to abandon them.

Permanent income theory helps us understand the mechanism and effects of automatic fiscal stabilizers. We have seen at PIH specified under REH that households may be supported by transitory incomes in keeping consumption at the permanent level. It had to be stressed when studying Friedman's (adaptively grounded) permanent income theory that transitory incomes may be also utilized in consumption expenditures. Transitory incomes were tacitly assumed to emerge as factor incomes–but it is not a necessity. When factor incomes start contracting, the consumption path appertaining to the permanent income may be underpinned even by transitory incomes paid as transfers. Takings from automatic fiscal stabilizers should be regarded exactly as such transitory incomes, since incomes from these stabilizers gradually fade away when an unfavourable macroeconomic situation ends, in other words, when factor incomes start rising again. For example, it is an interesting remark that *governments are able to smooth individual incomes and consumption paths much more efficiently than agents, since governments face fewer liquidity constraints and lower risk premiums than households* (cf. Noord 2000). According to PIH, households may be forced to borrow in hard times. However, we have seen that social segments that are mostly affected by the unfavourable cyclical impacts may be less likely to have access to credits–so, smoothing of income and consumption is undertaken by fiscal policy through automatic stabilizers. Government can obtain the required fiscal resources easier and cheaper.

When forming expectations, market participants may deliberately rely on these stabilizers, so they can calculate the path of permanent income by taking these transfer payments into account as well. Their impact does not seem to be dependent on the applied (or postulated) method of expectation formation. If expectations are rational, it is evident that lost factor income gets compensated by transfer payments from stabilizers as it has been expected by the agent in advance. In this case, takings from stabilizers support smoothing of consumption and, at the same time, the stability of the national economy. We have the same situation when the adaptive scheme is assumed. If factor incomes are contracting, market participants preserve consumption through spending out of transfer payments–and surprises within the adaptive scheme play no role in this mechanism. Although market participants forming adaptive expectations and considering only past data and developments may be surprised both by the emergence of a recession and by the launch of automatic stabilizers,[47] consuming out of transitory (transfer) incomes has no alternatives. Albeit the consequence of the direct manifestation of the theory would be that agents experiencing a drop in factor income (and the temporary unsustainability of permanent consumption) opt for borrowing so that simultaneously realized transfer incomes (as transitory incomes) can be saved, the result is the same: even if, speaking in formal terms, the representative consumer unit is assumed to be able to borrow and save at the same time, netting is inevitable. An act of actual borrowing or saving can happen only if the realized transfer income is not sufficient to preserve permanent consumption completely (there is an income gap) or, on the contrary, if transfer payments are not needed in toto for maintaining planned (permanent) consumption–so, transfer incomes from stabilizers can be saved partly.

5.9 Summary: Fiscal Policy in the New Classical Framework

New classical theories on fiscal policy were reviewed in this chapter. Lucas and his followers also extended their influence to Friedman's permanent income theory, though new classicals tried to *specify* the circumstances of the inefficiency of fiscal policy rather on the basis of the Ricardian equivalence theorem. Immense arguments are listed in this chapter to underlie the statement: a limited validity of the equivalence theorem can hardly be precluded under life-world conditions, however, equivalence should not be expected to hold in that pure form as it was specified by

[47] Keep the case of the Friedmanian Phillips curve in mind: agents operating within the adaptive scheme have no information on either the current state of the macroeconomic environment or economic policy responses.

new classical macroeconomics assuming artificial-theoretical environment. We could see that finite and uncertain lifetime, differences between lifetime and maturity period, the character of capital markets and the structure of the rates of interest constitute such circumstances that can undermine the equivalence theorem even individually.

Considerations regarding the underlying assumptions of the equivalence theorem proved to be useful. Theory assumes, among others, the revenues and expenditures of public sector to be kept intact in order that debt-for-tax-swap could be held with no real effects. We could draw interesting conclusions through assuming this condition away. As long as government sector does not modify the qualitative parameters of its own operative actions, tax-cuts implemented to enhance aggregate demand can hardly be expected to trigger the desired effects–since, following the theoretical considerations, it requires some shifts exactly in these qualitative features in a way that demand-enhancing interventions could be accompanied by a perceived rise in permanent income, i.e. a positive turn in expectations. As we have seen, *new classical macroeconomics got quite close to the earlier message of Keynes by exerting the Ricardian equivalence and PIH specified under REH*. In addition to the limited potential of monetary policy, fiscal policy can be successful again in stimulating aggregate demand within the new classical framework, if, eventually, business cycles introduced during the monetary analysis deviate a given macroeconomic system from the full-employment equilibrium state. *Of course, the criteria of success has become tighter*. Although, for Keynes, a fiscal stimulus emerging as a tax-cut promised to enhance aggregate demand successfully, agents as new classicals assumed try to discover the forthcoming steps of economic policy as well. If the public finds fiscal policy not to be credible, real effects of demand-enhancing actions will probably fail to emerge or, at least, will become considerably weaker.

References

Aarle B, Garretsen H, Huart F (2003) Monetary and fiscal policy rules in the EMU. Ger Econ Rev 5(4):407–434

Abel AB (1985) Precautionary saving and accidental bequests. Am Econ Rev 75(4):777–791

Afonso A (2001) Non-Keynesian effects of fiscal policy in the EU-15. ISEG, Lisbon

Almond GA, Powell GB (1992) Comparative politics today–a world view. Harper Collins, New York. Hungarian edition: Almond GA, Powell GB (1996) Összehasonlító politológia (trans: Béndek P). Osiris, Budapest

Andorka R (1997) Bevezetés a szociológiába. Osiris, Budapest

Báger G (2006) Az állami megrendelések (keresleti) hatása a gazdaságra. In: Vigvári A (ed) Vissza az alapokhoz! Tanulmányok a közpénzügyi rendszer reformjáról. Új Mandátum, Budapest, pp 111–131

Báger G, Vigvári A (2007) Államreform, közpénzügyi reform – nemzetközi trendek és hazai kihívások. ÁSZ FEMI, Budapest

Báger G, Pulay G, Vigvári A (2010) The main characteristics of the operation of the Hungarian state and the directions of necessary changes. Public Financ Q 55(2):223–250

Ball L, Elmendorf DW, Mankiw GN (1995) The deficit gamble. NBER working paper No. 5015. National Bureau of Economic Research, Cambridge, MA

Barr N (2004) Economics of the welfare state, 4th edn. Oxford University Press, Oxford

Barro RJ (1974) Are government bonds net wealth? J Polit Econ 82(6):1095–1117

Barro RJ (1989) The Ricardian approach to budget deficits. J Econ Perspect 3(2):37–54

Begg DKH (1982) The rational expectations revolution in macroeconomics–theories and evidence. Philip Allan, Oxford

Benczes I (2009) Expanziós költségvetési politika és hatásai az EU-ban. Külgazdaság 53(7–8): 4–22

Benczes I, Kutasi G (2010a) Költségvetési pénzügyek – hiány, államadósság és fenntarthatóság. Akadémiai, Budapest

Benczes I, Kutasi G (2010b) A költségvetési politika fenntarthatósága és a szabályalapú gazdaságpolitika. Köz-Gazdaság 5(2):59–74

Benczes I, Ürögdi A (2008) The new theory and empirics of political business cycles. Public Financ Q 53(2):269–282

Benk S, Jakab MZ, Vadas G (2005) Potential output estimations for Hungary–a survey of different approaches. Central Bank of Hungary, Budapest

Bernheim BD, Bagwell K (1988) Is everything neutral? J Polit Econ 96(2):308–338

Bernheim BD, Shleifer A, Summers LH (1985) The strategic bequest motive. J Polit Econ 93(6): 1045–1076

Bilson JFO (1980) The rational expectations approach to the consumption function–a multicountry study. Eur Econ Rev 13(3):273–299

Bird RC, Bodkin RG (1965) The national service life insurance dividends of 1950 and consumption–a further test of the 'strict' permanent income hypothesis. J Polit Econ 73(5): 499–515

Blanchard OJ (1985) Debt, deficits, and finite horizons. J Polit Econ 93(2):223–247

Blanchard OJ (1990) Suggestions for a new set of fiscal indicators. OECD, Paris

Bokros L (2009) A reformok kritikus tömege. In: Muraközy L (ed) A jelen a jövő múltja. Járatlan utak – járt úttalanságok. Akadémiai, Budapest, pp 194–240

Boulding KE (1966) Economic analysis–microeconomics. Harper & Row, New York

Bowden RJ (1973) Some implications of the permanent-income hypothesis. Rev Econ Stud 40(1): 33–37

Brown TM (1952) Habit persistence and lags in consumer behaviour. Econometrica 20(3): 355–371

Brunila A, Buti M, in't Veld J (2002) Fiscal policy in Europe–how effective are automatic stabilisers? European Commission, Brussels

Buchanan JM (1976) Barro on the Ricardian equivalence theorem. J Polit Econ 84(2):337–342

Buti M, Noord P (2004) Fiscal policy in EMU–rules, discretion and political incentives. European Commission, Brussels

Capet S (2004) The efficiency of fiscal policies–a survey of the literature. CEPII, Paris

Carroll CD (2001) A theory of the consumption function, with and without liquidity constraints. NBER working paper No. 8387. National Bureau of Economic Research, Cambridge, MA

Christiano LJ, Eichenbaum M, Marshall D (1991) The permanent income hypothesis revisited. Econometrica 59(2):397–423

Csaba L (2008) The new kind of macroeconomic populism. Public Financ Q 53(4):601–616

Csillag I (2007) Notoriously repeated budget policy. Public Financ Q 52(2):293–306

Cullis J, Jones P (1998) Public finance and public choice, 2nd edn. Oxford University Press, Oxford

Demény Á (2008) The central budget management system and the impact thereof. Public Financ Q 53(4):633–644

Erdős T (1998) Infláció. Akadémiai, Budapest

Erdős T (2000) A fenntartható növekedés egyensúlyi feltételei I. Közgazdasági Szemle 47(2): 101–115

Feldstein M (1974) Social security, induced retirement, and aggregate capital accumulation. J Polit Econ 82(5):905–926

Feldstein M (1988) The effects of fiscal policies when incomes are uncertain–a contradiction to Ricardian equivalence. Am Econ Rev 78(1):14–23

Fisher I (1982) The rate of interest. Garland, New York/London

Flemming JS (1973) The consumption function when capital markets are imperfect–the permanent income hypothesis reconsidered. Oxford Econ Pap 25(2):160–172

Friedman M (1948) A monetary and fiscal framework for economic stability. Am Econ Rev 38(3): 245–264

Friedman M (1957) A theory of the consumption function. Princeton University Press, Princeton

Friedman M (1968) The role of monetary policy. Am Econ Rev 58(1):1–17

Gallaway LE, Smith PE (1961) Real balances and the permanent income hypothesis. Q J Econ 75(2):302–313

Giavazzi F, Pagano M (1995) Non-Keynesian effects of fiscal policy changes–international evidence and the Swedish experience. NBER working paper No. 5332. National Bureau of Economic Research, Cambridge, MA

Giavazzi F, Tullio J, Pagano M (1999) Searching for non-Keynesian effects of fiscal policy. CSEF, Fisciano

Giddens A (1993) Sociology, 2nd edn. Polity Press, Cambridge. Hungarian edition: Giddens A (1997) Szociológia (trans: Acsádi J, Babarczy E, Nagy M, Nagy Z, Tóth L). Osiris, Budapest

Hall RE (1978) Stochastic implications of the life cycle–permanent income hypothesis: theory and evidence. J Polit Econ 86(6):971–987

Hegedüs A (1970) A bürokratizmus, mint a szakigazgatás szociálpatológiája. Kortárs 14(7): 1101–1114

Heller WP, Starr RM (1979) Capital market imperfection, the consumption function, and the effectiveness of fiscal policy. Q J Econ 93(3):455–463

Hodrick RJ, Prescott EC (1997) Postwar U.S. business cycles–an empirical investigation. J Money Credit Bank 29(1):1–16

Holbrook R, Stafford F (1971) The propensity to consume separate types of income–a generalized permanent income hypothesis. Econometrica 39(1):1–21

Holmes JM (1970) A direct test of Friedman's permanent income theory. J Am Stat Assoc 65(9): 1159–1162

Holmes JM (1971) A condition for independence of permanent and transitory components of a series. J Am Stat Assoc 66(3):13–15

Houthakker HS (1958) The permanent income hypothesis. Am Econ Rev 48(3):396–404

Hubbard RG, Judd KL (1986) Liquidity constraints, fiscal policy, and consumption. Brookings Pap Econ Act 17(1):1–59

IMF (2009) Fiscal implications of the global economic and financial crisis. International Monetary Fund, Washington, DC

Ize A (1983) Measurement of fiscal performance in IMF-supported programs. In: Blejer MI, Cheasty A (eds) How to measure the fiscal deficit. International Monetary Fund, Washington, DC, pp 52–84

Keynes JM (1936) The general theory of employment, interest and money. Palgrave Macmillan, London. Hungarian edition: Keynes JM (1965) A foglalkoztatás, a kamat és a pénz általános elmélete (trans: Erdős P). KJK, Budapest

Kim C (1996) Measuring deviations from the permanent income hypothesis. Int Econ Rev 37(1): 205–225

P. Kiss G (1998) A költségvetés tervezése és a fiskális átláthatóság problémái. Central Bank of Hungary, Budapest

P. Kiss G, Vadas G (2005) Légy résen! Az államháztartási egyenleg ciklikus igazítása. Közgazdasági Szemle 52(2):109–129

P. Kiss G, Karádi P, Krekó J (2004) Az euró bevezetésével járó strukturális politikai kihívások – költségvetési politika. Central Bank of Hungary, Budapest

Kiss MN, Mák I (2009) Szuverén kötvénykibocsátások alakulása a kelet-közép-európai régióban a Lehman-csőd óta. MNB-Szemle 4(4):13–23
Kopányi M (2003) Mikroökonómia. KJK–KERSZÖV, Budapest
Kornai J (1984) Bürokratikus és piaci koordináció. Akadémiai, Budapest
Kőrösi I (2009) Az állam gazdaságfejlesztő és jóléti szerepe. A globális válság – hatások, gazdaságpolitikai válaszok és kilátások 4. MTA VKI, Budapest
Kotlikoff LJ, Razin A, Rosenthal RW (1990) A strategic altruism model in which Ricardian equivalence does not hold. Econ J 100(12):1261–1268
Kraan DJ (2007) The political economy of budgeting in Hungary. Public Financ Q 52(2):307–319
Kutasi G (2007) Budget deficit and the achievement of fiscal equilibrium – institutional approach. Public Financ Q 52(1):37–52
László C (2001) Vargabetűk az államháztartási reform tízéves történetében (1988–1997). Közgazdasági Szemle 48(10):844–864
Laumas PS (1969) A test of the permanent income hypothesis. J Polit Econ 77(5):857–861
Laumas PS, Mohabbat KA (1972) The permanent income hypothesis–evidence from time-series data. Am Econ Rev 62(4):730–734
Levhari D, Mirman LJ (1977) Savings and consumption with an uncertain horizon. J Polit Econ 85(2):265–281
Mackenzie GA (1993) Are all summary indicators of the stance of fiscal policy misleading? In: Blejer MI, Cheasty A (eds) How to measure the fiscal deficit. International Monetary Fund, Washington, DC, pp 21–51
Maddala GS (2001) Introduction to econometrics, 3rd edn. Wiley, New York. Hungarian edition: Maddala GS (2004) Bevezetés az ökonometriába (trans: Békés G, Cseres-Gergely Z, Csorba G, Reiff Á, Szentpéteri Á, Szilágyi K). Nemzeti Tankönyvkiadó, Budapest
McElhattan R (1982) On federal deficits and their economic impact. Econ Rev 12(2):6–17
MNB (2010) Inflation report, Nov 2010. Central Bank of Hungary, Budapest
Modigliani F (1961) Long-run implications of alternative fiscal policies and the burden of the national debt. Econ J 71(12):730–755
Muller P, Price RW (1984) Structural budget deficits and fiscal stance. OECD, Paris
Noord P (2000) The size and the role of automatic fiscal stabilizers in the 1990s and beyond. OECD, Paris
O'Driscoll GP (1977) The Ricardian nonequivalence theorem. J Polit Econ 85(1):207–210
OECD (1999) Economic outlook, No. 66. OECD, Paris
OECD (2001) OECD economic surveys 2000–2001: euro area. OECD, Paris
OECD (2009) Economic outlook, No. 85. OECD, Paris
Patinkin D (1956) Money, interest, and prices. MIT Press, Cambridge, MA
Perrow C (1986) Complex organizations, 3rd edn. McGraw-Hill, New York. Hungarian edition: Perrow C (1997) Szervezetszociológia (trans: Hercsik A, Kósa K, Szalay I). Osiris, Budapest
Pistaferri L (2001) Superior information, income shocks, and the permanent income hypothesis. Rev Econ Stat 83(3):465–476
Rzońca A, Ciżkowicz P (2005) Non-Keynesian effects of fiscal contraction in new member states. European Central Bank, Frankfurt am Main
Samuelson PA (1958) An exact consumption-loan model of interest with or without the contrivance of money. J Polit Econ 66(6):467–482
Sargent TJ (1986) Rational expectations and inflation, 1st edn. Harper & Row, New York. Hungarian edition: Sargent TJ (2005) Infláció és racionális várakozások (trans: Szabó-Bakos E). Nemzeti Tankönyvkiadó, Budapest
Seater JJ (1993) Ricardian equivalence. J Econ Lit 31(1):142–190
Shaw GK (1984) Rational expectations–an elementary exposition. Harvester, Brighton
Shea J (1995) Union contracts and the life-cycle/permanent-income hypothesis. Am Econ Rev 85(1):186–200
Singh B, Drost H (1971) An alternative econometric approach to the permanent income hypothesis–an international comparison. Rev Econ Stat 53(4):326–334

Singh B, Ullah A (1976) The consumption function–the permanent income versus the habit persistence hypothesis. Rev Econ Stat 58(1):96–103

Solomon R (1964) A note on the full employment budget surplus. Rev Econ Stat 46(1):105–108

Spilimbergo A, Symansky S, Blanchard O, Cottarelli C (2008) Fiscal policy for the crisis. International Monetary Fund, Washington, DC

Stiglitz JE (1988) Economics of the public sector, 2nd edn. W.W. Norton, New York. Hungarian edition: Stiglitz JE (2000) A kormányzati szektor gazdaságtana (Mezei G). KJK–KERSZÖV, Budapest

Tanzi V (1993) Fiscal deficit measurement–basic issues. In: Blejer MI, Cheasty A (eds) How to measure the fiscal deficit. International Monetary Fund, Washington, DC, pp 13–20

Tarafás I (2001) A monetáris politika a nagy válságtól az ingatag pénzpiacokig. Aula, Budapest

Thurston TB (1977) The permanent income hypothesis and monetary influences on consumption. J Money Credit Bank 9(4):586–596

Tobin J (1981) The monetarist counter-revolution today–an appraisal. Econ J 91(361):29–42

Tóth GC (2010) Boosting demand by the state and automatic stabilizers. Public Financ Q 55(1): 51–69

Varian HR (1999) Intermediate microeconomics. W.W. Norton, New York/London. Hungarian edition: Varian HR (2001) Mikroökonómia középfokon–egy modern megközelítés (trans: Bara Z, Temesi J). KJK–KERSZÖV, Budapest

Vigvári A (2002) Közpénzügyek, önkormányzati pénzügyek. KJK–KERSZÖV, Budapest

Vigvári A (2006a) A possible path to implement public finance reforms. Public Financ Q 51(2): 131–151

Vigvári A (2006b) A magyar gazdaság versenyképességét szolgáló közszektorhoz vezető út egy lehetséges forgatókönyve. In: Vigvári A (ed) Vissza az alapokhoz! Tanulmányok a közpénzügyi rendszer reformjáról. Új Mandátum, Budapest, pp 229–247

Weber M (1923) Wirtschaftsgeschichte: Abriss der universalen Sozial- und Wirtschaftsgeschichte. Duncker & Humblot, Munich. English edition: Weber M (1961) General economic history (trans: Knight FH). Collier, New York

Weeks J (1989) A critique of neoclassical macroeconomics. Macmillan, London. Hungarian edition: Weeks J (1998) A neoklasszikus közgazdaságtan kritikája (trans: Szalai Z). Aula, Budapest

Yaari ME (1965) Uncertain lifetime, life insurance, and the theory of the consumer. Rev Econ Stud 32(2):137–150

Chapter 6
Economic Policy Consequences

[...] economists are always in a state of total intellectual disarray and confusion and are too busy arguing with each other to be able to advise politicians who are usually rather ignorant of economics.
(Jonathan Lynn—Antony Jay: Yes Minister—The Diaries of a Cabinet Minister)

Always to be right, always to trample forward, and never to doubt, are not these the great qualities with which dullness takes the lead in the world?
(William Makepeace Thackeray: Vanity Fair)

6.1 Introduction

Up to this point, we have mainly considered what high economic theory formulizes and what has been taught in the past few decades with regard to the efficiency of economic policy. A special but *not* a general or ever-pervasive nihilism of economic policy has stood out by now: economic theory was mainly engaged in defining precisely the scope of fiscal and monetary policy–what they can and what they cannot do. As far as monetary policy is concerned, new classical macroeconomics has moved within the usual compass, endowing monetary policy with only limited potential to exert countercyclical control. With rational expectations plus unforeseeable (i.e. random) macroeconomic shocks and complete price flexibility, striving for a monetary policy taken as a set of countercyclical instruments would not be only ineffective but also *impossible*. Since monetary policy and market participants posses the same set of information, real economic shocks are unforeseen and unpredictable for monetary policy makers as well, so no offsetting-countercyclical measures can be taken in advance. If a shock has already been developed and is actually perceived both by agents and the monetary authority, a corrective-neutralizing-countercyclical response is only a useless attempt, since monetary policy actions do not exert any influence in the current period. As shocks within the new classical framework are white noises, there are no long-living cycles, that is to say, the direction and intensity of the shock in period $t + 1$ cannot be inferred on the basis of the period t shock. Although monetary policy is able to make effects, its actions could only be of random character (*surprising*, as it was designated above). However, it cannot be known previously whether these actions can contribute to the

P. Galbács, *The Theory of New Classical Macroeconomics*, Contributions to Economics, DOI 10.1007/978-3-319-17578-2_6

elimination of a shock or may make the macroeconomic situation even worse later, when the actual effects develop. And since the necessary monetary policy actions cannot be planned in advance because of information deficiencies (i.e. because of the lack of foresight), *surprising* policy actions must be *random* as well. Under these circumstances, monetary policy should dissociate itself from any kind of countercyclical efforts. However, we could have drawn interesting conclusions when some of the assumptions of the base-line model had got set aside. If long-living cycles are in the game, monetary policy is not forced anymore to take random actions: steps that were regarded as surprises in the beginning are still able to produce effects even after they have become known, so monetary policy can be exerted in a countercyclical way now. Of course, this potential is limited: monetary policy is still not an efficient or high-priority device of macroeconomic control (in real terms), but it can obviously contribute to cyclical smoothing.

However strange it may sound, the most gripping theoretical conclusions in new classical macroeconomics were drawn with regard to fiscal policy. For instance, even the Keynesian case of flexibly adjusting consumption can be inferred from PIH specified under REH, assuming the appropriate policy actions. Moreover, Ricardian equivalence revealed that a countercyclical control over aggregate demand is only possible if it is accompanied by a process of fundamental structural reform. But if all the favourable conditions can be assumed, countercyclical fiscal policy promises to be efficient, even with anticipated actions. On the one hand, these capacities are limited, and, on the other hand, this limited potential of economic policy highlights the relation between operative public sector actions and public expectations. Obviously, structural reforms are not aimed at countercyclical stabilization of aggregate demand–however, regaining the possibility of macroeconomic stabilization proves to be a favourable by-product of such reforms.

However, we have seen that structural reforms resulting in qualitative changes tend to be late and to lose actual content. *Through these arguments, new classical macroeconomics hardly restores the possibility and the potential of countercyclical policies built on activist demand management.* Success is possible, but the playing-field of fiscal policy may be very narrow. If market participants entertain negative beliefs about public sector operations, countercyclical efforts may be unfruitful. *Possibilities of activist economic policy have critically dwindled.* Fiscal policy can trigger reactions of aggregate demand that follow the Keynesian pattern, but the success of this action is not pre-guaranteed anymore: the set of necessary conditions for actual effects has considerably broadened compared to the prior (Keynesian) considerations.

However, such an intellectual wilding in economic theory can be viable only temporarily, since economic policy, as well as its purposes and instruments, have been a hot topic of everyday political discourses from times immemorial. *Politics cannot exist without an activist economic policy*–since it is true, for example, that one of the most important performance-indicators of governments and their current opposition is related to economic considerations.[1] Political thinking has included

[1] This is obviously true of the *rational choice theory* (Downs 1957), while other theoretical approaches of voting behaviour (e.g. *partisan identification model*, Campbell et al. 1960; or the

both economic thoughts and considerations about economic policies since the very beginning (cf. e.g. Bayer 1998; or see Plato's Republic; it can be argued that the formation of state for Plato served economic interests as well–cf. Maróth 2002; Plato 2006). The inseparability of politics and economic policy manifests itself for instance, in the fact that the economic sphere definitely outlines the playing-field for political actions (cf. Bihari and Pokol 1998). So, accordingly, the purpose of economic policy[2] is to find a proper setting for this playing-field, to broaden it by pushing its border-lines out. Countercyclical economic policy built on the unconstrained and undisturbed operation of automatic stabilizers is obviously useful in terms of fiscal sustainability, but forces political decision making to play only a minor and highly subordinated role. Properly set stabilizers almost withdraw the possibility of economic policy interventions from the scope of politics. As it was true of some authors that the theory of Keynes had already served the interest of politics and economic policy as well (cf. e.g. Deane 1978), *similarly*, new classical theory could not dare to urge to *question the possibility and potential of activist economic policy actions either*. It is argued here that, after all, *this* is the nihilism of economic policy itself. If economic policy cannot exert essential influences on anything real in an activist way, or it can do so only with disturbances at most, then, on the one hand, economic policy makers should not even strive for it and, on the other hand, the whole issue should be cancelled from the key topics of political discourses and campaigns. So, new classicals were forced to struggle with the problem of laying the foundations and clarifying the transmission mechanism of a different type of economic policy, keeping the probable failure of systematic countercyclical economic policies in mind. That's why we are concentrating now, after the chapters on economic theory, mainly on operative considerations, scrutinizing how new classical thoughts have appeared in this sphere. Since the scope of economic policy of Keynes got considerably curtailed by new classical theory, now we are interested in, first of all, what strategies Lucas and his followers suggested in order to replace the economic policy built on activist, i.e. demand-side control. Vulgarized Keynesian theory got limited and simplified to justify countercyclical-related budget deficits–however, Keynes himself argued only for expansionary but not deficit-oriented fiscal policies. For him, it is not the *amount of* fiscal balance but the *change in* it that matters. Expansion can be carried out by spending out of the surplus as well. If we study the relation between demand-

sociological model of voting behaviour, Harrop and Miller 1987) do not regard this relation as of primary importance (Körösényi et al. 2005).

[2] After all, macroeconomic theories discuss relations between (macro-)variables. On this ground, economic policy *for us* is the efforts of exerting influences on a given macroeconomic system with regard to some desired purposes. In other words, a macroeconomic system is stimulated in order that a required level of inflation, employment, exchanges rates, etc. could evolve. Of course, the concept of economic policy can be interpreted in a wider sense (e.g. Veress 2007), when the definition process of goals and targets, or establishing the social acceptance of an economic policy program are all regarded as responsibilities of the economic policy. However, here, it is rather a wider social policy task, since a broader definition and interpretation of economic policy may distract public and political attention from the mechanisms of macroeconomic systems.

enhancement and the dynamics of budget balance, it should be noted that only the shifts in the balance count as far as the demand-impacts are considered, while the balance itself is related to sustainability (cf. P. Kiss 1998). However, it is undoubted that the fiscal policy principles distilled from the theory of Keynes did not pay considerable attention to the balance-impacts of expansions and to the resulting effects on government debts. To put it in other words: Keynes concentrated on the possible roles a government can play in demand-enhancement and attributed only marginal importance to the impacts of these countercyclical and welfare efforts on budget balance and government debts. A huge success of new classical macroeconomics was exactly the findings that *demand-enhancement and the dynamics of budget deficit and government debt were not independent any more. Fiscal expansion is not sufficient on its own any more, public sector operations have risen to be relevant determinants of the final effects.* It should be realized that how much more refined and sensitive the new classical framework is compared to the theory of Friedman–or to those oversimplifying views that started burgeoning as well as on new classical soil. *There is no predestined inefficiency, fiscal policy can be preserved to be an efficient countercyclical device under particular circumstances* studied above.

It seems as if new classical macroeconomics gets rehabilitated somehow in Chap. 6 after the previous attacks. This work was primarily aimed at exerting a critique on new classical doctrines so that, at the same time, they could be prevented from a complete destruction. In other words, we tried to separate *grounded* and *groundless* components definitely. What conclusions can be drawn if some of the standard new classical assumptions are assumed away? Are these findings inconsistent with the genuine purposes of new classical theory? And what was this purpose at all? Is it sure that a simple and almost insipid doctrine lay behind the well-ordered and robust mathematical apparatus such as the impotency of fiscal and monetary policy? Hitherto, it has been argued here that surely *not*. According to the present interpretation and our constructive critique, *new classical macroeconomics have made efforts to define and specify the criteria of inefficiency*. However, if these criteria do not prevail, i.e. if they cannot be held as assumptions, since they are set aside or life-world conditions can only approximate them more or less, then radically different conclusions should be drawn concerning fiscal and monetary policy. Activist and discretional economic policies can obtain their objectives only under specific and very tight conditions and only in a specific way. *It is argued that this and not the simple inefficiency-impossibility theorem was the most important component and message of new classical macroeconomics.*

Chapters 2–5 were unambiguously aimed at expressing a constructive critique, following the methodological guidelines established in the introduction. It is worth introducing the things to be done in Chap. 6 through the theory of the Phillips curve. Short- and long-term considerations were equally applied in the Phillips curves given by A.W. Phillips, Friedman and Lucas (or Phelps). Whether there is a vertical Phillips curve at all and, if any, in which time-horizon does it exist–broadly speaking, the whole evolution of the Phillips curve was uniquely determined by these questions. Actually, the reverse relation between prices/wages and unemployment was not doubted long ago–this *short-run* relationship could not have been denied even by new classicals. The only question arising constantly is *whether the mechanism (in other word: the explanation) is right on the basis of which macroeconomic models aim at describing a given phenomenon*. It is exactly the problem to which Chaps. 2–5 are devoted: it was argued that the mechanisms given by Friedman and Lucas are fraught with serious logical inconsistencies and other mistakes at some points. Such mistakes were found mainly in the different interpretations of the Phillips curve, but the revealed incompatibility of long-living real business cycles and the inefficiency of monetary policy emerged as similar inconsistency (it must be admitted that this latter inconsistency is not perplexing at all, but, on the contrary, it is a credit to the theory on grounds of the conclusions drawn while analyzing fiscal policy considerations). Discussion about fiscal policy did not have to take trouble over such logical confusions anymore, so we could more easily pay attention to conclusions that can be drawn by setting some assumptions aside.

The next question is why we can declare in this chapter that monetarism seems to be more easily defendable on the basis of its long-run considerations, in spite of the previously exerted critical remarks. Either orthodox monetarist or new classical long-term considerations are weighed, it can be seen that these findings, eventually, assert not more than that macroeconomic systems tend to move towards equilibrium from time to time in the long-run. It is such a clear and genial thought that it cannot even be debated–and it would be a serious mistake to phrase, for instance, that Keyes also would have opposed this doctrine if he had had the opportunity. If we reconsider and perhaps vulgarize all the things new classicals thought about long-run macroeconomic mechanisms and processes, we can get to simple and clear sentences. *Macrosystems tend towards stabilization after and in spite of short-run fluctuations, since it is in the common interest of market agents*. Or: *macroeconomic systems cannot be kept influenced through economic policy on the long-run, since agents adjust themselves to the environment which is determined by these policy actions among others, so policy acts lose their effects sooner or later* (by the way, this is the plainest phrasing of the Lucas critique). Or: *as far as the problem of economic growth is considered, supply-side stimuli may also play a special and important role*. However, it should not be forgotten that we face a twofold problem again: *what* and *how* new classicals teach. To put it in other words: we have to scrutinize separately the things new classicals thought about short- and long-run mechanisms. Long-run lessons as to supply-side economics appear as the consequences of short-run theories. Of course, not as logical consequences, in a sense as scientific theorems follow from basic assumptions. It is argued here that long-run messages were needed in new classical macroeconomics in order that the potential of economic policy could be restored after it was seriously maltreated by some short-run concerns. There is no direct and inferential logical connection between the doctrines of supply-side economics and the short-run countercyclical inefficiency-theorem. The potential of countercyclical policy is limited in the short-run and real economy tends towards equilibrium in the long-run, so it seems to be reasonable for new classicals to put the emphasis on supply-side stimulations, suggesting some quantitative improvements in this equilibrium state. In other words, a rise in the level of potential GDP is advocated. *However, it should be realized that supply-side economics holds independently of new classical cycle-theory*. Moreover, even those Keynesians are ready to accept the long-run potential of economic policy who, otherwise, are still entertaining an undiminished faith in short-run countercyclical policies and who do not entrench the possibilities of such policies with various limiting factors as new classicals did. If we denied the merits of supply-side economics, it would implicitly attack and undermine even the theory of Keynes, since this case is nothing more than the admittance of a theory proved to be successful in operative economic policy as a guidance for operative economic policy. Before it could arise at all, *we must refute the accusation in advance that charges us to filch the F-twist, criticized consistently and vehemently hitherto, back stealthy in the reasoning through admitting the merits of supply-side economics*. Because of the logical independency mentioned above, we should not think of supply side economics as to be accepted in a Friedmanian way: it is out of question that supply-side economics works well in practical and empirical terms and, as far as these merits are considered, the set of its theoretical assumptions concerning short-run mechanisms is only of secondary importance, as Friedman would argue. However, supply-side doctrines can be regarded as fundamentally right without restoring the short-run theory of new classicals in a Friedmanian style, i.e. disregarding the logical order or how correct the presumptions are and paying attention to empirical validity only. We are not forced to discharge new classical macroeconomics on its apparent failures. So we cannot talk about an implicit and comprehensive rehabilitation of new classical theories–it is argued only that successes and merits of new classicals should not be disputed. "Give the emperor what belongs to the emperor, and give God what belongs to God" (Matthew 22:21).

So now, we turn our attention to the long-run lessons of the theory. Stressing the importance of long-run effects in new classical macroeconomics has a twofold root. Both monetary and fiscal policy have only limited potential to exert countercyclical effects–though for different reasons. If the success of these policies is uncertain, it may be worth finding an area where economic policy interventions can be expected to achieve the desired outcomes. This area is supply-side economics.

6.2 Supply-Side Economics

The answer required by both economic policy and politics was the emergence of supply-side economics. We can refer to the fact even without tiresome wanderings taken in the history of economic thought that the followers of supply-side economics tried to separate themselves from the (alleged) fallacies of the Keynesian systematic countercyclical economic policy and ideology. By the 1970s, economists started focussing increasingly on the factors through which the potential level of aggregate supply can be raised, while the idea of the stimulation of the aggregate demand was left to fall into oblivion. So, as a result, the questions of capital accumulation, the relationship between capital accumulation and taxation (especially the considerations on how taxation can influence capital accumulation– cf. Lucas 1990), technological advancement, individual incentives or improving labour force could get in the lime-light (Feldstein 1986) [→ **growth theory**].

Actually, through curtailing the systematic potential of monetary and fiscal policy in affecting real variables (real output/income and unemployment), new classical macroeconomics and REH suggested that we should allow these variables to take their natural values that, *generally*, promises to be the best *macroeconomic* policy. So, the purpose of macroeconomic policy should be *mainly* to minimise the shocks that (can) impede adjustment processes and that come from variables otherwise subjected to the control of this macroeconomic policy.[3] This suggestion is usually identified by the constant growth rate rule that was advocated even by Lucas as well. And if economic policy, if any, is capable of exerting a systematic control over real magnitudes only under very tight conditions, it would rather focus on the control (the stabilization) of nominal variables, regarding e.g. disinflation as its ultimate goal (cf. Shaw 1984). Moreover, according to some interpretations, only a monetary policy being capable of breaking inflation down and establishing price stability is the best way of stimulating aggregate demand, since erratic price fluctuations may lead to serious disturbances in the real system (King 2005). Controlling real economy may be predestined to fail by two reasons: though some deliberately generated disturbances can deviate real system from the natural state, but the re-arrangement process will be painful; and, moreover, the instruments available for the authorities bear a (hardly stable) time-relation with macroeconomic variables so countercyclical interventions may even exaggerate the prevailing disturbances (cf. Kaldor 1970).

The limited potential of macroeconomic policies was argued explicitly above. While fiscal and monetary policy make efforts to stabilize some nominal magnitudes,[4] it may be possible through successful and efficient *microeconomic* policies *not* to divert a macroeconomic system from the natural level of output and employment, but to extend these natural levels in a positive direction, *in the long-run*. [→ **growth theory**] The differences in the effects of these two mechanisms under idealized conditions can be summarized schematically in Fig. 6.1.

With a rigid aggregate supply, a rise in aggregate consumption from AD1 to AD2 triggers a shift only in the price level ($A \rightarrow B$), real income/output and employment do not change even in the short-run. Of course, all this may be realised only if there are no long-lasting cycles, i.e. if aggregate supply is inflexible and does not even adjust to the policy-induced increase in aggregate demand. We have seen that if long-living cycles are assumed, if expectations are not fully rational or if

[3] When we use the words "generally" or "mainly", we want to highlight again that the countercyclical efforts of economic policies are not predestined by new classical theory to fail–though the necessary conditions for a macroeconomic success usually make it hard to follow an efficient and successful policy.

[4] The stimulation of aggregate demand was theorized to have the capacities only under tight conditions to take the effects envisioned by Keynes.

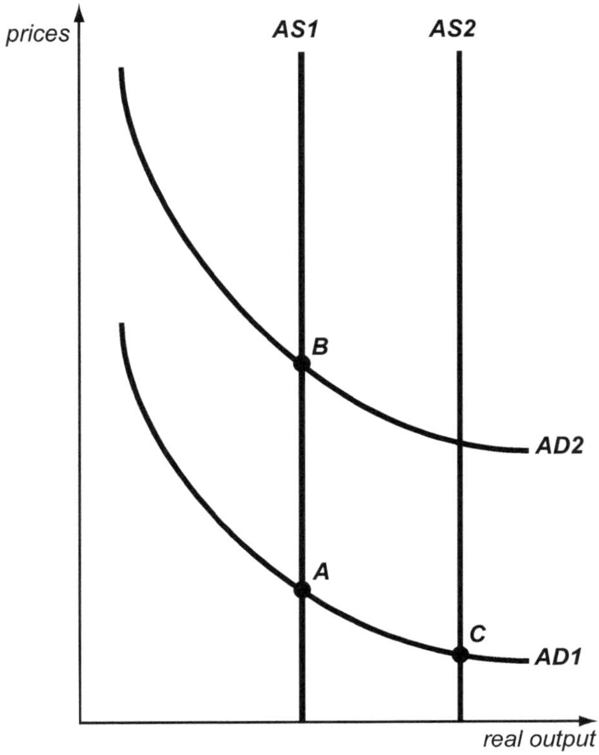

Fig. 6.1 A slight comparison of the effects from macro- and microeconomic policies (Shaw 1984)

there are surprising structural reforms (or a mixture of them) in the game, the big picture may radically change.

If the dynamics of aggregate supply is free from cyclical fluctuations, hence manipulating the aggregate demand is completely useless, ineffective and expensive, it is still possible through microeconomic policies to relocate aggregate supply from AS1 to AS2, ending up in a higher real output and a lower price level ($A \rightarrow C$). Putting aside the circumstance that this theory in its simplicity could easily be attacked by highlighting the grotesque effects of the quantity theory on prices,[5] we will pay our attention only to the intervention mechanisms and effects of macro- and microeconomic policies.

Controlling unemployment compensations or minimum wages *may* be capable of triggering these favourable supply-side effects–while setting them into inappropriate levels may induce negative-unfavourable changes in the natural level of unemployment. Suggesting reductions in the unemployment benefits as a means to enhance working intensity has been a recurrent topic in the theory of social and economic theory (cf. e.g. Erhard 1957). Recently, even arguments for the welfare

[5] I.e. a higher level of real output leads to a drop in the price level.

state got some flavour of antipathy stressing that unemployment compensations hardly contribute to finding new jobs, though these payments can really help the unemployed going through hard times[6] (Barr 2004). In order to eliminate these adverse effects, activation policies have emerged so that they can give support to people seeking for and accepting new job opportunities rather than subsidize the time of being unemployed. One of the most important techniques on which these policies were built is corrective adjustments in the tax system, creating the system of employment/working tax credits (such schemes are successfully applied both in the United Kingdom and in the United States). We may even say that negative effects of unemployment benefits can be eliminated effectively through setting these payments into a low level and, at the same time, encouraging job-seeking activities. Of course, it should be mentioned as well that the micro-foundations are far from unequivocally approving employment-enhancement based on tax incentives. In this context, it is quite enough to recall the analysis carried out in standard microeconomics on the backward bending labour supply curve, according to which an increase in the wage-unit, beyond a certain level, does not incite additional employment (Varian 1999). However, it is also true that these microeconomic investigations are quite sceptic regarding minimum-wage controls so the lack of such controls promises to be more beneficial than a robust system of such measures: setting the minimum wage above the equilibrium level may trigger adverse employment effects, while defining a lower level is a completely irrelevant circumstance– cf. e.g. Kopányi (2003). So, it is not surprising that supply-side economics rather urged avoid the harmful consequences of missetting.

Growth Theory Macroeconomics takes models and theories of cyclical fluctuations and of economic growth as separate problems (Weeks 1989).[7] Mainstream economics rooted in the neoclassical theory has its special growth conceptions and models that have determined the framework of thinking of real economic growth for a long time.

[6] In accordance with these considerations, high levels of unemployment tends to be increasingly attributed to the welfare state (unemployment compensations and minimum-wage control) and not to an insufficient level of aggregate demand. According to these findings, these welfare policies are able to hold labour supply back.

[7] This discrepancy also appears in the topic of courses in elementary macroeconomics. Keeping the capital input fixed in traditional two-factor production functions is a related circumstance. In this case, only the fluctuations in output are relevant (if the model framework we apply allows such fluctuations–if not, it is exactly the stability of production that makes the main message of deductions and demonstrations). Accordingly, growth models are those that allow stock of productive capital to change, since if the real system settles in the stage of its own general equilibrium (represented by the equilibrium of the labour market in the basic models) and if there is no increase in population (that could trigger utilized equilibrium labour input to increase) and technological change, then output cannot change between periods, i.e. economic growth is precluded. Under such circumstances, the issue (and purpose) of economic growth is boiled down to the problem of capital stock dynamics.

After all, neoclassical growth models regarded economic growth as an equilibrium concept. Accordingly, each macroeconomic system has a long-term equilibrium growth path influenced by various factors that can be identified, eventually, with its growth potential. In these models, the most important underlying factor as a determinant is technological change that is taken as exogenously given.[8] In addition, per capita investments, triggering temporary growth processes, can also enhance real output, but without resulting in a persistently lasting additional growth.

Neoclassical growth models considerably contributed to the explanation of the regional differences in economic growth. If a region, e.g. a national economy or a macro-region within an integration, accumulates higher per-capita capital stock or is capable of a faster technological change, then its economic growth rate will also be higher. However, it is an important circumstance that regions should be regarded as open macrosystems, so a significant allocation mechanism built in neoclassical theory contributes to the equalization of regional incomes. Within the realm of neoclassical economics, production factors are reallocated among the regions in accordance with the differences in factor incomes, so factor mobility has to be taken into account. The regions in which capital-to-labour ratio is low produce higher levels of capital income, while wages are minor. This condition appeals to capital, while labour force tends to wander away. As a result, capital-to-labour ratio increases and labour incomes also start rising. On the contrary, regions with high capital-to-labour ratio and, hence, higher levels of labour incomes lose from their stocks of capital, but become attractive to the employees. So, neoclassical growth model postulates an automatic drop in (and, after all, a complete elimination of) income differences. Actually, the concept of *convergence* roots somewhere here, in the neoclassical model of economic growth. Investigations inspired by neoclassical growth theory were aimed at testing the theoretical prediction that lower capital-to-labour ratios tend to be accompanied by higher levels of economic growth. If the theory is right, we can expect regional income differences to lessen without any government or supra-national interventions (Tondl 2001).

Of course, the extent to which tax-cuts can be taken as the *only* (meaning: *exclusive*) instrument of supply-side economics is dubious. Keynesian economics also had a penchant for tax-cuts. For Keynesians, it was a device of exerting control over aggregate demand (cf. Kalecki 1935), since boosts in aggregate demand can be carried out not only through government purchases. Of course, it is not worth stressing traditional instruments of economic policy to be compatible with both theories. In this context, tax-cut is a special case, since e.g. the various forms of unemployment compensations were reckoned by new classicals as definitely harmful, while, taken as an automatic stabilizer by believers of "conventional" economic

[8] Stressing exogenity, i.e. the independence of technological change of economic conditions implies this technological change to be similarly independent of economic policy (as well). Perhaps it was the characteristic for which neoclassical growth models were criticized the most (cf. Erdős 2003). This conclusion, on second thoughts, seems to be not surprising at all: as neoclassical economics refuted the possibility and potential of economic policy in other contexts as well, a higher degree of tractability regarding technological change would have been very strange.

polices striving for demand control, this set is an efficient instrument.[9] Moreover, it should not be forgotten either that some explanations of non-Keynesian effects argue the credible and permanent changes in public sector operations to exert influences through the supply-side (Rzońca and Ciżkowicz 2005; Aarle et al. 2003). Thanks to these changes, enterprises' costs fall, resulting in a rise in business profitability and in improvements in competitiveness on international markets (Neicheva 2007).

However, in Chap. 5 we spent a long time discussing the Ricardian equivalence related to the questions of the impotency of fiscal policy. There, we could see that the theory can hold only under very tight conditions. Putting it in other words, we can say: new classical macroeconomics separated *efficiency* and *inefficiency* (plus their cases and conditions) of fiscal policy precisely (and, to a lesser extent, of monetary policy) regarding the control of aggregate demand.[10] As a result, we got a list of special conditions that may end up in this inefficiency. In this context, first of all we should not refer to the theses by which the significance of public sector operations got highlighted above. Analyzing the equivalence-theorem, we have already mentioned the results with regard to the case of altruism, differentiating rates of interest or the complications originating from lifespan that stressed the limited life-world functionality of the equivalence. So, we are justified to say: *much less is needed in order that the conventional mechanisms of fiscal policy could be in operation* (as Keynes assumed and elaborated it) *than that its potential would be infinitely curtailed.*

One may get deeply confused if required to evaluate the theories and the results of new classical macroeconomics. Most of the difficulties were raised by PIH specified under REH and the Barro-Ricardian equivalence theorem. We have seen that equivalence can hold in its pure form only under very tight and special conditions–so, under life-world conditions, we had better expect the theorem to get violated. It was argued above that new classicals themselves also called attention to this fact, which appeared mainly as the failure of the demonstration of the inefficiency. The importance of structural reforms adds to the complexity. Although the perfect manifestation of the equivalence cannot be expected under life-world

[9] In 1940s, even Friedman himself did not regard these stabilizers as evidently wrong–cf. Vigvári 2008b; as we have seen, even new classical theory did not preclude the efficiency of stabilizers.

[10] There is a wide consensus on the lack of a *long-run* ability of monetary policy in stimulating real economy (Horváth 2006). Keynes himself was also sceptical about the efficiency of monetary policy, since he attributed significance *not* to variables in the emergence of recessions that are under monetary policy control. Moreover, for him, the case of the liquidity trap limited the potential of monetary authority further. In the monetarist context, central banks can cause only disturbances at most if they desire more than just assisting, even if we can find model frameworks in which money is non-neutral. However, *we should not infer the short-run inefficiency in output control from the case of the long-run inefficiency*. By the way, intervention mechanisms of inflation targeting regimes are built on this short- and medium-term efficiency of monetary policy. So, it is argued here that new classical macroeconomics, through persuasively constraining the potential of fiscal policy, placed the attitude and instruments, problems and responses of economic policy on radically new foundations.

conditions, unsustainable processes and operations performed by the government sector make a factor that can *still* narrow the playing-field of stabilizing-countercyclical fiscal policy, even if the demand control ability of fiscal policy does not get fully eliminated because of the lack of conditions being necessary for the perfect equivalence. *It has to be highlighted that the results and findings of new classical macroeconomics were aimed only partly at the demonstration of the inefficiency of economic policy*. The real purpose was rather highlighting a *conditional efficiency*, since Lucas and his followers consistently stressed the tight conditions on which new classical models were built. In other words: the demonstration of the *conditionality* of the equivalence theorem was definitely sought after by new classicals. But anyway, in these terms, they were not too successful, since numerous critical remarks called attention to *further* factors that limit perfect equivalence. However, new classical theory deliberately underlined the qualitative parameters of government sector and their importance. *The genuine and paramount purpose was to highlight: equivalence holds if and only if government is not capable of changing its own processes and operations.*[11] It was the demonstration of the conditional character of equivalence. So, though the effort was extant, new classical macroeconomics could not persuasively take the wind out of the sails of the believers of systematic countercyclical fiscal policy. Notwithstanding, new classical theory and its burgeoning reconsiderations suggest that albeit some potential of countercyclical fiscal policy should be assumed, it can hardly be regarded as viable and probable because of the countless limiting conditions (and assumptions) and the recurrent narrowing of the fiscal playing-field. So, new classicals recommended instruments to operative economic policy that were expected to have effects mainly *in the long-run*.[12] The success of new classical economic policy is considerable in these terms as well, since, following this line of reasoning, it could lead to a field which was not covered by Keynes and the Keynesians whose main focus was on short-run processes and effects.[13]

6.3 Economic Policy and Credibility

We have seen that fiscal policy plays a special role in the theory of new classical macroeconomics. The assumption of Ricardian equivalence makes this sphere of economic policy ineffective, though orthodox monetarism has already looked at

[11] Critiques supplemented this theorem by calling attention to further limitations of perfect equivalence.

[12] By stressing long-run impacts, new classical macroeconomics was highly benevolent to political elites. If we admit that economic policy performance of different political forces is also a significant factor in the circulation of elites (Pareto), denying the short-run potential notably reduces the accountability of elites.

[13] It is typical that the history of economic thought attributes the considerations of some long-run consequences uncovered by Keynes' theory to Joan Robinson, among others.

fiscal policy with a critical eye.[14] Milton Friedman consistently refuted that fiscal policy instruments could produce effects on the level of real income or output (cf. Tobin 1972). Friedman (1948) suggested the monetization of budget deficit, therefore the only effect of deficit is additional inflation at most. Compared to the original Friedmanian conception, James Tobin (1981) gave a more elaborated explanation to this issue when evaluating the theory of monetarism, recognizing that the inflationary consequences of deficits are not evident or inevitable and the governments of countries with well-developed financial markets are not forced to finance deficits by money issuance, even if Friedman himself urged it.[15]

New classicals overwent orthodoxy: by introducing the Ricardian equivalence, they denied not only the possible effects on anything real but all the inflationary consequences were assumed away, at least in that case when a government can complete the debt service without being forced to monetize. Sargent and Wallace (1981), citing results from their model analyses, state that the actions of a fiscal authority financing deficits through new bond debts should be expected to end up in inflation in the end. It could be achieved by an assumption according to which the stock of bonds the public holds is stuck to a constant level after a given time, and monetary policy being subjected to this requirement decides on the growth of the monetary base that serves as the money in circulation. Since budget deficit can be financed through either increasing the money supply or new bonds, a more tight current monetary policy must be accompanied by a more intense bond issuance if the fiscal authority fixes the level of budget deficit for each period. This story can already be interpreted as there is an upper threshold for the stock of bonds that cannot be enhanced further, so government cannot be unwilling to increase money supply anymore in order to preserve the balance (Sargent 1986). Thus, inflation is inevitable and even non-deferrable, since if a money demand function is assumed according to which the demand for money is dependent on expected inflation as well, price level and inflation depend both on the current level and the expected path of the monetary base. So, after all, the currently tight monetary conditions can lead to a higher level of current inflation (cf. Sargent and Wallace 1981).

So, the final and overall effect of a deficit is dependent both on the fiscal-financial strategy and, in relation to it, on the commitment and credibility of economic policy as well. The most important lesson of the Sargent-Wallace model mentioned in the previous paragraph was that monetary policy should be adjusted to fiscal policy, if it is the fiscal authority that moves first in the game. As it was mentioned previously, the prohibition of financing the budget deficit through newly issued money has advanced to be a common political and economic policy norm by now–and this development makes the analysis even more gripping. If

[14] We have already mentioned in Chap. 5 that debt-for-tax swap is inefficient only under very tight conditions. A change in the level or composition of government expenditures undeniably affects real economy.

[15] This extraordinary suggestion can be well explained by Friedman's view that this procedure can completely make budget surpluses and deficits appear as the consequences of real economic performance.

Ricardian equivalence works, neither any inflation nor (temporary) real effects should be taken into consideration. However, it requires the public to link the current tax-cut with (the necessity of) a subsequent tax increase. It could be an interesting and special situation when this relation is missing or is eroded–and the case is often mentioned in which the public evaluates the initial tax-cut as a rise in permanent income, so demand impacts have to be expected.

We can find a situation in economic history that is a good example of an economic policy the fiscal elements of which are inconsistent with each other (in other words: there is no credibility) and, simultaneously, there is another contradiction lying between the paths of this fiscal and the monetary policy. U.S. economic policy of the 1980s under the Reagan administration contained several incongruities of this kind–and the effects and results of this economic policy (the so-called *Reaganomics*) generated huge and dividing debates even in the mid 1990s.

The fundamental characteristic of the situation is that the fiscal policy guidelines announced by the government could not be followed simultaneously. According to the diagnosis given by Reaganomics, the roots of the problems could be found in the government expenditures having got out of the previous tight control while, at the same time, tax burden reached the limits of the society's tolerance. The consequence was increasing budget deficit and, moreover, President Reagan also placed the rising inflation to this account. Cutting back government expenditures was expected to settle the problem, but, over and above these plans, there were some recommendations for cuts in personal income tax, accompanied by various measures to stimulate investments. It was the operative foundation of the new tax act voted in August 1981.[16] Additionally, these policies were complemented by rigorous monetary measures[17] (ERoP 1982). In the end, inflation could be restrained with serious social costs: unemployment increased considerably. However, this success should have been attributed rather to monetary policy,[18] since there were no advances in cutting off government expenditures (there was a long-standing and considerable budget deficit under the reign of Reagan administration). Although non-defence, non-social-security spending (e.g. on education) were

[16] Economic Recovery Tax Act (ERTA), or, in other words, the Kemp-Roth Act. According to the plans, this act would have reduced budget revenues by $787 billion between 1982 and 1988; from which individual tax cut would have amounted to more than $500 billion, and corporate income tax cut to $200 billion. These numbers were modified even in 1982: TEFRA (Tax Equity and Fiscal Responsibility Act of 1982) reduced these cuts by $244 billion. Accordingly, actual total tax-cuts dropped by $72 billion for individuals and by $131 billion for corporations (Boskin 1988a).

[17] So, the program did not form a coherent view on the best option to break inflation down. Cutting expenditures back and the restrictive monetary policy were both expected to serve this purpose.

[18] The official policy of FED between October 1979 and October 1982 was aimed at targeting aggregate M1 (FRBoSF 2003), however, the actual policy was carried out rather as nominal GDP targeting. The way of breaking inflation down was the cooling of the growth in nominal GDP, and this purpose was achieved according to the schedule. At the same time, monetary targets were systematically missed (Modigliani 1988).

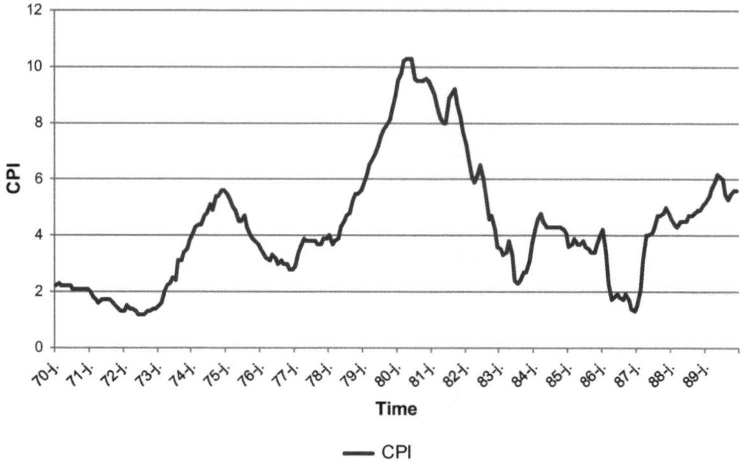

Fig. 6.2 Seasonally adjusted monthly CPI inflation, change to the same period of the previous year, United States of America, January 1970–December 1989 (*Source* St. Louis FED)

reduced (Kamlet et al. 1988) and these cuts could largely offset the drops in budget revenues, defence spending increased to a serious extent. So, the *actual* economic policy was only *partly* compatible with the suggestions of supply-side economics emerging recently: to stimulate the economy through tax-cuts and to ease budget pressures through cutting off non-defence spending (Fig. 6.2).

Budget deficit turned out to be long-living. This tendency was strengthened further by the circumstance that the society did not cry out against this situation. Major macroeconomic indices were found correct, inflation and unemployment were dropping–though, as far as the latter is concerned, the success was only modest. So, for the public, a fiscal intervention seemed to be completely unnecessary (Fig. 6.3).

The satisfying solution was deferred until the middle and the second half of the 1980s. By the early 1980s, the government had misjudged its own playing-field, and policy makers expected non-defence spending to promise further possibilities of contractions. The Reagan administration also had a mistaken belief that tax-cut-induced economic growth could compensate for the lost tax revenues–however, this optimism was completely ungrounded (Feldstein 1994). A tax increase was still precluded–actually, President Reagan did his best to prevent the taxes-to-GDP ratio from rising (Taylor 1995). In October 1986, further significant tax changes were announced,[19] but some slight increases had happened previously both in 1983[20] and in 1984.[21] Since, in political terms, any measures, due to which any social segment would have been considerably worse-off, would have been unacceptable, corporate

[19] Tax Reform Act of 1986.

[20] Social Security Amendments of 1983.

[21] Deficit Reduction Act of 1984.

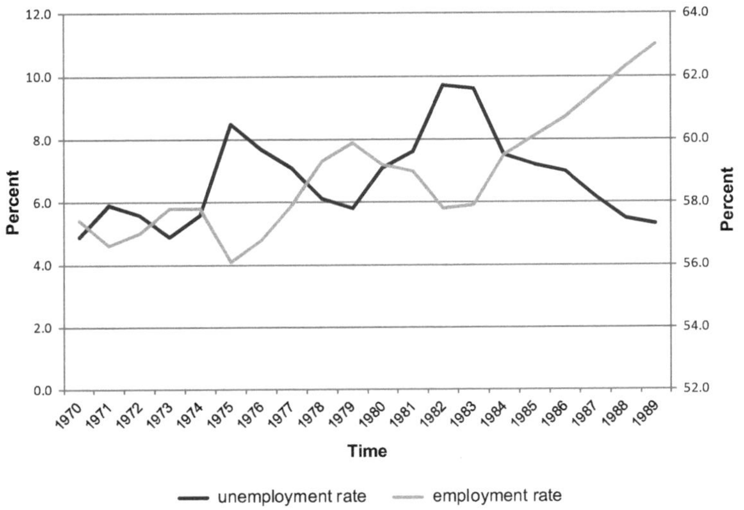

Fig. 6.3 Unemployment as the percentage of labour force (*left axis*), employees as the percentage of population (*right axis*), United States of America, 1970–1989 (*Source* U.S. Department of Labor, Bureau of Labor Statistics)

income taxes were increased in 1986, ending up in the ceasing of a number of incentives. Simultaneously, monetization of government debt was not carried out and, moreover, the growth rate of M1 was rather diminishing during the second half of the 1980s (Fig. 6.4).

After the Reagan administration had passed its time, additional tax increases came in 1990 and in 1993 as well (Taylor 1995) that left marks also on the dynamics of real GDP (Fig. 6.5).

This situation can well be interpreted on grounds of new classical concepts and theses. The policy of tax-cuts was inconsistent with the rising federal deficit, therefore, considering the lack of any intention to monetize, the public might have expected the story to end up in a tax increase finally. Of course, all this eroded the predictions of supply-side economics, and real economic growth would be (would have been) much more moderated than expected. The outcomes were strengthened further by the disinflationary policy of FED which could hardly be regarded as beneficial for the expansion of employment and investment. On these grounds, we could expect Ricardian equivalence to hold and the *failed* boost of real economy to create an opportunity to establish the balance through increasing the tax burden. The only problem was that *fiscal policy had been able to resist budget pressures for long and, hence, to defer successfully the necessary tax increases.* Thus, the public tended to regard the tax-cut as their own additional disposable income, so consumption expenditures could rise. Market participants had no reason to expect any drops in disposable income to occur in the near future, so household consumption rather followed a Keynesian path (cf. Boskin 1988b). From this phase on, the macroeconomic situation could have been settled only at the expense of the

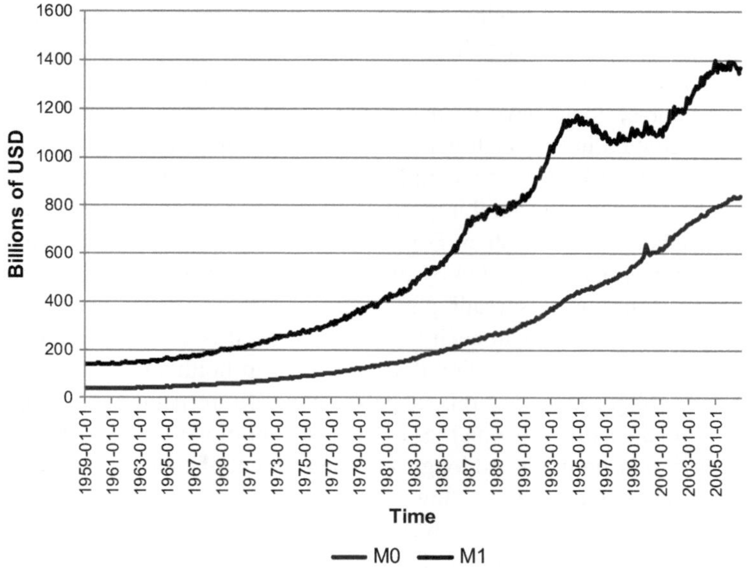

Fig. 6.4 Monthly data of aggregates M0 and M1, 1959–2006, United States of America, billions of 2000 USD (seasonally non-adjusted) (*Source* St. Louis FED)

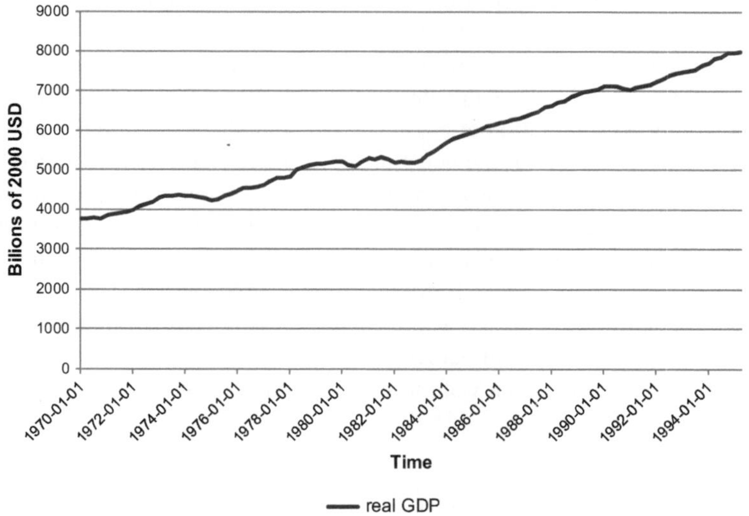

Fig. 6.5 Quarterly data of real GDP, 1970–1995, United States of America, billions of 2000 USD (seasonally adjusted) (*Source* St. Louis FED)

society.[22] A partial or full monetization of government debt would have cut a threatening inflationary process loose and a tax increase would have taken negative effects on real economy. There was no painless therapy to choose. Economic policy, of course, could have been feasible, if the government had been able to stop debt accumulation through cuts in defence spending–since a serious government debt could hardly have been built up without the star wars ambitions.

It may be a lesson of general importance that monetary and fiscal policy should be in accordance with each other. For instance, central bank credibility in inflation targeting regimes may be eroded by expansionary fiscal policies generating inflationary pressures. A wide-spreading government sector and low levels of inflation can hardly be reconciled with each other. However, it is also important to realize that the impotency of fiscal policy does not seem to hold: shifts in revenues and expenditures trigger modifications in public expenditure pattern–however, as we have seen, the final structure emerges as the resultant of various impacts. Short-run stimulation of consumption and long-run enhancement of capital accumulation can harmonize well with each other. However, credibility constitutes a serious limiting factor for these policies. If the government comes forward with an inconsistent fiscal policy, i.e. the enhancement of consumption and capital accumulation promises to be only temporary, the ability to exert a control over real variables may be easily a double-edged weapon in the hand of the economic government: as allowances encourage macroeconomies to grow, contractions hold them back. The prohibition of monetizing government debt does not end up automatically in the Ricardian equivalence. A fiscal authority rolling the repayments over (hence, behaving erratically and incredibly) may trigger serious fluctuations in real economic performance. As the equivalence theorem when violated offers room for fiscal policy in the stimulation of aggregate demand, these possibilities may easily become sources of danger, if the boost in aggregate demand may be traced back to the lack of economic policy credibility. The underlying mechanisms of non-Keynesian effects have to be highlighted here again. As it was stressed above (in Chap. 5), quantitative corrections lacking structural reforms cannot modify the demand of agents forming rational expectations, since the public, under these conditions, knows that the expansion is only of temporary character. Of course, impacts change without REH. Government follows a mistaken (damaging) strategy if relies on the lack of rationality: though it can succeed in stimulating

[22] Boskin (1988a) also takes the circumstance into account that the intensified willingness to spend in a highly open economy runs up the rates of interest, that strengthens the currency and, through this development, negatively affects export. This latter can offset the positive effects of household consumption. So, the composition of national income changes, the emphasis relocates from interest rate dependent activities to household consumption, undermining the plausibility of scenarios supply-side economics outlined. It does not have to be mentioned to what extent economic growth and development process built on home consumption is fragile and unstable. The big picture can be made more complicated and sophisticated by taking into consideration the fact that household consumption may be highly dependent on the available credits and the rates of interest.

aggregate demand in the short-run, but since we *cannot* talk about a structural reform carried out but only about a simple quantitative correction intended to enhance consumption, budget pressures remain extant and the resource gap generated by the initial tax-cut has to be offset by a subsequent tax increase. In this case, due to government intervention, real economy suffers serious disturbances, the seriousness of which gets further intensified by the success of the initial tax-cut in demand-enhancement. The initial boost in aggregate demand is later followed by a decline, due to the tax-cut. Now, there is nothing more than generation of destructing fluctuations. So, we are here again: *fiscal demand-enhancement is not predestined to be inefficient, only the criteria have become tighter and more rigorous*. Without credibility, without improvements in the quality of government sector processes and mechanisms, demand-side impacts will either fail (if REH works) or a boost will be followed by a decline (without REH), so the outcome of a fiscal demand-enhancing action is a trigger of real economic fluctuations.

6.4 Symbolic Economic Policies[23]

This section offers a few remarks on the so-called heterodoxy commentaries of recent times (e.g. Bod 2013; Csaba 2011). It will be highlighted that, in accordance with the growing popularity of unusual economic policy actions, a set of "tools" is emerging that aims to exert its effects breaking with usual, common, instrumental actions. Outlining a framework of the history of mainstream economics and treating fiscal and monetary policy separately, it will be argued that economic policy only gradually has become capable of applying this system. The point is that though the economic effects of expectations (on output, unemployment, business cycles, etc.) were highlighted in the theory of Keynes (and even in those of Marx and Mises), the models vulgarizing the theory of Keynes completely have forgotten about it. Moreover, some modern theories have regarded public beliefs (in other words, expectations) as a power that deprives economic policy of its instruments, causing further delays in making deliberate efforts by economic policy to influence these public beliefs. In our view, both the emergence of symbolic economic policies mentioned above and the rise of heterodoxy are on the same level, since certain governments, not having much leeway, can only operate through giving signals to markets. Although it is not the time to formulate ultimate and eternal generalised statements, it may perhaps be stated that symbolic economic policies, as part of the intersection of *heterodox* and *unorthodox* interventions, cut away from instrumental policies, can make some room for manoeuvring available as a last resort for economic policies. Our theoretical framework will be filled with some episodes from the last few years of the economic policy of Hungary. Of course, it is not suggested that the Hungarian case analysed here is the only example of symbolic

[23] This section is the amplified version of the forthcoming paper Galbács (2015).

economic policies. The possibility of symbolic economic policies follows from the theoretical foundations of new classical macroeconomics (as it will be detailed below), however, it cannot be judged at this point how widely this set of tools will be utilized later–the set of tools that is suggested here on theoretical grounds only as a mere option.[24]

6.4.1 Preliminary Remarks

Recent times have witnessed the emergence and evolution of quite a new set of economic policy tools that, now, aims at deliberately breaking with the traditional and common mechanisms of economic policy. Episodes to study and analyse have been offered by the recent economic policies in Hungary, so our generalised remarks will be supported by these observations. Of course, these tools did not appear out of nothing. Somehow, mainstream economic theory has already anticipated the appearance and spread of these *symbolic economic policies*, as we designate them.

In the literature, these unusual economic policies have two labels. One of them, on first appearance is *heterodoxy*, which means diversity and abundance of forms. Of course, it also has a slightly pejorative connotation, since this diversity often refers to the erratic and hectic nature of economic policy. For Bod (2013), the former governor of the Central Bank of Hungary, heterodoxy is a reapplication of old economic policy actions (e.g. protectionist policies–cf. Botos 2009), revival of forgotten transmission mechanisms. Of course, meeting the requirements of explicitly orthodox recommendations can be part of heterodoxy as well, if a government otherwise takes rather confusing and chaotic measures. The point is the deviation from instruments that are *accepted* and *recommended* in a given period (i.e. from the mainstream proposals named orthodoxy), and naturally this deviation covers contradicting and logically confusing interventions as well. To distinguish the special Hungarian *unorthodoxy* from this heterodoxy is not an easy task since these two concepts significantly overlap. In his review, Veress (2013) mentions a

[24] It is somehow similar to the case of the liquidity trap that follows from the theory of Keynes (1936). Economists, though the case of the Attlee-administration seems to be the only uncontroversial instance, doubted that Keynes had discussed a real risk embedded in the routine of economic policy only by virtue of their theoretical mindset (in Chap. 3 it was suggested that monetarist theory completely excludes the possibility of liquidity trap). Although history offered only one example, the case of the liquidity trap for Keynesians was a true and accepted part of the theory. Similar subtle and ambiguous examples can be found in other sciences as well. Michael Polanyi (1964) analyses in detail those consequences that are deduced from the theory of physics and the empirical confirmations of which must be waited for long. As theoretical physics has become more and more sophisticated and complicated, the consequences are increasing in number the empirical tests of which seem to be desperately impossible. Here one may think of the black holes that are smaller than a nucleus while weighing billion tons–these are the direct but unobservable consequences of the accepted theory of physics–cf. Hawking (1993).

number of measures inconsistent with the old and well-known logic of markets (e.g. crisis taxes), but one can find here actions that contradict recommendations by theorists. Here, mention should be made of the Hungarian flat rate income tax, from which, according to theorists, one could not expect higher household consumptions. However, *the main point of unorthodoxy seems to be the explicit punishment of certain market participants* because of their alleged contribution to the development of the crisis. We can hardly draw a lesson from all of this, but it seems that *unorthodoxy is an extreme manifestation* (or form, if you like) *of heterodoxy*. Both of them come into the lime-light when the possibilities of orthodox policies fade away. In the following parts, it will be argued that the possibilities of both heterodox and unorthodox measures (just as those of all kinds of economic policy interventions) are limited, so when economic policy, for one reason or another, reaches the end of its possibilities, a switch to symbolic policies occurs naturally.

6.4.2 Conceptual Framework

It is hardly an exaggeration to claim that the economic theories of the twentieth century contributed to the broadening of our knowledge of economics to the highest degree by recognizing the importance of expectations and public beliefs for economic policy and by mapping their operational mechanism. Beyond the fact that our theoretical knowledge of the functioning of macroeconomic units has grown considerably thanks to this research, new insights reached far into other areas of economic theories. Mention should be made of the basic issues of experimental economics since it investigated, on the one hand, the degree of rationality and the ways it is developing, and, on the other hand, the micro level mechanisms of expectations and public beliefs (cf. Smith et al. 1988; Porter and Smith 2003), thus doubling back hereby to the theoretical framework where hypothesis for rational expectations emerged for the first time (Muth 1961).

The essence of conventional, realist-instrumentalist economic policy is to plan and apply such interventions that generate desired macroeconomic outcomes through tuning (or maintaining the previous level of) some variables (e.g. interest rate, budget deficit, the level and growth rate of national debt, certain government spending programs–the list can be expanded almost infinitely) controlled by the government in an economic environment in which some degree of rational expectations of the actors[25] (i.e. those affected by the economic policy) can be presupposed.[26] Probably, it is not an overstatement to say that mainstream

[25] Here, for the sake of a simpler phrasing, actors will be referred to as "the public".

[26] Expectations have become a very popular concept of economic theory during the last four decades. As it is just one of the widely used concepts, its accurate denotation has paled, has become diluted and somehow diverse–so, it is not an easy task to find a clear and effective definition. It is not a huge mistake to allude to expectations as forward-looking (public) beliefs directed towards future events and developments which serve and support the public's utility

economics has devoted its past 50–60 years to identify each intervention mechanism and, eventually, to formulate economic policy recommendations based on the current macroeconomic knowledge. These models, oversimplifying the abundance of forms of theories, are always about the questions of instruments of economic policy and the effects they produce. Usually, this development is described as a narrowing process of the scope of discretional policies. The ultimate framework for all kinds of economic policies is a combination of inputs and outputs that are interconnected by a theoretical model (i.e. a *supposed* theoretical model–that means this model is not verified in the strict sense, only a non-falsified model, in Popperian sense). Guided by a given economic model a government (at least in better cases) tells policy makers which instruments they should use in order to achieve certain goals of economic policy, and *vice versa*: what effects will be produced by a given intervention (cf. e.g. Dobias 1980; Bod 2003). However, saying that each economic policy intervention and routine has its own theoretical background does not mean that it is the so-called high theory background. This can be some kind of "theoretical", fully heterodox or unorthodox consideration (in this case, of course, one may say, that it is only *feigned* theorizing). A theoretical basis, however, definitely exists since economic policy makers can plan actions only on the basis of a theory (perhaps, of an *ad hoc* theory) abstracted from recurrent events–these ideas connect *ex ante* the actions based on the use of instruments with expected and desired outcomes. For economic governance, this background can be an empirical, generalizing observation that completely lacks theoretical explanations ("if–then"). Suffice it to say that the Friedmanian programme of positive economics (Friedman 1953) favoured these generalizing observations (quasi-rules).

The (relevant) economic theory followed by economic policy and commonly mentioned by orthodoxy always contains behavioural equations (or, in other words, structural equations), on the one hand, that describe the actions and reactions of private sector agents and, on the other hand, the system of rules applied by economic policy–with the latter being, formally, a behavioural equation as well or a bunch of such equations. These systems are built upon a set of endogenous and exogenous variables. The economic policy, in this approach, aims at triggering shifts in the dependent variables of economic policy by regulating the variables which are under its control and are the inputs to private sector agents i.e. tuning them into beneficial levels. This is the way how desired policy outcomes emerge. Actually, all the variables controlled by the economic policy are endogenous indeed (setting aside the extremity of the rule of a constant money growth rate which,

maximizing behaviour. Literature (e.g. Keech 1991), using the term of "economic policy" in our sense, directs us back to the principles of Keynes' theory as a source–that's why it is important to stress that discussing forward-looking beliefs in a macroeconomic theory was a momentous (though not stressed by its commentators) part of the masterwork of Keynes (1936). Of course, these expectations could not be designated as "rational", though he had presupposed a careful judgement-making process on the part of the public when describing short- and long-term expectations that pointed right in this direction.

therefore, is independent of the conjunctural position of the economy), so their numerical levels are determined by and in the interdependency of behavioural equations: policy responds to the behaviour of private sector agents through applying economic policy rules, just as private sector agents respond to economic policy. These two sectors (i.e. private sector and government) are in interdependency (that can be easily turned into the language of mathematics–and has been willingly done by mainstream economics). Of course, these rules of economic policy are not necessarily automatic–the circle is wider, since, e.g. the loose, non-specified and non-formalized behavioural rule can be mentioned here according to which a government starts an expansionary fiscal policy in periods of recession on a discretional (so, on a non-rule) basis.

It may be noticed that hardly anything has been said here about the applied economic policy; both rules and discretionality (or their combination[27]) are options–according to whether an equilibrium or a disequilibrium setting is assumed.[28] For us, the only important aspect is that in orthodoxy these *realist-instrumentalist economic policies* aim at producing real economy effects through tuning real[29] instruments. Using a present tense in this case is proper, since this system of instruments[30] is still used actively, though not exclusively. Taking the risk of some oversimplification, one may say that this routine of economic policy means simply "twiddling" adjusting knobs (although that needs plenty of improvisation skills and artistic intuition) where these adjusting knobs as concrete and touchable setting devices mean a very proper analogy to the instruments used in (realist) economic policy interventions (even though nobody holds a market interest

[27] This combination may be needed when an economy supported by an economic policy based on automatisms and rules cannot leave the consequences of a shock behind–or can, but only during a long adjustment period (see Galbács 2011; Goldfeld 1982).

[28] Of course, this is not only a question of macro-modelling or philosophy of economics–but of economic policy as well that was answered by ideologies, as long as ideologies existed at all.

[29] It is mentioned here only for the sake of good order that the *symbolic–realist* distinction has nothing to do with the common opposite of the terms "real" and "nominal". What is usually called as a nominal change, in our terminology, even that counts as a realist (that is, actual) intervention of economic policy. In this context, changing and controlling the market (that is, the nominal) interest rate proves itself to be a realist intervention as well (phrasing correctly, an economic policy intervention based on an actual change of an instrument)–as opposed to all the cases of symbolic economic policies, as labelled here.

[30] By this phrasing, it is not intended to suggest that this system of instruments is well-defined, well-identified and unchangeable in terms of time. Indeed, high theory after Keynes has been searching, in the first place, for the instruments that can produce effects in real economy. Suffice it to mention all the complications raised by the quantity theory of money, i.e. the (alleged) existence of the neutrality of money and of the homogeneity postulate. It is not an overstatement to say that the most important dividing line among the schools of economic thought existed by the assumptions defining the variables supposed to have the ability to exert influence on real economy. For the sake of good order, it can be added (i.e. the problem of determinants of dividing lines) that schools classify different macroeconomic phenomena as cases needed to be treated by economic policy instruments. It is a cliché in the history of economic thought that inflation has been judged in completely different ways by Keynesians and monetarists.

rate in his or her hands–however, any of us can touch a thermostat). Of course, it should be stressed that, in the case of realist economic policies, mentioning the existence of real instruments, in the strictest sense, is mere tautology. Given the theoretical basis for the conceptual distinction, it is better to talk about realist economic policies as the ones that are built upon the use of *well-known and commonly applied* instruments–as opposed to symbolic economic policies that, actually, *operate no instruments at all*. So, *by a symbolic economic policy action, we mean a quasi-intervention* with a meaning that points far beyond the action itself, in the case of which the deeper sense (the true and proper content) makes the point. The only goal of these actions is to convey the message to the public–while the directly observable action is only the carrier of this message.[31] On the contrary, for a realist-instrumentalist policy everything is what it is. There are no hidden meanings, messages, or content–there is only an expectation that the public will respond in accordance with the aim of the policy action. In this comparison, unlike realist-instrumental policies, a symbolic action (similarly to heterodox and unorthodox interventions) can easily get a pejorative connotation. If an economic policy is symbolic, it can readily be described as "only" symbolic, as only a substitute for instrumental actions by its activities that are empty and tenuous compared to realistic policies. In comparison to symbolic actions, realist-instrumentalist policies, if they can be, for instance, credible (we will return to this problem soon), seems to be to the point and, what is more important, *actual*. The contrast often cited is made between orthodoxy and heterodoxy–unorthodoxy. The majority of these policies, however, can be subsumed under instrumentalist actions since economic policy mainly works in the same (i.e. realist) way, whether it follows mainstream recommendations or not. *So, at this point of the discussion, we can conclude that heterodox and unorthodox policies have common elements which are already outside the pale of realist-instrumentalist interventions: they are labelled here as symbolic policies.*

[31] Using this concept, we follow the interpretation of symbolism conventional in literary studies and arts according to which the aim of symbolism is to describe the esoteric spheres beyond and above phenomena, the perceptible surfaces (cf. e.g. Komlós 1965; Sőtér 1965). This is the manifestation of the same Platonic tradition (Enn. V.9, 5–6; see Plotinus 250 AD) which, via physics, as it was highlighted in Chaps. 1 and 3, influenced profoundly even the methodology of economics. When a physicist says that he or she looks for the ideas of God (Hawking 1993), then, eventually, he or she talks about the intelligible laws being far beyond the phenomena governing the world. Economists do the same, when they seek the ultimate laws of human behaviour lurking beneath the surface.

6.4.3 On the Way Towards Symbolic Policies: The First Attempt

It should be stressed that the theoretical demarcation between realist and symbolic economic policy interventions cannot be based on an economic policy's utilization of expectations and beliefs–although this option is apparently tempting and, moreover, seems to be simple. In this case, one would say that symbolic economic policies are those that exert their influence through the channel of expectations. However, this only appears on the surface. For a long time now, the discussion has been about an active and deliberate utilization of expectations and beliefs on the part of realist-instrumentalist economic policies. Eventually, this means the completion of the transmission mechanisms related to the instruments of economic policy (cf. Horváth 2006). Among the case studies on operative monetary policy there is a classic episode elaborated in which the monetary authority, eventually, can produce the effect of an interest rate cut by maintaining the previous level of the interest rate while the market expects an interest hike (cf. King 2005), and *vice versa*. Although, in this example, the effect emerged via the utilization of expectations, and although the central bank did not modify the value of any of its potential instruments, it is still the case of a realist economic policy that is, naturally, related to an instrument and that reckons with the consequences of expectations.

It is undeniable, as we have seen, that *the boundaries around economic policies have become tighter and the possibilities of macroeconomic regulations have been tightened thanks to the recognition of the economic effects of expectations and beliefs*. One of the most important lessons from theories of expectations is that all realist economic policy interventions also have effects produced through beliefs and expectations, as well as the influences connected directly to instruments, that *must* be considered during the planning and realization phase of economic policy making. The chances and effects of interventions (it must be stressed that, here, we are talking about realist economic policy interventions) mainly depend on public beliefs: a fiscal or monetary expansion is in vain if it (i.e. its effects, especially on the macroeconomic environment, or its sustainability) is judged pessimistically by the markets.[32] Beliefs are relevant circumstances in every case when, in the

[32] Typically, it is a question of credibility (cf. Schelling 1982) examples of which were abundantly offered even by the Hungarian economic policy based on unorthodoxy. Firstly, the public must believe and have confidence in the fact that instruments the government wants to apply are adequate for (or, in other words, proper to) the desired goals (macroeconomic outcomes). On the way to economic policy success, the public acceptance of government's theoretical framework is a necessary but, of course, not sufficient condition. If there is no acceptance, the real problem is that the government starts a large-scale social experiment. In this case, the main question is whether we can talk about extraordinary and novel ideas, about unusual answers and form-breaking prescriptions to well-known problems with the possibility of solution–or about only an economic policy the instrument set of which has been known for a long while but the elements of this set and the transmission mechanisms they work through are deleted from the suggested

environment of economic policy, some kind of forward-looking behaviour (i.e. not necessarily rational expectations, in the strong sense) of the public should be taken into consideration (cf. Blinder 1999). The channel of expectations of monetary policy transmission mechanisms and the theoretical description of central bank credibility, for example, illustrate this secondary (but really and always only secondary?) mechanism.

Credibility is rooted in actions, whether one talks about private sector agents (Sobel 1985) or an (economic) government.[33] Anyway, every economic policy intervention, using the new classical terminology[34] (Shaw 1984; Blinder and Fischer 1979), has a *signal extraction process* as a concomitant phenomenon in the course of which the public (as a receiver) makes a judgement about the soundness and presumable intentions of the government (as a sender) on the basis

---

instruments and ways, i.e. they are among the bad examples. In other words, the conflict is that whether (economic) policy is really new and revolutionary, or forgets (or, at least, tries to forget) about facts that were thoroughly considered (and refused) in the literature on economic theory and economic policy decades ago. In the latter case, economic policy tries to achieve an unachievable state, since it expects the success from instruments and mechanisms proven improper (and, if this is the case, it cannot be credible). By and large, the answer to this questions determines that whether government's social and economic policy experiment can be a challenger (or, perhaps, an alternative) of the Western-type and democratically founded (and heavily criticized) capitalist order–or it will be an inevitably finite process carrying the promise of failure, on which the way back will be as painful as was the road towards.

[33] McCallum (1988) defines credibility of economic policy referring to public beliefs. According to this, a policy is said to be credible if it suits the public's expectations–or, formally: when the public's expectations of a policy parameter is equal to the actual value of that parameter (the public knowledge on a Taylor-rule can be mentioned as a good example). After all markets make judgements about credibility–on the basis of, for example, whether announced policy actions can be rationally expected to be executed in the promised way (cf. Black 2002; Schultz 1996). Of course, this line of reasoning sets the value of policy announcements pretty high. Here, however, it is argued that *public* (or, in other word, *market*) *judgment on the adequacy of the macroeconomic model followed by government is a separate source of credibility* as well (cf. Blackburn and Christensen 1989). If it is missing, even actions executed in a preannounced way do not create credibility. But anyway, credibility *cannot* be interpreted in itself, only in the context of the public (markets) and government. There is a slight but unambiguous hint towards value theories of economics–since, as a substantial form of value (as a property that an object possesses in itself) independent of the perceiving subject (i.e. the conscious mind) cannot be assumed (e.g. Marx 1867), there is no kind of credibility independent of the perceiving subjects, or of the community they form. Even if there could be such a form of credibility, it cannot root in a statement phrased by the government itself ("We are credible!"). Maybe strange to say, but *something is credible if markets take it so*. However, this reasoning definitely relates the problem of credibility in economics to an epistemological framework. Even if one can presume the existence of the world-in-itself (something external to the mind), one cannot access the "things-in-themselves", that is, we are not able to transcend the bounds of our own mind. So, everything we perceive from reality is a product of the (human) understanding, that is, comes from the interaction of things-in-themselves and our own minds (cf. Boros 2009; Kant 1787; Russel 1980; Turay 1984).

[34] Eventually, Lucas (1972, 1973a) used the concept of signal extraction problem in the same sense, even if his context differs, since he discussed the comparison of expectations and perceived facts (with the latter providing only imperfect information). However, this difference does not affect the point.

of, firstly, whether its executed actions are in accordance with the promises it has made (we can label it as a conventional concept of credibility–see e.g. KVA 2004) and, secondly, whether its applied (or prospective) instruments seem to be proper for the macroeconomic problem in point. Successful episodes do matter and they can persuade the originally sceptical public of the adequacy of the applied instruments and mechanisms. Of course, these successes are less likely (but of course, no one can rule them out) in a sceptical–oppositional macroeconomic environment. Even a realist intervention is always more than a mere mechanical regulation–and, accordingly, it *also* has a meaning that points far beyond the action itself.

Thanks to the rapid broadening of our knowledge of the economic significance of expectations, now we have the possibility of interpreting and planning symbolic interventions–i.e. of economic policy actions that exert their influence *without* operating any economic policy instruments. In the simplest realistic case, policy produces effects through tuning its instruments in an environment with stable parameters. This mechanism works in the case of any kind of economic policy rules and their automatic application. However, modifying behavioural equations is necessarily attached to all kinds of policies that take the effects of expectations into consideration (being the most impressive lesson from the Lucas-critique). Nevertheless, it should be noted that a symbolic policy, though it does not count on the potential instruments of economic policy, still produces its effects through behavioural equations–but the *essence* of its mechanism is the modification of these equations. This way it becomes possible that unchanged values of economic policy instruments can create different outcomes–similarly to the case of a monetary policy taking expectations into account, as we have seen.

The evolution of the Phillips curve accurately reflects that economic theory and operative economic policy hardly considered the effects of expectations for long, though Keynes, vulgarized by Phillips, regarded some psychological factors as fundamentally significant ones when elaborating his theory of economics. IS–LM-model, which is another attempt to give a useful, easily explainable and manageable (moreover, a very simple) technical guidance to operative economic policy, does not take into account these factors either. Ironically, models attaching outstanding importance to the effects of expectations were so over-formalized and so meaningless in their simplified messages that fiscal policy could hardly make any attempt to put these theories into practise. It is not an overstatement to say that operative fiscal policy would have liked to exert influence even when high theory predicted the complete failure of activist policies. Moreover, fiscal policy that was regarded as only a second-rate topic by the neoclassical mainstream could not get over the fact that Friedmanian theory, for a long time, had virtually put it into the category of monetary policy.

It was a different case with the monetary policy. It has been a living tradition right from the monetarist revolution that the most illustrious representatives of high theory have appeared in the world of operative monetary policy again and again – and *vice versa*, it has been typical that leading central bankers have been economic theorists of great learning.[35] It is quite easy to argue for this as it is enough to skim through the research reports of modern central banks: we can find that all these empirical research programs are linked firmly to academic circles. However, one has to say that the vulgarized theory and operative economic policy (fiscal policy, in particular) recognized the possibilities of influencing expectations only with some delay. Today, of course, expectations constituting an almost independent transmission channel should be regarded as an established instrument of economic policy. Therefore, symbolic economic policies could only appear as a late stage of the development process– seemingly, only at that time when instrumental polices (relying on expectations *as well*) as intermediate forms ran out of steam, and, therefore, when governments having no leeway have to merely confine themselves to giving signals.

So, it seems that two independent but parallel phenomena have given way to symbolic policies. *Firstly, a late aftermath of the post-2007 crisis in which (orthodox, heterodox and unorthodox) the realist-instrumentalist actions of unsustainable fiscal policies became more and more impossible or, at least, difficult and expensive due to the loss of confidence of markets. Secondly, the trickling down of the knowledge of expectations and beliefs from high theory that happened with a considerable time lag, so the fiscal policy elite who were less familiar with academic thinking could discover these elements of the set of tools only at this time*. It is questionable whether heterodox-unorthodox economic policies (under which, as a subset, symbolic interventions are subsumed) multiplied because the confidence in the then- (and, mainly, even now-) prevailing theory of professionals was shattered. One may rather say that a simple exigency presented itself in which the economic policy could not apply the instruments and follow the established practices (both of them root in high theory and in daily operative routine, simultaneously) available in "peacetime"(cf. Csaba 2011). Everyone did what he or she could and had to do–the extraordinarily diffuse, picturesque and diverse episodes in economic policy have emerged thanks to this. Where usual, realist-instrumentalist interventions have some leeway, there the economic policy was not forced to take heterodox-unorthodox actions.

[35] Perhaps, Milton Friedman and Stanley Fischer are the best-known from those who got into the field of operative monetary policy as economic theorists. Alan Greenspan or Ben Bernanke are well-known and acknowledged even by theorists. In Hungary, Péter Bod and Júlia Király (the former Governor and Deputy-Governor of the Central Bank of Hungary, respectively) interconnected academic sphere and economic policy. In the case of fiscal policy, at least in Hungary, researchers are rather academic consultants who, according to the accepted cast, give advice to fiscal policy makers that remain usually untaken.

6.4.4 On the Way Towards Symbolic Policies: Economic Policies Relying on Expectations

As it has become possible to make symbolic economic policy interventions, the issue of judging traditional-usual ones has evidently become problematic and more complex. Both rational expectations theory and new classical macroeconomics have already demonstrated that an economic policy by actively and deliberately influencing public beliefs can attain its desired macroeconomic goals at much lower expenses. It is less expensive to keep inflation under control through a credible monetary and fiscal policy than to achieve the same end via successive interest rate increases and, therefore, via sacrificing growth (cf. Backus and Driffill 1985; Blackburn and Christensen 1989). Expectations and beliefs function as a special filter: every realist intervention (as it has been known for a long time now) exerts its influence (filtered) through them. These effects can be affected and influenced by beliefs in a number of ways: they can be intensified, cushioned, or even completely altered by them. The essence of symbolic policy making is that it intends to produce its effects only through influencing beliefs–i.e. without any kind of realist interventions, cut away from (real) instruments of economic policy. Keynes (1936) was the first to stress this possibility in his theory of the rate of interest: here, the efficacy of monetary policy depended on what the public view was about the safe level of the rate of interest. However, this belief can become independent of monetary policy. Although Keynes in his *General Theory* did not unambiguously suggest an economic policy that would definitely exert its effects through influencing expectations, he dropped several hints that, later, could become the foundation of the theoretical development under discussion. In one of these hints, in Chap. 15, discussing the liquidity function, he made a distinction between the changes in the rate of interest which are due to changes in the size of the money supply (that can be controlled by the central bank quite well) and those which are primarily due to modification of the liquidity function (or, eventually, in the liquidity-preference) fully affected by public beliefs (that is, by psychological factors). Among the attempts of Keynes pointing in this direction, however, the theory of the liquidity trap is of the utmost importance, in which the controlling potential possessed by the monetary authority is unambiguously dependent on public beliefs and expectations. *All stable relationships between the instruments governed by the central bank and the macroeconomic outcomes associated with them cease to exist*: if agents prefer treasuring money instead of holding bonds because of a fall (regarded as excessive by the public) in the rate of interest, the liquidity trap emerges and the intervention mechanism of economic policy working trustworthily hitherto dwindles to nothing. Of course, Keynes did not suggest that monetary policy should try to shape public beliefs if it wanted to avoid the liquidity trap. His stress was rather on the role of expectations (able to become independent of economic policy) that limit the scope of economic policy and filter its effects.

Having based the functioning of macroeconomic systems on short-and long-term expectations regarded as beliefs rooted in public attitudes, as it is well-known today, Keynes definitely anticipated the most essential message of later expectations theories. Keynes' original intention was rather to stress the role of psychological factors, by which, primarily, he tried to oppose the supposedly stable equations of neoclassical economics. Doing this, he maintained the neoclassical doctrine according to which all things in an economy are interconnected, while completely denied the faith in a precise and deterministic control of economy. By attaching great importance to public beliefs in the functioning of macroeconomic system,[36] Keynes tried to support a notion of an economy lacking calculability and controllability (in short: stable, deterministic relations), in contrast with the neoclassical idea of a stable and machine-like economy.[37]

The same filtering effect already mentioned in connection with Keynes stands in the centre of the basic interpretation of the Lucas-critique (Lucas 1973b), since, according to it, public beliefs eliminate the effects of all realist interventions, i.e. completely filter and neutralize them because of the adjustments in behaviour in advance (cf. Lucas 1995). However, all this does not mean the complete ineffectiveness of economic policy, since, according to the equivalence theorem, *there can exist interventions the effectiveness of which is ensured just by their impacts on public beliefs*. Doing the debt-for-tax swap, standing in the centre of the equivalence theorem, has no effects on the real economy only if, in the meanwhile, the operation of the public sector remains unchanged (stressing this stipulation made it possible to judge the equivalence theorem in Chap. 5 in a more refined way). Of course, in vulgarized economics, new classicals are usually associated with rigid economic policy rules–but, although it is not intended to put forward more arguments for and against rules in this section, the discussion here can be supported by some related considerations. The essence of following the rules in an economic policy is not to cross the path of the automatic mechanisms of an economy but to allow expectations based on economic policy rules (as well) to realize and work, exerting unboundedly their influence on real economy. It must be stressed that symbolic policies, of course, can be properly adjusted to this goal. This case, somehow, can be ranked as the so-called *market-fetishist view* (cf. Vigvári 2008a) because of the faith in the efficacy of market expectations/beliefs and, generally speaking, of market operations. Undoubtedly as a border-line case, *besides* applying instruments, governments always have the chance and potential

[36] In the literature, even today, authors willingly regard expectations as essential determinants of cyclical movements, not citing Keynes, of course–see e.g. Magas (2009).

[37] Using the term "idea" in its Platonic sense is overly proper. Neoclassical economics as a pure theory describes specifically the functioning of a perfect (idealistic) macroeconomic system. Real macrosystems *can only near* its stability, calculability and (last but not least) its tendency towards equilibrium and the ability to remain in this state more or less. We cannot be sure of what view neoclassical economics held on this idea elaborated by itself. Certain texts imply that neoclassical theorists were completely aware of the fact that their theories and models described the functioning of non-real economic systems–nevertheless, relying on the results of their pure theory, they tried to answer some questions raised by real economies and it makes this problem far more complex. However, it is evident that Keynes broke with the description of ideal economies and, instead he tried to suggest directly applicable solutions to real functional problems of real macrosystems.

to give signals[38] to markets *quasi* recognizing their superiority on each occasion.[39] Of course, adapting to market processes is only an option; the intention of giving up declared goals of an economic policy can arise at any time. Naturally, the government gives signals even at this time but a discrepancy (or even a conflict) occurs between government actions and public beliefs. In such a case, policy is based on the inherently implicit assumption (that can be easily made explicit), or *conviction* that public beliefs are wrong and market needs a shock to get out of its current (presumably suboptimal) position, and to do this, an economic policy regulator is needed that is wiser than markets themselves.[40] Expectations of market participants always serve the interests of their own utility maximizing behaviour (e.g. Hahn 1984). Agents think that if their expectations are fulfilled, their realized profit or utility will be maximized. If one prevents market expectations from exerting their influence (i.e. it is achieved by a shock, or even by a symbolic intervention that expected and actual outcomes diverge) it means that either we are smarter than the market itself (of course, such faith should always be judged sceptically), or at least we think we are (it is more likely), or, simply, we want to damage the market. This problem is even more complex if expectations cannot be regarded as single valued, i.e. if public beliefs/expectations are heterogeneous–however, this case can be reduced to the convergence of expectations, since there is always one expectation that can turn out to be correct, and agents with mistaken belief try to take over it (i.e. the correct expectation), or its formation manner (the expectations equation). Whichever road an economic policy takes (i.e. whether it moves towards fetishizing

[38] Giving signals can be a part of economic policies that suit orthodox norms. If, for example, we look at the institutional components of Washington Consensus (that commonly regarded as orthodox), it is evident that establishing them gives signals that can fundamentally influence market beliefs on a national economy. In our opinion, these actions should not be put under the category of symbolic policies, since they seem to be very real interventions that influence market beliefs, too, as a secondary channel. Well, it seems to be the aforementioned case of expectations channel of realist economic policies.

[39] Using the most neutral phrasing is intended. These lines here do not aim a normative analysis, i.e. we are not arguing for the superiority of markets–we do not suggest more than that there exists a faith in the superiority of markets.

[40] Some supplementary remarks may be useful at this point. From Hayek (1988), to mention the economic policy regulator as a kind of central planner can raise some scepticism, even if the past few decades of operative economic policy supported theoretically by the convincing arguments listed by Marx and particularly Keynes questioned the theorem of impossibility of centralized economic policy at a minimum. We have seen that the success of centralized discretional policy cannot be ruled out even in the new classical/monetarist theory if we are ready to take some non-extreme assumptions (Chap. 4). It is more important to see that a government contradicting market beliefs can act as an anchor if it meets some serious credibility criteria, in which case the change of public beliefs means the solution of the discrepancy. The role of credibility in this situation cannot be overvalued: without it, this effort remains only a risky experiment as the negotiation process of the Hungarian economic policy with the international institutions that was communicated to be opened at the end of 2011 and that met a weird end later, and, not independently of it, the government commentaries on markets developments showed it.

or overriding the market), its symbolic actions surely imply the market model and philosophy (in its everyday sense) on which its economic policy is built.

One should bear in mind that in the above-mentioned case the change in beliefs was triggered by a realist-instrumental action (one may suppose some technical-institutional features of that). For example, a comprehensive public sector reform that can surely support the recovery of the regulating potential of economic policy through influencing expectations may be regarded as realist (actual–i.e. not merely as symbolic) at least. In this story, economic policy makes realist interventions, and the fact that their effects evolve is due to changes in beliefs resulted from real actions. To put it simply one may say that beliefs (or the changes in beliefs) were the necessary but not sufficient conditions of the above consequences. Here, changes in expectations and beliefs were not suggested to have the potential to support the achievement of economic policy goals on their own.

6.4.5 Symbolic Policies as Signals

Special attention should be paid to the case mentioned above, in which the government tries to "override" market beliefs. Although it is stated *reductio ad absurdum* that government in such cases is smarter than markets (at least it thinks it is), an information asymmetry favouring the government has been a very common assumption in the literature since the seminal paper of George Akerlof (1970). According to the monetarist Robert Barro (1976), the success of activist economic policies depends on whether a government possesses an informational advantage over the public that can be exploited by the government during the game it plays *against* (sic!) the public. Stanley Fischer (1977) also quoted information asymmetry favouring the government when he stressed the potential for the success of systematic economic policies. In each example, putting it simply, the government has more knowledge than the public. In the literature, however, the case in which information asymmetry favours the public[41] is completely disregarded–even if this case is deeply rooted in economic thought, though, presupposing rational agents, the formulation of this assumption is fraught with difficulties.

If *all* agents rationally form expectations, the public cannot know more than the government, since everything that is known to the public is also known to the government being economically rational. If something is known to everyone that is

[41] Of course, it follows from the nature of the case in point. If the question is whether economic policy deceives the public to succeed in the short-run or not, an information asymmetry favouring the public is not a relevant starting point, since whether market is better informed than the government or both have the same information set, the results are the same: government cannot carry out systematic economic policies (if, of course, all further conditions necessary to it are held–see Chaps. 3 and 4 here).

known to the government as well.[42] *However, if the principle of economic rationality is violated on the part of the government, then the public knows more than the institutions responsible for the economic policy*. When the economic rationality of a government fails, then, evidently, it cannot be expected that this lack of rationality will be recognized and eliminated by the government (not including the case when the effacement of economic rationality is voluntarily–one may say, done because of *political* rationality). Several examples of this case can be found in the literature: it is enough to think of the planning risks that are typical in budget planning and often emerge deliberately, many times definitely related to political goals (for further considerations, see Báger and Pulay 2008). So, if the public knows more than the government, a credibility problem directly and immediately occurs. In this case, a government may follow a policy of which the public knows in advance that it is wrong or false or, one may say, unsustainable. Such an economic policy cannot be credible, and the mechanism of self-fulfilling prophecies starts working and interventions lacking credibility are doomed to failure.

The question that which party knows more, i.e. who can judge correctly the economic policy of a government, is a very complex problem and here one can suggest only a few viewpoints to consider. Public judgement of the macroeconomic model followed by the government is a directly relevant aspect. It is not too hard to realize that how limited the potential of an economic policy that follows a logic contradicting economic common sense is. In this case, *government gets measured in terms of economic rationality*. Studying the contradictions of political and economic rationality has a long and remarkable tradition in the literature (for a great summary see Kutasi 2007), and, according to this, it seems that economic considerations probably fade into the background political rationality. If a government chooses a macroeconomic model questioned by the public to guide its economic policy, it can be not only the result of a mistake (a wrong decision), but also that of deliberate considerations. Both political and economic actions, quoting Max Weber (1964), can be pure purposively/instrumentally rational acts, their purposes may, however, contradict. In such a case, economic instruments may be subordinated to the achievement of political goals, i.e. to pure instrumentally rational political actions, so economic policy interventions performed become the instruments of getting and preserving government power (cf. e.g. Bihari and Pokol 1998). This is a situation in which political rationality involves giving up economic rationality (Benzes and Ürögdi 2008). We can draw a direct parallel with the case that put a problem in the lime-light of economic thought: the problem of sacrificing long-run goals for short-run successes (Kydland and Prescott 1977). Naturally, the judgement of the contradiction between short-run successes and long-run goals is not always simple, since one can find an economic policy decision that, in addition

[42] For Hayek (1988), this is evident that public knowledge outweighs government's information set, but this feature does not follow from the inability to rationality, rather from the shortage of information process. Government cannot efficiently collect and process pieces of information dispersed in the market.

to forgoing economic rationality, also contradicts political rationality (cf. e.g. Papp 2008).

Of course, it is a separate problem whether markets are right. A government can always regard a failed bond auction or a downgrade (or a risk of it) as a speculative attack or, simply, as a misunderstanding, or mistake, if not outright hostility. It is a matter of fact that the institutions that analyse and monitor the fiscal and growth processes of a national economy may be at a distance of thousands of kilometres from the given country and, their analysts do not often speak the native language either. So mapping actual economic and fiscal processes and making proper judgement about fiscal and real economic risks may present a serious challenge for them. Therefore, these opinions (and even credit ratings themselves) are always disputable. A government with transparent and credibility-building actions can do a lot to support the true judgement and judgeability of the situation, of course. However, all this is a minor problem but not necessarily of secondary importance from the point of view of the economic policy and its attitude towards the markets. Whether markets are right (in the government's opinion) or not, investors and credit rating agencies (together with a number of other institutions) constitute the market itself. If the public's concerns about fiscal and real economic processes grow, i.e. the expectations about them deteriorate, there is always a possibility to (1) dispute or (2) dispel these beliefs. A government can dispute public beliefs, but it might end up in a delicate situation in which the economic government relies on those (i.e. the market) in further financing national debt under unchanged circumstances or, for example, in (re)accelerating inflow of active capital whose value judgement is questioned at the same time. If the international judgement of the sustainability of fiscal balance deteriorates and the market *as such* lays down requirements as conditions for keeping on financing the budget, then economic policy cannot count on the market until it meets them. However frustrating it is for economic governments, markets together with certain international institutions make judgement about the economic policy. Of course, it may considerably limit the leeway for economic policies–at least for the ones that do not live their lives in a vacuum without any external relations. But such policies are hard to find. It is not necessary for us to move towards the extremity of market-fetishizing views to regard the tuning of the attitudes of market participants as an elementary need. *The primary goal of symbolic policies is just to give orienting-tuning signals to markets.* It can be disputed whether the market is omnipotent, whether endless deregulation and liberalization, or whether the excessively broadening of the extent and limits of the market is the best policy. All this is primarily a question of ideology (or, if you like, that of faith) in which economics has not too much to do. However, it is a different issue to accept that markets are not to be overridden but to be calmed down if there is dependency upon them.[43] Governments are always different in terms of

[43] Once the problem of unorthodoxy has already been mentioned above, the case of the IMF-agreement of Hungary emerged after November 2011 should be put right under the symbolic economic policy actions. The importance of these steps were probably recognized even by the

their ability to oppose markets, naturally. Just in the wake of the crisis, we could witness episodes in which some governments exploiting advantages from early reactions and from the monetary sector's needs for rescue packages gained superiority over the market due to shifting the responsibility (Bod 2013). In such cases, governments can even have more leeway in the economic policy and it can be one of the sources of above-mentioned heterodoxy. Governments are ready to take extraordinary steps in extraordinary situations.[44]

When we discussed the power of markets above, an international scene was often mentioned on which governments of national states have to operate. Indeed, there is a view that connects the power of the market and the reduction in the possibilities of economic policies with the realization of globalization (e.g. Veress 2007). It cannot be doubted that there are some arguments for this view, and though this problem shall not be analysed here, some remarks concerning our topic are needed. It is doubtful whether the appearance of markets as powers limiting economic policies can actually be regarded as a consequence of globalization. Talking about the dominance of markets is an expressly modern development. However, the process of globalization already started by the geographical explorations of the fifteenth and the sixteenth centuries (e.g. Giddens 1993), though it is a commonplace that globalization is considerably expanding in our days (e.g. Szentes 1999). To avoid contradictions, it is enough to link the dominance of the market to the emergence of *total* globalization (Veress 2006) which is, however, a recent development. Actually, the situation is that *the description of markets as filters of*

government, since it explains why the government tried to keep negotiations on the agenda for such a long time, while, actually, contradicting and deferring statements were born regarding the concrete agreement. It is also mentioned by the non-official label (the so-called peacock dance) widely used in common talk and originated in the highest level of government that the goal was to give a firm signal to the markets. Taking all this into consideration, we can see that this action perfectly fits into the set of symbolical economic policies, since the signal, as an action, intended to the markets lacked all kinds of realist-instrumentalist element. Of course, this "peacock dance" can only be regarded as a mere symbolic action if the intention of a concrete agreement is questioned automatically and evidently. One can, however, only infer it from the details of the story, but cannot prove. If this agreement had been concluded, then, today, we could talk about a realist-instrumental action. It would be the case as well if the negotiation process had fallen through in spite of an actual intention to contract, creating realist-instrumentalist action that had only signalling effects. However, the government explicitly aimed at reducing national debt (in accordance with foreign requirements) and this undertaking simultaneously narrowed the leeway of economic policy. Right because of this undertaking one should think that government did not attempt to enter into this agreement (and played only the so-called Turkish card).

[44] It was a different story in Hungary. These extraordinary actions were occasioned right by the excessive and tragic narrowing (not the broadening) of the economic policy playing-field, and the government, quite originally, did not rely on its communicational readiness or on the dependency of the monetary sector, but on its parliamentary power and on public communication consumed huge amounts of financial resources in taking its actions. Of course, what a government can do inland and what it can do facing globalized financial markets is always a good question. It definitely seems that only the (partly self-constraining) inability to act effectively and the defencelessness against markets (and the inevitable acceptance of it) forced governments to take symbolic actions.

this kind can easily be derived even from models of a closed economy. Therefore, the dominance of markets is *not* the consequence of globalization, though, undoubtedly, it is a substantial concomitant phenomenon of that. Even in terms of economic policy, of course, globalization is much more than the dominance of markets and the reduction in economic policy instruments, since globalization might even increase the number of economic policy instruments. Actually, *the dominance of markets*, or, in other words, their powerful ability to control, can be derived from all theories that take the effects of expectations into consideration (Keynes, Friedman and Lucas–to suggest only those theories that can be easily regarded as similar ones[45]). Markets are always mechanisms that limit and shape the scope of the economic policy (even assuming a closed economy) when they are considered as having beliefs and expectations that determine macroeconomic outcomes. It is argued here that the narrowing of the scope of economic policy is not the result of globalization, but that of the advancement in market behaviour (i.e. rationality) and that of unsustainable economic policies that concentrate on short-run interests. These factors together put market participants in the social and economic environment of governments into a position of substantial power. The fact that there are globalized agents among them as well is a consequence of long-term (centuries-long) social processes.

Now we hardly live in a period when we can successfully argue for the efficient and reliable information processing of markets (or, if you like, for their wisdom). All significant crises that have shocked the world economy can be described as ones that have been triggered (or, at least, they have considerably contributed to their deepening) by information inefficiencies, overheating, or, simply and generally speaking, the irrationality of markets. Comprehensive summaries of crises (e.g. Lámfalussy 2000; Gál 2010) can always be interpreted as a list of market crimes, though there have been authors driven by their attitudes towards markets who blamed governments for the same mistakes (e.g. Schwartz and Friedman 1963). Perhaps it is the most correct (at least the wisest) to regard crises as the result of interferences between markets and governments[46] (e.g. Csaba 2008). The question of who is better informed and whose actions can be assumed as proper *a priori* is unanswerable–even if we want to draw conclusions on the case in point *a posteriori* by analysing certain episodes, the chance of a consensus remains small. We can still make certain assumptions regarding this problem, but the solutions and

[45] Mentioning Keynes may be striking. Although Keynes can easily be accused of the idea of central planning, Keynes actually never wanted to suggest substitutes for markets but to complement them by state intervention, exactly in the same way as literature of market failures recommends today. Our examples mentioned from the theory of Keynes stress that markets are important filters and controls of economic policy for him.

[46] The answer to this question fundamentally depends on whether macroeconomic systems are considered as systems with tendencies towards equilibrium. If yes, government interventions are naturally refused as shocks disturbing equilibrium. If no, government interventions can always be justified–and this is true as well if interventions are needed not because of the lack of self-regulating nature of markets but, simply, the missing markets.

the inferences that can be drawn from these models will be directly offered just by the apodictic nature of mathematics. Here, it is an extraordinarily relevant thought that was stressed by Kant: within the realm of aprioristic analyses, we can only draw inferences that *we* placed into our theories through our concepts (Kant 1787; Cassirer 1921; Tengelyi 1995). If we consider markets as the pledge of proper and efficient economic behaviour, then government interventions will always turn out to be harmful. If market is an imperfect medium and in the need of regulations and corrections, government interventions adhering to certain principles will always be justified and even efficient. Discussing it, however, we can only sail the oceans of ultimate faiths.

6.4.6 Conclusions

But anyway, in the period of extremely intensified financial and general globalization, symbolic economic policies may have special importance. To direct the effects of markets into the desired channel by legislative power is only one option. However, this legislative control always lags behind market innovations–moreover, this regulation problem is not quite relevant regarding the case in point. Some effects of markets may be considered as harmful either in advance, i.e. independent of experience or relying on empirical evidence. These effects can be ruled out by laws and measures (cf. Király 2009). As far as symbolic economic policy actions are concerned, it should be accepted that the market has the final say–and this does not mean that markets are necessarily right. Of course, the statement "markets are right" can be interpreted in numerous ways and can be concluded from various concerns. In the period of the Keynesian economic policy[47] one could argue that a sufficiently persistent fiscal stimulus, sooner or later, would produce the desired effects via the multiplier mechanism. Even if the expansion was deficit financed,[48] the fiscal position and the accumulated national debt did not remarkably limit the

[47] This phrasing is from Hicks (1974). It was stressed by him that the economic policy regarded as Keynesian rather bore a spontaneous resemblance to the system suggested by Keynes himself than was a conscious adaptation of his recommendations. However, elsewhere in the literature it is underlined that operative economic policy sometimes explicitly supported the theory of Keynes (cf. e.g. Andor 1998).

[48] If we deny that Keynes had a significant role in the prolonged boom of the post-war period as Hicks did, then the recommendations of the Keynesian economic policy cannot be blamed either for national debts naturally expanded by broadened government activities (see e.g. Sivák and Vigvári 2012). By the way, Keynes himself had quite definite ideas on deficit financing, though he paid little attention to this topic in his General theory. For him, the effect of current revenues and expenditures on the budget show only cyclical patterns, i.e. a deficit is accepted only in hard times. This deficit is settled by the revenues rising automatically during recovery (regarding longer periods). Keynes did not suggest discretionary tax changes, so the modern concept of fiscal balance management based on automatisms was definitely present in his theory. Of course, deficit financed public investments for capital goods were allowed and were suggested to repay during

scope of the economic policy. Now we know that the world has drastically changed after Keynes. It is also a commonplace that the greater market openness and the interdependency of national economies make the success of government actions evidently ambiguous today. Shortly, budgets do not have enough resources to start expansions the beneficial effects of which emerge partly abroad. It seems to be sure that we can find other arguments, but these two are completely enough to give rise to a suspicion: the time of traditional economic policies wobbling on expansions and restrictions is over. Economic policies need new tools that help them sidestep budget constraints and that preserve the possibility of controlling economies without expensive actions having dubious potential for success. As we see now, budgets rather have to struggle with taking operative actions, finding the resources necessary for them and maintaining the benevolence of markets. This benevolence is needed not only in accordance with budget operations but also with economic growth and development, meanwhile one should bear in mind that governments try to share the burdens of development and growth processes. Realist economic policies have much less leeway and symbolic actions can lower the limits faced by governments. It seems that applying these symbolic polices will be an urgent necessity for governments that suffer this loss of room for manoeuvre. Of course, we do not know now whether these symbolic actions will be degraded on their own to be mere emblematic surrenders to market power since during an exposure to the market it is not easy (and not feasible) to defy it. Without a bilateral communication process between governments on the one hand and markets on the other, i.e. governments have to face a global economy having and preserving superiority, these symbolic actions can be regarded by the other side as simple acts of capitulation. The question is that whether the tools[49] of symbolic policies remain a last resort with no other means and methods available or these symbolic policies will have an etiquette and acceptance (it is hardly a mistake to say that in the cases above markets did not like the actions we analysed) that make them be an equal channel of interaction between actors of the global economy.

Theory has taught us how important expectations are regarding the functioning of the real economy, how much is up to credibility, or how gentle interventions are needed to anchor expectations and to fine-tune them slowly. Stressing the importance of symbolic economic policies is not more than another step on the *same* path. When international capital flows reach a massive scale we have seen recently or when governments must increasingly rely on home and foreign markets to finance their budgets, it is unavoidable to get a sophisticated set of tools that can efficiently influence the market climate without disturbing the functioning of the real

their lifetime out of tax revenues (Brown-Collier and Collier 1995). This is, practically, the suggestion of meeting the requirement of the Golden Rule of fiscal policy (cf. Vigvári 2008b).

[49] I do not dare to say that such a set of tools really exists. By analysing examples, it seems that anything having a meaning that refers to something being far beyond the action itself can be a symbolic action that can give a signal to market agents standing on the other side of the communication channel.

economy. However, the question of symbolic economic policies can only be raised here; the answers will be offered by the (near) future.

6.5 Some Further Remarks on Inflation Targeting

Inflation targeting tends to be an increasingly general and common framework of modern economic policy. International literature puts large emphasis on the benefits of this regime. Typically, the pros stress the improving macroeconomic performance, the flexibility of transitions to this regime or the compatibility with the characteristics of emerging markets (Allen et al. 2006; Batini and Laxton 2006). It has to be stressed that inflation targeting is often referred to as the synonym for price level targeting: the confusion probably comes from the fact that price stability may be interpreted as the *stability of* either *price level* or *inflation*. However, the consequences of following these two regimes are radically different. It is not too hard to realize that a return to the target under inflation targeting requires *disinflation*, while *deflation* under price level targeting. Deflation can hardly be a desired target for monetary policy, since deflation would increase the real value of debts, crushing the debtors–in other words, deflation implies a depression (cf. Tarafás 2001). It is argued here that the higher degree of real economic volatility implied by price level targeting evidently settles which of these regimes is the more desirable or beneficial for macroeconomic systems. In spite of all these, Svensson (1996) still argues for price level targeting, highlighting the circumstance that inflation may show a lower variability under price level targeting. However, his results are very sensitive to the assumed dynamics of output. Sweden in the 1930s was the only country that made a trial of price level targeting–but, in a climate that was penetrated by a fear of deflation (cf. Berg and Jonung 1998). In modern times, price level targeting still shows up as an efficient nominal anchor against deflationary expectations, with direct implications for the economy of Japan. However, its application in this case as well is directed towards only an initial period that is set by the successful anchoring of expectations (cf. Svensson 2001). Once the targeted price level is achieved, inflation targeting can replace the regime built on targeting the price level (Svensson 2003a).

We have seen in the theoretical chapters that the assumption of rational expectations is not enough to ground the short-run inefficiency of monetary policy–in this context, the cardinal issue is the assumption of perfect wage and price flexibility. In this time-horizon, monetary authority can influence the level of output and employment and, *through this*, inflation dynamics, striving for the deliberate utilization of this ability in inflation targeting regimes. Actually, it seems as if the notion on the transmission mechanisms of monetary policy turned back to the Keynesian idea. While inflation in the Friedmanian theory was a mere monetary phenomenon, then Keynes related inflation to real economic performance. *It seems* as if inflation targeting placed this Keynesian tradition back in the centre of operative monetary policy, as far as short-run implications are considered at least, since, after all, expectations and real economic performance, through interest rate control, contribute *together* to sticking the rate of inflation to a desired path. According to the widespread view, *inflation in the long-run is still a monetary phenomenon, standing under the influence of the quantity of money and monetary policy*. By the way, stressing the relationship between output and inflation is exposed to the danger that a stable, function-like relation between inflation and unemployment strikes root in common knowledge again (cf. King 2002). *In the long-run, central bank has to set its key policy rate according to the neutral real rate of interest*, i.e. short-term real rate of interest on the average should be equal to the average neutral real rate of interest.[50] If the central bank tries to fix the short-term real rate of interest below this neutral rate for long, then aggregate demand will persistently exceed supply, and the given macroeconomic system is to be expected to show the symptoms of overheating, creating serious inflationary pressures. Moreover, higher levels of inflation are usually accompanied by higher degrees of volatility of inflation, taking adverse effects on real economic performance (cf. Svensson 2003a). This is exactly the special feature because of which it would be harmful to refer to the Keynesian tradition, since a higher level of real economic performance in the theory of Keynes was assumed to be permanently maintainable at a given and stable level of inflation–though it was not the monetary policy that played the main role in this story, as it was mentioned above. However, it should be obvious on the grounds of our pervious discussion that monetary policy is not capable of it and it is not too wise to link the purpose of long-run real economic control with monetary policy. So, we are not saying that the expense of a higher level of employment is a persistently higher and stable level of inflation and that adjusting these magnitudes is a decision variable lying under the control of monetary policy. We should say rather that meeting the inflation target sometimes requires temporary excess demand. However, it is obvious that, in the short-run, the pulling effect of excess demand on inflation prevails. It is argued here that highlighting the efficiency of monetary policy and distinguishing the short- and long-run aspects of this efficiency are definite merits of monetarism. By the way, *it should not be forgotten that there were two time-horizons mentioned in the discussion.* The response-time of real economy has to be separated from the time-horizon of the stimulation or from the time-horizon of the *ability* of stimulation. Real economy can hardly be expected to response to monetary policy actions in the short-run, so, in this context, typically long term considerations come in the lime-light. However, stimulation is possible only in the short-run, i.e. only temporarily. So, we can say that *monetary authority can exert short-living influences in the long-run only*; in other words: with a considerable response-time.

What is more, all this is not independent of the fact that monetary authorities can take responsibility only for controlling the "sustained" inflation, since prices may

[50] *Neutral real rate of interest* is the interest rate being consistent with the output equal to the potential level. According to some theoretical considerations, this neutral rate is determined by *non*-monetary factors.

be considerably deviated from the desired path by both monetary and non-monetary shocks. It is not the usual practice to judge the performance of inflation targeting central banks on the basis of the monthly inflation data. A more appropriate measure is a moving average computed from these high-frequency data that disregards dysfunctional short-run fluctuations. Of course, it may be a misleading measure, since it is possible that the targeted rate of inflation stands in the centre of fluctuations of actual inflation (cf. Rasche and Williams 2005).

This framework of monetary policy, while gaining popularity, reformed the previous regimes in various aspects. It is completely enough to mention the more intense and clear communication between monetary policy and the public (markets) or the increased discipline and accountability (responsibility) of monetary policy. Inflation targeting has also brought about fundamental changes in the way monetary policies are followed. As opposed to earlier times, when monetary policy in practical terms was not more than (efforts towards) controlling the monetary base, operative monetary policy under inflation targeting is built on manoeuvres with the key policy rate. [→ **the goals of monetary policy**] This change can be traced back to technical or, in other words, *behavioural* causes. According to some insights, *demand for money* was often completely uncontrollable by *shifts in the money supply*, thus central banks try to influence both the real rate of interest and the monetary aggregates through changes in the (short-term) key policy instrument (cf. King 2005).

6.5.1 Expectations and Constrained Discretion

Inflation targeting is *not* an ideal manifestation of the new classical monetary policy recommendations, but we can still say that some assumptions and findings phrased by new classical macroeconomics are realized within the framework of inflation targeting. Due to the fact that monetary policy makers within inflation targeting strongly rely on the effectiveness of discretional economic policy actions, new classical high theory is attacked exactly at the point where it was the least convincing.

Let us start with the differences. We could see that new classical macroeconomics urged an intervention-free monetary policy, which is based on a constant money growth rate.[51] Inflation targeting abandons this principle, with the difference the bottom-line of which is not the fact that the rate of interest was advanced to be the

[51] *However, following a rule does not necessary mean the acceptance of the constant money growth rate rule*. We have seen that Sargent and Wallace (1974) favoured a monetary policy rule that systematically responds to real economic performance. Broadly speaking, the essence of rules is not about the issue of the constant money growth rate but abstaining from surprises–Kydland and Prescott (1977) stressed at least that surprises may lead to serious welfare losses.

key instrument. High theory literature still echoes with the *rules-rather-than-discretion* debate.[52] The bottom-line of the problem can be expressed well through a simple model (cf. Debelle 1999). The relevant mechanisms in a macroeconomic system are described by the following equations:

$$\pi_t = \pi_{t-1} + \alpha\left(y_{t-1} - y^*_{t-1}\right) + \varepsilon_t; \tag{6.1}$$

$$y_t = y^*_t + \beta\left(y_{t-1} - y^*_{t-1}\right) - \gamma\left(r_{t-1} - r^*\right) + \eta_t, \tag{6.2}$$

where π is the rate of inflation, y is real output, y^* is the level of potential output, r is the short-term real rate of interest functioning as the key policy rate and r^* is the neutral real rate of interest, while ε and η are random disturbances with zero means. This system has several special features. For instance, it follows from the first equation embodying the Phillips curve that changes in the rate of inflation can be traced back to a twofold reason. First, lagged output above the potential level enhances inflation; and second, inflation of the previous period builds in the mechanism and the impacts may be long-living. If output in period $t-1$ being above the potential level increases the inflation of period t, this period t inflation leads to a higher rate of inflation in period $t+1$, even without a positive output gap. So, the higher rate of inflation keeps prevailing until the output drops, or until a reduction in the inflation rate occurs as a result of the corresponding random term. Current inflation depends not only on the output gap of the previous period but also on the inflation rate of the previous period. It means actually that inflation is higher today in part (!) because it was higher previously. The significance of this thought is incredibly high: *the cause of inflation* (at least partly) *is inflation itself.* It is exactly the message that should be regarded as the major merit of new classical macroeconomics. A further characteristic is that monetary policy can influence the output of the next period only, while inflation can be affected only indirectly (through output), with a two-period lag.[53] Parameters are constant, that is, this system does not concede the consequences of the Lucas critique: policy interventions do not modify the reactions of market participants (cf. Ball 1997).

This simple system can be criticized for its excessive rigidity and for its insufficient elaboration. The bottom-line of monetary policy transmission or the ways the different channels are strengthening and supporting each other are well-known. However, it has to be stressed that this model is not aimed at describing life-world conditions and mechanisms in a *realistic* way, but considering some consequences drawn on grounds of an assumed model framework. In this context, we have to refer back to the principle already phrased in Chap. 1. Mainstream

---

[52] These rigid rules can be regarded as the case being closest to the new classical monetary policy recommendations.

[53] Since the rate of interest of period t affects output of period $t+1$, and period $t+1$ output exerts influences on inflation of period $t+2$. The circumstance that the interest rate of period t determines the output of period $t+1$, as it will be seen, does not mean that output responds to shifts in the rate of interest with a time lag.

economics is a *pure theory*, that is to say, it does not give a realistic description of reality, even if new classical macroeconomics, driven by a conceit, sought it. This algebraic deduction is not intended to highlight all the transmission channels of modern monetary policy or the interactions between these channels. Actually, this simple mathematical structure is too rigid–as opposed to life-world conditions. *Real-life economic mechanisms do not show up as outputs from equation systems. It* is exactly one of the major problems of mathematical economics and, in general terms, of mathematization: once we make efforts to describe directly experienced reality through mathematical equations, we can get to such structures only that are infinitely rigid and *artificial* compared to reality. As far as this simple (i.e. oversimplified) model is considered, we have nothing else to do but to take the consequences of the system defined *somehow* (i.e. in *a certain* way among the number of possible options) into account. As a critical remark, we can stress that inflation under inflation targeting is *also* affected by the exchange-rate effects of shifts in the key policy rate or that effects do not emerge following a rigid sequential order. Beyond these circumstances, we have to highlight only that Eqs. (6.1) and (6.2) reach their ultimate purpose: they could accentuate the fact of lagged reaction ability of monetary transmission, even if this model offers only a schematic analogy of reality. Deficient equations are not able to embrace the infinite complexity of reality.

The rule monetary policy follows in this model can be easily distilled. Central bank may target to minimize the variance of real output or of inflation, or (evidently), under inflation targeting, to stabilize these two variables compromisingly. [→ **inflation and output gap**] Monetary authority, by setting r_t, is able to strive for the direct control of the output of period $t+1$ only, since inflation for this period (π_{t+1}) is already determined by current output and inflation. Monetary authority can influence only period $t+2$ inflation by this interest rate decision. Output expected for the next period can be formulized as

$$E(y_{t+1}) = E(y^*_{t+1}) + \beta(y_t - y^*_t) - \gamma(r_t - r^*) \tag{6.3}$$

on the basis of the second equation since $E(\eta_{t+1}) = 0$, so the problem of monetary authority is to control and set y_{t+1}. Of course, it does not follow that inflation does not affect this decision–it is only that central bank takes inflation expected for period $t+1$ given, since its interest rate decision has no effect on next period inflation, as

$$E(\pi_{t+1}) = \pi_t + \alpha(y_t - y^*_t), \tag{6.4}$$

and $E(\varepsilon_{t+1}) = 0$. In other words, inflation expected for the next period is determined exclusively by current values of the relevant variables (of course, not disregarding that r_{t-1} also contributes to setting y_t).

To review this mechanism, we need to consider two simple (but extreme) cases. Assume that our two-equation macroeconomic system was in equilibrium for a long time, i.e. there was no disturbance that could deviate either output or inflation from the potential level or from the target the monetary authority set, respectively. For the sake of simplicity, consider a constant level of potential output and a time-invariant inflation target. However, assume that $\eta_{t-1} > 0$ and, as a consequence, $y_{t-1} > y^*$, while we know that $y_{t-2} = y^*$ and $\varepsilon_{t-1} = 0$. On these grounds, $E(\pi_t) > \pi_{t-1} = \pi^*$ (where π^* is central bank inflation target). In this example, central bank can choose from two opposing extremities. It can target output stabilization,[54] where inflation dynamics (variance) has no impact on monetary policy. In this case, monetary policy strives for $y_{t+1} = y^*$ to become true.[55] The central bank response can be interpreted on grounds of the equation describing output dynamics:

$$E(y_{t+1}) = y^* + \beta(y_t - y^*) - \gamma(r_t - r^*), \tag{6.5}$$

since $E(\eta_{t+1}) = 0$. We have to assume that the actual value of y_t had already been known when the monetary authority made its interest rate decision for period t (in other words, when the central bank set r_t). In other words, interest rate decisions are always due at the end of periods (e.g. of calendar years). Therefore, it is not true that output responds to shifts in the rate of interest with a one-period lag–but that *in the next period*. Because of the intention to set output into the potential level as soon as possible, central bank seeks after $y_{t+1} = y^*$, from which it follows that the sum of the other terms on the right is zero, that is:

$$\beta(y_t - y^*) = \gamma(r_t - r^*), \tag{6.6}$$

and, solving it for r_t, we get

$$r_t = r^* + \frac{\beta}{\gamma}(y_t - y^*). \tag{6.7}$$

A central bank seeking after only real output stabilization makes its interest rate decision at the end of period t on the basis of the output gap actually observed in this period; and, as it is expected, if actual output exceeds the potential level, central bank responds through an interest hike.

[54] This is definitely a short-run stabilization of real output. In the long-run, as we have already seen, monetary policy could keep real economic performance above the potential level only with considerable costs. Output fluctuates around the potential level in the long-run and monetary policy plays no role in setting this potential output. The long-term target of monetary policy should be a control over the rate of inflation, while the long-term *real economic* target can be nothing more than to support real output in sticking to the potential level as close as possible. Naturally, breaking inflation down raises real economic costs, that's why monetary policy should not disregard the real economic consequences of its disinflationary policy.

[55] However, monetary authority cannot seek after $y_t = y^*$ to emerge, since output, according to the interest rate decision made for period t, is going to respond only in period $t + 1$ after the output swing of period $t - 1$.

As an alternative setting, the central bank can make efforts to turn inflation back to the target as soon as possible.[56] Because of the period $t-1$ real economic shock it is true that $\pi_t > \pi^*$ (but $\pi_{t-1} = \pi^*$), and if there is no offsetting disturbance in period t, it still holds that $\pi_{t+1} > \pi^*$. Central bank, through its interest rate decision made in period t, tries to ensure $\pi_{t+2} = \pi^*$ to be true (it is not possible sooner since there is only an indirect control over inflation), the only way of which is a reduction in output. On the basis of lagging the equation that describes inflation dynamics, we get

$$E(\pi_{t+2}) = \left[\pi^* + \alpha\left(y_{t-1} - y^*\right) + \alpha\left(y_t - y^*\right)\right] + \alpha\left(E\left(y_{t+1}\right) - y^*\right). \tag{6.8}$$

It can be seen that three different lags of the output gap are included. Period $t-1$ data is factual, while period t output gap is uncontrollable because of the time-pattern of real economic responses. Now, the purpose of monetary policy is ensuring $\pi_{t+2} = \pi^*$ to occur, on the basis of which we can say that

$$\alpha\left(E\left(y_{t+1}\right) - y^*\right) = -\alpha\left(y_{t-1} - y^*\right) - \alpha\left(y_t - y^*\right); \tag{6.9}$$

$$E\left(y_{t+1}\right) - y^* = \left(y^* - y_{t-1}\right) + \left(y^* - y_t\right). \tag{6.10}$$

By combining the output equation with Eq. (6.10), we get

$$\left(y^* - y_{t-1}\right) + \left(y^* - y_t\right) = \beta\left(y_t - y^*\right) - \gamma\left(r_t - r^*\right). \tag{6.11}$$

Solving it for r_t, the solution is

$$r_t = r^* + \frac{(1+\beta)\left(y_t - y^*\right)}{\gamma} + \frac{y_{t-1} - y^*}{\gamma}. \tag{6.12}$$

However, since

$$y_{t-1} - y^* = \frac{\pi_t - \pi_{t-1}}{\alpha} \text{ and } \pi_{t-1} = \pi^*, \tag{6.13}$$

we can also say that

$$r_t = r^* + \frac{(1+\beta)\left(y_t - y^*\right)}{\gamma} + \frac{\pi_t - \pi^*}{\alpha\gamma}. \tag{6.14}$$

In case of this second reaction function, central bank responds to both inflation and output deviations, since both are needed in order that inflation could be maintained at the target level. Either inflation exceeds the target or real output eclipses the potential level, both cases (or their combination) necessitate an interest hike to

[56] Here, it should be noted that this model was formulized assuming inflation persistence, i.e. a *single* real economic shock has a persistent inflation effect.

break inflation down. However, it is important to stress: central bank, when it applies the reaction function of Eq. (6.14), does not weigh the effects of the interest hike, carried out in order to ease the inflationary pressures, on output dynamics (output variance).

However, these two cases designate only the endpoints of the possibilities available for the central bank. First of all, what we can expect is that the monetary authority takes the extents of both the output derailment and the missing of the inflation target into consideration, weighing them somehow, when making its interest rate decisions. No matter how the central bank preferences are defined, the way of controlling is to affect $E(y_{t+1})$ through the rate of interest set for period t. According to the structural equations, central bank is not capable of controlling either π_t or π_{t+1} through r_t, that is to say, both are exogenously given–but as the central bank can expect real economy to return to the potential level after the period $t-1$ shock (so, on these grounds, no intervention is needed), it makes its decision on the basis of the expected inflation and its own preferences. Monetary authority tries to set period $t+1$ output gap in order to respond somehow to the missing of the inflation target:

$$E(y_{t+1}) - y^* = -q\left[E(\pi_{t+1}) - \pi^*\right], \tag{6.15}$$

which is equivalent to

$$E(y_{t+1}) - y^* = -q\left[E(\pi_t) + \alpha(y_t - y^*) - \pi^*\right], \tag{6.16}$$

where $q > 0$. Setting this parameter q is the central bank's responsibility: monetary policy preferences for the stabilization of output and that of inflation are manifested through the value chosen for q. It can be demonstrated in algebraic terms that q cannot be set arbitrarily–as we will see it below. However, it is worth phrasing a preliminary remark that the higher weight stabilization of inflation gets by the monetary authority, the higher the value of q is–and *vice versa*: the more important output stabilization is for the central bank, the lower q is. Moreover, it comes from the equation describing output dynamics that

$$E(y_{t+1}) - y^* = \beta(y_t - y^*) - \gamma(r_t - r^*). \tag{6.17}$$

By combining these two latter equations, we get

$$\beta(y_t - y^*) - \gamma(r_t - r^*) = -q\left[E(\pi_t) - \pi^* + \alpha(E(y_t) - y^*)\right], \tag{6.18}$$

on the basis of which (solved for r_t) the Taylor-rule of monetary policy (instrument rule)[57] is available as

[57] It is not accidental that the rule is formulized for the real rate of interest; on the one hand, market agents are assumed to consider the real interest rate when making decisions, and, on the other hand, monetary authority tries to influence the real rate of interest through setting the nominal key policy rate.

$$r_t = r^* + \frac{\beta + \alpha q}{\gamma}\left(y_t - y^*\right) + \frac{q}{\gamma}\left(\pi_t - \pi^*\right). \tag{6.19}$$

During the discussion, two alternative monetary policy reaction functions got deduced according to two extreme preference systems–and a third one as the general case. *The relation among these three functions constrains the possible values of q*. The output-stabilizing equation can be derived from the Taylor-rule, since it can be revealed through comparing them that

$$\frac{\beta}{\gamma} = \frac{\beta + \alpha q}{\gamma} \text{ and } \frac{q}{\gamma} = 0, \tag{6.20}$$

where, of course, $\gamma \neq 0$ and $\alpha \neq 0$. It also follows that $q = 0$ if output stabilization is the only goal of monetary policy. It is also true that the Taylor-rule can easily be re-cut for the case in which a rapid disinflationary process is the only purpose. Comparing the Taylor-rule and the inflation-oriented reaction function, we get

$$\frac{\beta + \alpha q}{\gamma} = \frac{1 + \beta}{\gamma} \text{ and } \frac{q}{\gamma} = \frac{1}{\alpha\gamma} \tag{6.21}$$

where $\gamma \neq 0$ and $\alpha \neq 0$. It follows that

$$q = \frac{1}{\alpha}. \tag{6.22}$$

Ball (1997) shows that a value can be chosen for q, depending on the parameters of the structural equations and on the central bank's preference system,[58] that can simultaneously minimize the variances of both target variables.

Actually, Taylor-rules establish monetary policy actions in advance, i.e. they link given central bank responses to given macroeconomic situations. It is exactly the characteristic that should be highlighted when contrasting Taylor-rules with (the possibility of) discretional policy actions. Referring back to the issues discussed at the head of this section, the Taylor-rule derived here is a *rule* in a literal sense, so monetary policy does not seek after surprises through it. However, opponents of these rigid rules support discretional policy steps yet not on these grounds, but stressing that these rules are not capable of dealing with unforeseen shocks or changes in the structure of the economy (cf. Bernanke and Mishkin 1997). In the context of the rules, we should not stress the lack of surprises, since it would suggest that every regime that gives some leeway for discretional actions makes efforts to surprise market participants, but rather the fact of *automatic operation* (cf. Bernanke et al. 1999) as their main character. We will see that flexible inflation targeting can be the best characterized by the concept of "constrained discretion" (Bernanke 2003), since this policy is regular (analogously to rules) as it strips

[58] As far as the latter is considered, as a function of the relative weights associated to the alternative targets.

monetary authority of the possibility of following time-inconsistent policies,[59] while, at the same time, it gives some room for taking some discretional steps, and monetary policy actions are not grounded on automatisms but on comprehensive analyses *as well as* value judgements.

In case of a Taylor-rule, the weights attached to the variances of inflation and output determine how the central bank responds to fluctuations in these target variables–*however, no discretional steps are possible*. According to some authors, it can hardly be referred to as the most efficient way of monetary policy. We have seen that Taylor-rules make it possible to weigh the output and the inflation target, but, in every other aspect, such a rule operates as an automatism. Inflation targeting, as an alternative to rigid rules, does not set a reaction function of this kind, but relies rather on the value judgements and information processing activities of central bank apparatus and experts, admitting the existence of pieces of information that are relevant with regard to the controlling of the macroeconomic system but that cannot be taken into account in an automatic (i.e. discretion-free) regime (Bernanke and Mishkin 1997). [→ **inflation and output gap**].

Inflation and output gap The trade-off between efforts towards output- and inflation-stabilization (that is, the possibility of weighing) is also given in case of inflation targeting. Svensson (2003a) mentions Strict Output-gap Targeting (SOT) and Strict Inflation Targeting (SIT) as endpoints. In this terminology, Flexible Inflation Targeting (FIT) makes the intermediate case, in which central bank can exert a simultaneous (compromise) control over real and nominal variables. Literature portrays flexible inflation targeting as the most efficient monetary policy framework (Debelle 1999). In this case, central bank is more tolerant of missing the inflation target and does not try to break inflation down as soon as possible.[60] Under FIT, monetary policy allows 2–3 years to achieve the inflation target,

[59] In other words, central bank does not seek after real economic interventions which are going to prove to be detrimental in the long-run but are or seem to be beneficial only in the short-run, since central bank is constrained by its own commitment to keep inflation at a low level.

[60] The trade-off is available between the variances of output gap and inflation. It means that a tight control of inflation (keeping it around the target) can be carried out only at the expense of higher fluctuations in output, and vice versa. Taylor (1979) was the first to suggest the long-term reverse relationship between the variances of output and inflation, labelled as a "second order" Phillips curve, highlighting that the "first order" Phillips curve (describing the relation between inflation and the level of output) is vertical in the long-run. The explanation of this relationship does not require a detailed discussion, the basic elements are easy to reveal intuitively: because of the shocks affecting both the price level (inflation) and real economic performance, stabilization of the one variable is possible only at the expense of higher fluctuations in the other. To realize it, it is enough to see that inflation, at a given output gap, is affected by various independent factors (production costs, prices of imported raw materials and semi-finished goods built in the products, inflation expectations). If any of these factors produce inflationary pressures, these pressures can be eliminated only by cutting output (growth) back (if these inflationary tendencies promise to be long-living), not forgetting the significance of expectations either, since a right interest rate decision influences expectations as well, strengthening the effects of interest rate policy. So, under these circumstances, stability of inflation is achieved at the sacrifice of steadiness of output. In the opposite case, central bank does not make attempts to stimulate real economic performance (so it allows output to settle at the natural level), so inflation should be expected to show higher fluctuations.

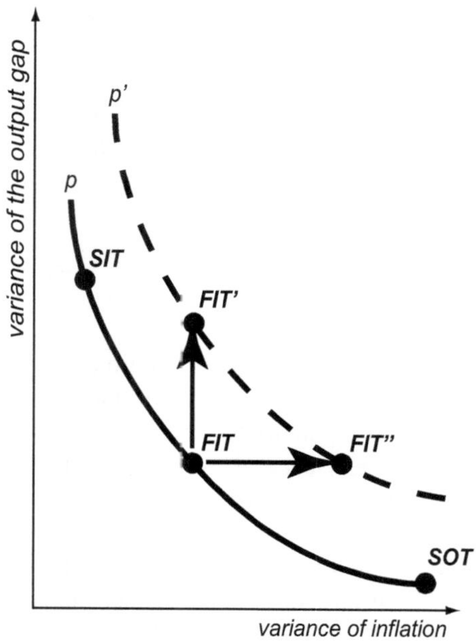

Fig. 6.6 Targeting strategies of monetary policy on the basis of the trade-off between the fluctuations of inflation and output around the target and potential level, respectively

and missing is tolerable within this time-horizon–it is exactly the reason why drastic real economic interventions are not needed. To put it in other words: meeting the target is only a medium-term requirement which makes the monetary authority entitled to seek after also (short-run) output stabilization *at the expense* of short-run fluctuations in inflation.[61] So, it has to be stressed that central bank does not make any efforts to push output above the potential level, just tolerates some kinds of output dynamics within certain limits: if these output swings do not seriously endanger meeting the inflation target.

Flexible inflation targeting does not only mean that central bank takes an intermediate position between strict inflation targeting and strict output-gap targeting–but as well as that it can reconsider its own position, depending on the macroeconomic situation (Fig. 6.6).

If a macroeconomic system is affected by an exogenous shock (e.g. a cost shock), the previous trade-off between the variances of inflation and output gap is not available for the central bank anymore. So, curve *p* characterizing the initial options shifts to *p'*. Keeping inflation at the target level under these circumstances could be achieved only at the expense of an output-gap variance higher than previously, i.e. of more serious real economic losses (FIT'). However, central bank under flexible inflation targeting can opt for allowing temporarily larger fluctuations in inflation (in other words: allowing inflation to deviate

[61] The higher variability of inflation under strict output-gap targeting should not be confused with the (empirical) relationship between the variability of inflation and relative prices. The former, as we have already seen, means that inflation responds to efforts of different strength to close the output gap through fluctuations of various extents. However, the latter means that various degrees of variability of individual prices appertain to various rates of inflation. To put it in other words: relative prices show different extents of variability at different levels of inflation. For further considerations, see Tommasi (1992), Debelle and Lamont (1997), Choi (2009, 2010).

from the target for a longer period and/or in a larger extent) in order to exempt real economy from the burdens of a stricter inflation targeting[62] (FIT", or takes an intermediate position between FIT' and FIT" as a compromise).

The bottom-line of the mechanism is that the monetary authority defines and declares an explicit loss function in the following form (Debelle 1999), for example,

$$L_t = E_t \sum_{s=t}^{\infty} \delta^{s-t} \left[(1-\lambda)\left(\pi_s - \pi^*\right)^2 + \lambda\left(y_s - y^*\right)^2 \right], \tag{6.23}$$

that[63] penalizes if actual inflation and output miss the target. It is quite evident on the basis of this loss function that monetary authority does not seek real output to exceed the potential level. Although theoretical literature urges central banks to specify precisely and communicate transparently such loss functions, it is still not a general routine in operative monetary policy. For instance, though the Central Bank of Hungary stresses that price stability is of primary importance, this preference is not revealed numerically through proclaiming an explicit loss function (cf. László 2002). It has to be realized that the (relative) importance of central bank targets cannot be clear for the public without revealing this loss function. It may lead to credibility problems, since it cannot be expected in all probability that the central bank will always refute to induce short-run economic growth at the expense of inflation. It evidently undermines the intention to anchor expectations. Specifying a loss function does not make a considerable excess burden for a central bank, since only a decision on λ is needed over and above to choose a target level of inflation–the output-target is exogenously given. However, finding a numerical value for the potential level (or potential growth rate) of output is fraught with methodological difficulties and there has not been a generally accepted or salutary solution to the problem so far. [→ **output gap**] In this context, one of the dimensions of central bank transparency, i.e. *political transparency* has to be mentioned that means openness concerning policy objectives. Its highest degree is when the central bank declares its quantitative output- and inflation-targets, plus their relative weights to reveal their importance and the precise form of its loss function (Eijffinger and Geraats 2006). According to the general routine, announcing the inflation target is not problematic, since it is inherently implied by inflation targeting. For Cukierman (2000), this partial refusal may be caused by the circumstance that central banks actually show some inclination to tolerate a temporarily higher rate of inflation in order to avoid considerable fluctuations in real economic

---

[62] As an example, during the 2007–2009 crisis, the Central Bank of Hungary was highly criticized for its alleged rigour–however, this rigour was definitely aimed at maintaining financial stability, though, simultaneously, it could contribute to completing the inflation target (see Neményi 2009).

[63] Where λ determines the relative weights of output and inflation as targets and δ is the relevant discount rate. Under strict inflation targeting, of course, $\lambda = 0$, while $\lambda = 1$ under strict output-gap targeting.

performance. So, inflationary policies involve a real danger, though this danger is not entertained by the possibility of dynamically time-inconsistent policies anymore–rather by careful central bank behaviour through which real economic consequences of inflation targeting are also considered. It can explain why central banks suppress loss functions, since, according to some models, expected inflation determines pricing behaviour of the near future, that is to say, central bank should avoid biasing beliefs of market agents towards higher rates of inflation.

We can make a critical remark by claiming that it is still possible to deceive market agents even with an explicit loss function. However, announcing an explicit loss function improves central bank transparency and credibility. Of course, it is true that even if there is such an explicit loss function, central banks can still opt for economic boost that is likely to lead to inflationary costs. Indeed, any central bank can give up its credibility at will. *Loss functions cannot guarantee central bank credibility.* Loss function is a device of a *credible central bank* to become transparent in order to anchor the public's inflation expectations as efficiently as possible. Announcing a loss function supports central bank in building credibility through supporting the public in considering the relation between actual central bank policy and the declared loss function from period to period. Without a loss function, interpreting the actions taken by the central bank and understanding its operative monetary policy impose a heavy burden on the public, since the central bank (or, precisely saying, decision-making and decision-preparing apparatus of the central bank) thinks something of the importance of output- and inflation-control for sure. So, *an implicit loss function necessarily exists.*[64] Constructing an implicit loss function on empirical grounds is very difficult for market participants, since the estimation process is hindered by extraordinary cases (i.e. by occasional deviations from the rule followed). A policy according to the loss function supports establishment of credibility. So, loss function is *not a pledge of credibility* (a central bank can be credible even without a loss function), but definitely supports building it up. A loss function makes the central bank more accountable, since this function serves as an explicit standard in judging the actions the central bank takes. Of course, the loss function can be set aside at any time–central bank is always allowed to opt for *rule-crossing* economic-enhancing interventions even with an explicit loss function. However, crossing the rules becomes evident immediately (or soon, at least) if an explicit loss function is announced, and the public can punish the monetary authority through withdrawing credibility. Rule-crossing actions and the efforts to boost economy at the expense of inflationary costs tend to remain hidden, since, in the lack of an explicit loss function, probably there is no standard underlying the evaluation. An uncertain empirical loss function that could be estimated only under pure model-conditions is completely unsuitable for this purpose. That's why it is argued here that an explicit loss function cannot guarantee

[64] Which, perhaps, cannot be explicitly formulized even by monetary policy makers, no matter how absurd it is; it is nothing more than the simple and common fact that the behavioural rules we accept and follow in our everyday life are often unknown to us (cf. Smith 2003).

credibility, but its lack raises credibility problems through drops in central bank accountability. Of course, the public's competence as regards the interpretation of this loss function or the comparison of actual monetary policy with the loss function may be limited and this circumstance may undermine the arguments for announcing an explicit loss function.

In case of a monetary policy carried out on the basis of an explicit loss function, central bank analysts, before every monetary policy decision, generate forecasts of the target variables conditional on different future paths of the key policy rate, setting the feasible combinations of inflation and the output gap by these forecasts. These forecasts are always produced with central bank notions about the transmission mechanism, views of the current state of the given macroeconomy and judgements on the relevant conditions and shocks in the background.[65] Monetary policy makers pick out that projection of them which is expected to minimize the loss function,[66] so to say, to bias inflation to the target and output to the potential level at an appropriate pace. Through announcing the inflation forecast, actually, central bank advances it to be an intermediate target, since inflation forecast bears quite a strong relationship with the ultimate goal (price stability), but this forecast can be successfully influenced through central bank instruments (this is the essence of *forecast targeting*) (cf. Horváth 2006; Svensson 2000). So, a central bank under flexible inflation targeting focuses on the ultimate goal when sets its intermediate target (inflation forecast) and takes actions in order to achieve this intermediate target. [→ **the goals of monetary policy]** In formal terms, setting an intermediate target means that the central bank makes efforts to achieve the target that can be written as

$$\pi_{t+\tau,t} = E\left[\pi_{t+\tau}\middle|i^t; I_t; z^t\right], \tag{6.24}$$

where $\tau \geq 0$, i^t is the instrument path (instrument-rate plan for period t) that can be phrased as $i^t = \{i_{t+\tau,t}\}_{\tau=0}^{\infty}$ (where $i_{t+\tau,t}$ designates the rate of interest planned for period $t+\tau$), I_t is the information set the central bank possesses in period t (i.e. when the interest rate decision is made), and z^t designates the subjective value judgements (cf. Svensson 2003a). As it has already been mentioned in Chap. 2: forecasts are not necessarily based on model predictions.

The goals of monetary policy According to the often cited concept of the system of monetary policy objectives, targets constitute a well ordered hierarchy (Fig. 6.7).

The intermediate target is aimed at intermediating between the ultimate goal, which is directly uncontrollable by monetary policy, and the operative target bearing a direct relationship with instruments.

---

[65] It has already been mentioned in Chap. 2 how important expert predictions are as they may completely overwrite model estimations in the forecasting routine.

[66] If deciding on the appropriate interest rate path and finding the optimal projection is the problem, applying a *targeting rule* can be an alternative. This is a simple rule that constrains the setting of targets. A targeting rule sets a requirement that has to be met by the forecast itself. In case of quarterly interest rate decisions, the simplest $\pi_{t+8,t} = \pi^*$ form can be regarded as such a rule. In other words, key policy rate is set in order that the two-year inflation forecast being conditional on a constant interest rate path could be at the target (Mizen et al. 2003).

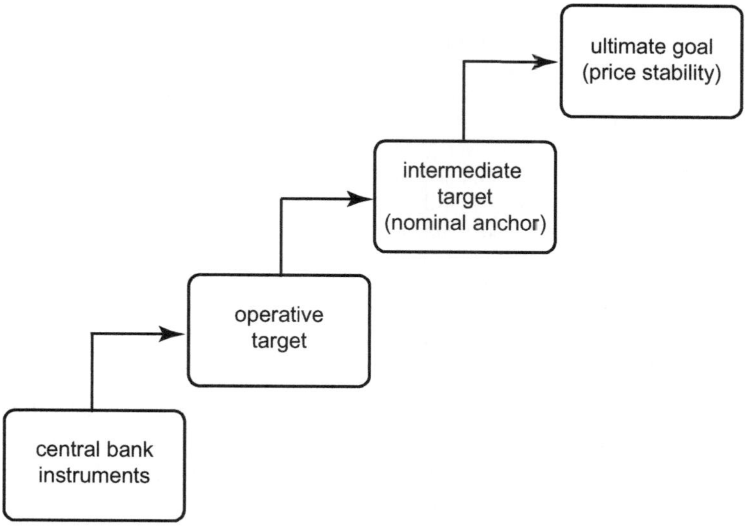

Fig. 6.7 The hierarchy of monetary policy targets (Horváth 2006; Vigvári 2008b)

It has to be stressed that actual achievements of the targets imply various strategies of monetary policy. As far as the ultimate goal is considered, there is a consensus among central banks: price stability as the ultimate goal has become commonly accepted by now, and monetary policy cannot be regarded as an efficient way of long-run real economic enhancement even by high theory. However, choosing an efficient intermediate target should not be independent of the fundamental structural characteristics of the given macroeconomy. For small, open economies, (wide-band) pegging of the exchange rate proved to be a well performing intermediate target. As long as the relatively stable relationship between the price level and one or another monetary aggregate existed (or, at least, was expected to exist), controlling the money supply was a popular intermediate target for larger, stronger and closer national economies with floating exchange rate regimes. The general interest rate level is a further possible intermediate target, exemplified by the U.S. monetary policy. Under inflation targeting, inflation forecasts published by the central banks are used as intermediate targets.[67]

So, on the one hand, monetary policy strategy interpreted as choosing the targets[68] means adjusting the intermediate target to the characteristics of the economy. On the other hand, strategic considerations are also needed in finding an operative target that suits the chosen intermediate target. In other words, there are incompatible intermediate and

[67] A further and recently highlighted aspect of defining the intermediate targets is whether a potential target can act as a nominal anchor. A nominal anchor is such a variable that can fix market expectations. It can be realized, for instance, that the pegging of the exchange rate or a narrow band is more efficient in settling market expectations (since these targets give proof of a higher-level commitment of the central bank) than a wider band (see Horváth 2006).

[68] For us, by choosing the targets is definitely meant that monetary policy picks some macroeconomic variables (e.g. quantity of money, level of the rate of interest, etc.) to act as intermediate and operative targets. The case should be clearly distinguished in which the monetary authority declares numerical values as targets for such variables–rather, it should be referred to as choosing the target(ed) value. The central bank chooses its target when starts using inflation forecast as an intermediate target–and sets the target value to 3 %. After all, it is the application of a targeting rule.

operative targets as replacing the intermediate target usually implies the modification of the operative target variable. When the general interest rate level was the intermediate target, FED's operative target was the interbank rate on free reserves of member banks. Targeting M1 used as an intermediate target was accompanied by imposition of member bank reserve requirements as an operative target, but when the system is based on M2/M3 applied as intermediate target, using the level of loans that the eligible financial institutions can receive from the central bank promised to be an efficient operative target (Szalkai 1995). Exchange rate dynamics can be easily influenced by changes in the quantity of high powered money. However, under inflation targeting, monetary authorities are seeking after affecting the short-term money market rates, i.e. the latter are regarded as operative targets. Whatever the intermediate target of a central bank is, it should be accompanied by an operative target by which this intermediate target (and, through it, the achievement of the ultimate goal) can be effectively influenced according to the view about monetary transmission mechanism.

Future "losses" are considered in the loss function at their expected values, since central bank interest rate decisions exert influence through the transmission channels with time lags. Evaluating the length of these lags is basically an empirical (econometric) problem. Because of the unforeseen and unpredictable shocks, and the lagged effects intermediated by the transmission channels, *there is no leeway for real economic fine tuning* and central banks cannot and should not seek after it.[69] With all this, central bank tries to meet the long-term inflation target, while tolerates short-run deviations from the target[70] (exactly this latter characteristic is one of the fundamental features of flexible inflation targeting). Following it, central bank reveals the optimal projections of the target variables and the interest rate path that is expected to realize these projections and sets its key policy rate according to this path. This sequence repeats before each monetary policy decision and the interest rate path gets modified if needed (Svensson 2003a).

As we could see, this mechanism (monetary policy framework) does not imply the automatic actions required by Taylor-rules.[71] Instead of the Taylor-rule-based central bank reaction function, it is justified to use the form

[69] One of the most important lessons that can be drawn on the basis of the model used above to derive the central bank reaction functions is to urge the central banks not to seek after powerful effects through their interest rate decisions, since the effect of such a decision can be modified *later* by an unforeseen real economic shock. We could draw similar conclusions in Chap. 5, when shocks were interpreted as white noises–according to the new classical doctrines. Eventually, flexible inflation targeting, through its tolerance of real economic fluctuations, allows the real economy to follow its own path within certain limits and does not make efforts to exert a tight control.

[70] Deviations from the target, as we have seen, appear as a kind of compromise, since output stabilization can be carried out at the expense of inflation, and vice versa. Central bank views and preferences on this compromise are revealed by the loss function.

[71] For some authors (Svensson 2003b), a Taylor-rule (whatever form it takes) cannot be regarded as optimal. Accordingly, following a targeting rule is beneficial in multiple aspects: monetary policy makers can rely on a broader set of information, there is room for subjective value judgements (of experts) plus shocks and model changes can be more easily managed within this framework. Moreover, instrument rules do not suggest any guidance for cases of deviations from the rule, though such deviations may be needed under certain conditions. It should not be forgotten either that central banks do not commit themselves to following instrument rules. By making some

$$i_t = F(I_t; z^t) \qquad (6.25)$$

under flexible inflation targeting, where i_t is the key policy rate (the privileged short-term nominal rate of interest) set in period t (in other words, i_t designates the current instrument-rate decision). It is not needed to explicitly specify this reaction function, since it is applied *without* mechanically following a rule.

The flexibility of *forecast targeting* (flexible inflation targeting) can hardly be overstated–and exactly this characteristic was highlighted here. Neither the generation of forecasts nor taking the policy actions is mechanical. The role of value judgements and individual situation evaluations is well exemplified in the case of the Bank of England in the early 2000s. According to the Inflation Report in February 2002, rate of inflation was below the 2.5 % target (that would have necessitated an instant intervention in case of an instrument rule), while, at the same time, real economic performance was reported to lag behind the trend (BoE 2002) (Fig. 6.8).

On the grounds of our previous considerations, we may have expected the Bank of England to respond to this development by a drop in the key policy rate–at least, a Taylor-rule would dictate it. However, actual events did not follow our way of thinking. As central banks under flexible inflation targeting have room for deliberation, decisions may also take market expectations into account. Figure 6.9 depicts the dynamics of the key policy rate and the shifts in market expectations during the relevant period.

According to the Report, private sector expectations in early 2002 envisioned a much more positive view regarding future developments than the actual probabilities. Optimism can easily be grasped even in interest rate expectations, since, in the first half of 2002, market agents forecast key policy rate to increase. Seemingly, market participants failed to recognize that home-consumption-based growth could hardly be sustained anymore. As the inflation fan chart reveals, missing the target seemed to be only temporary, projections showed inflation to return to the target at the forecast horizon (i.e. within 2 years). The Bank of England did not need to change the key policy rate to meet the inflation target *and* to stimulate the real economy,[72] since positive market expectations were sufficient on their own to underpin favourable economic developments. *Key policy rate was kept below the expected level and this instrument-rate decision proved to be sufficient on its own to*

compromise, Taylor-rules can be used effectively in regressions to describe central bank interest policies; as some results show, for example, FED's key policy rate (the federal funds rate) reacted to shifts in the output gap gradually rather than instantaneously (Judd and Rudebusch 1998). By the way, targeting rules are not commonly preferred. For instance, McCallum and Nelson (2004) highlight that a targeting rule closely suits the model constructed to describe private sector behaviour and this rule performs badly if this model is misspecified. Under these circumstances, however, instrument rules based on more generic properties of models may work better. According to King (2005), this is true of instrument rules as well.

[72] It has to be noted that a higher growth rate would have also been beneficial for inflation dynamics.

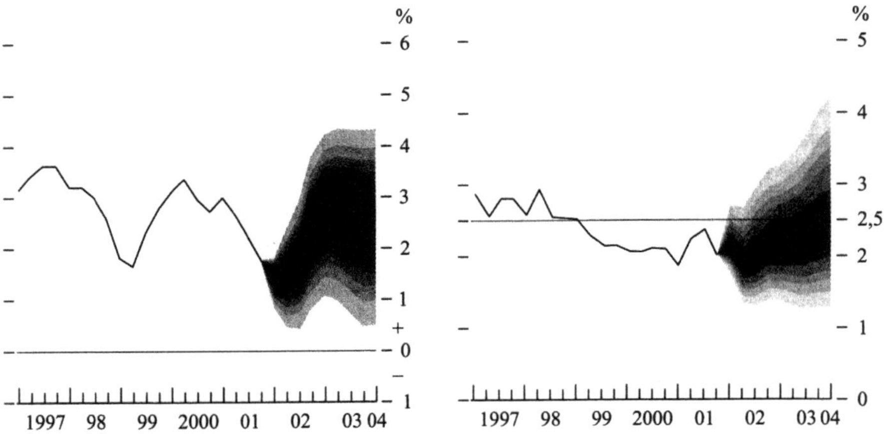

Fig. 6.8 Fan chart of output (*on the left*) and inflation (*on the right*). Percentage changes on a year earlier (*horizontal axis*: time; *vertical axis*: percent) [*Source* BoE (2002)]

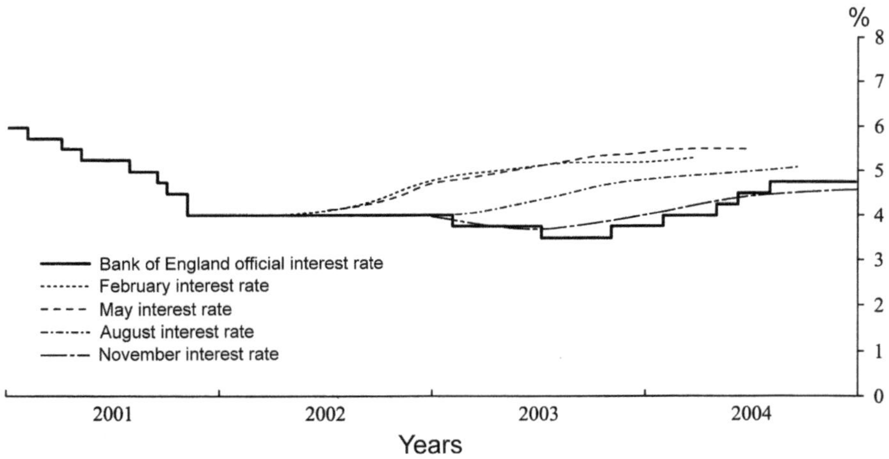

Fig. 6.9 Key policy rate (Bank of England official interest rate) and market expectations of future interest rates, 2001–2004, United Kingdom, as calculated in each of the four Inflations Reports published in 2002 (King 2005)

trigger boosting effects. The actual presence of discretion and deliberation is quite evident. The big picture can be more elaborated if further developments are also taken into account. Report from February 2003 mentioned a deterioration in private sector expectations (BoE 2003), reflected by the downward path of November forward interest rate (Fig. 6.10).

Because of the deteriorating macroeconomic outlook, central bank was forced to give up its previous policy. If private sector expected the rates of interest to drop, a constant official interest rate would have been highly detrimental, and the

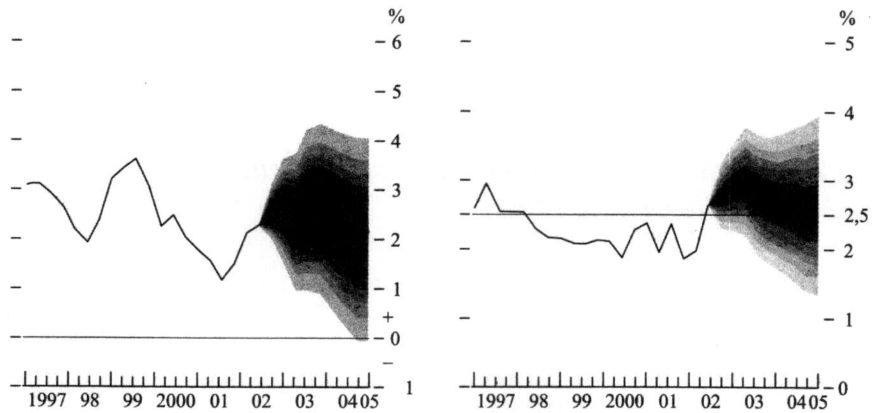

Fig. 6.10 Fan chart of output (*on the left*) and inflation (*on the right*). Percentage changes on a year earlier (*horizontal axis*: time; *vertical axis*: percent) [*Source* BoE (2003)]

prevailing unfavourable processes would have been strengthened further. The Bank of England took private sector expectations into account and optimized its decision accordingly. A Taylor-rule would have urged an interest hike on the grounds of the fan charts from the 2003 Report–however, a comparison with the actual events reveals significant deviations from this rule. The cut in the policy rate did not make it impossible to keep inflation on target, since inflation was expected to return to the target within 2 years, assuming a nominal interest rate at 3.75 % that underlain the inflation forecast as well.

Before stepping ahead, it has to be stressed that the Bank of England did not apply any delusion though its policy decision did not meet public expectations. On the grounds of these considerations, we can say that *a delusion applied by the monetary authority means an unforeseen and unexpected change in a revealed instrument rule*. In case of a monetary-aggregate-based intermediate target, an unexpected shift in the high powered money was such a delusion. A non-announced change in a policy-rate-based instrument rule or deviating from this rule all can be regarded as similar deceiving actions. We can hardly talk about such steps under flexible inflation targeting: transparent central bank tries to make optimal decisions by processing and utilizing as much information as possible in order that these decisions could influence (i.e. anchor) private sector expectations as well at the same time. Of course, *surprises are not ruled out of inflation targeting either* (cf. Vonnák 2007). If the public does not take the official monetary policy target plausible (or acceptable), central bank, through monetary shocks, may try to give signals to market agents that its target does not coincide with actual or forecast inflation.[73] However, these actions should not be regarded as typical instruments of an *appropriately* and *normally* operating central bank.[74]

6.5.2 *Inflation as an Expectation-Based Phenomenon and the Central Bank*

It was mentioned at the end of the previous section that monetary authority actively makes efforts to influence public expectations. The literature usually refers to these efforts as the anchoring of expectations, regarding an appropriately applied nominal anchor, i.e. the intermediate target (inflation forecast) as its most important tool. Papers describing the transmission mechanism of modern monetary policy mention the role of expectations as a separate case (channel) over and above the exchange rate, interest rate and credit channels.[75] Occasionally, expectations contribute to

---

[73] Shocks make only one option–a credible central bank can reach its goal through communication as well, however, in case of credibility problems, it can give clearer signals about monetary policy preferences through shocks.

[74] Some authors put large emphasis on the transparency of central banks (for them, transparency implies abstaining from surprises). In the case of FED, for instance, we can talk about a twofold goal: one means following an appropriate and efficient monetary policy, while the other declares that the central bank, when trying after these goals, should be transparent and open (Greenspan 2001). Moreover, in this context, we can meet the suggestion as well that the central bank, motivated by the desire to meet public expectations, tends to adjust its policy to market beliefs (Issing 2006). Market agents strongly lean on their (inflation) expectations when making decisions. When the central bank wants to make efforts to reach a target that does not harmonize with expectations, setting this target does not necessarily imply the lack of credibility, since the central bank, after all, can succeed in achieving this target through applying the available instruments. However, it may cause even serious real economic damage and, as a result, central bank policy may be heavily criticized. So, anchoring expectations may be expensive under certain conditions. Central bank can avoid it, if it sets the target in accordance with public expectations, through which, at the same time, it can guarantee the target to be met automatically. Of course, it is only possible when inflation is low and tolerable, being somewhere around the target. If deep structural processes do not compromise with the inflation target to be achieved, central bank can "succeed" in meeting this target with a completely passive policy. In this case, monetary authority allows inflation to go its own way in an economy being on the right track, since no intervention is needed. In other words, the best policy in such a case is setting the target according to market expectations.

[75] Sometimes (Mishkin 1996) asset price channel is suggested as a separate mechanism. We have to consider a threefold effect in this case. First: monetary policy affects share prices. If monetary policy is expansive, share prices appreciate, which, in turn, enhances going concern values compared to the replacement (capital) costs of fixed assets, resulting in a rise in investment expenditures. In contrast, if monetary policy is restrictive, then a reverse transmission should be considered. Second: as shares are popular instrument of store of value for households, an increase in share prices, through wealth effect, encourages consumption expenditures. And third: estate and property prices can also be taken into account within a more general framework. These instruments of store of value exert influences similarly to the previous two mechanisms: a rise in their prices triggers an increase in consumption expenditures and, hence, aggregate demand. Knowing by experience, in some cases, only the latter of this threefold mechanism plays an active role in

operations of the other channels.[76] However, expectations, as we have seen, affect also the pricing behaviour of market agents. Uncontrolled inflation expectations always raise a dilemma for the monetary authority, since inflation pressures can be eased, as *one* option, through an interest rate decision that may produce detrimental real economic effects in turn. However, a non-reaction may make inflation get out of control through expectations growing firm, since employees and producers may tend to raise wage claims and prices, respectively. So, in such a case, a monetary policy non-reaction may strengthen expectations on higher rates of inflation (Goodfriend 1993).

So, expectations have become an *autonomous* and *direct* source of inflation. This recognition should be regarded as a definite success of new classical macroeconomics. By the way, there are statements that attribute an ever-greater significance to the channel of expectations. Accordingly, a central bank is capable of affecting inflation in such an extent that it can influence public expectations (cf. King 2005). And if inflation is an expectation-based phenomenon (at least, partially), monetary authority has room to follow a more balanced interest rate policy through affecting expectations.[77]

In practical terms, anchoring expectations means that public inflation expectations scatter around the target–and if central bank is credible (in other words, if it is capable of setting actual inflation in accordance with the target through the available instruments), it implies that the criterion given by Muth (1961) concerning the rationality of expectations is met. Predictions from the (quasi-)relevant model describing the mechanisms and processes of a given macroeconomic system coincide with public expectations. Missing the target does not imply the non-rationality of expectations: the relevant model indicates the detour and, moreover, all of this is going to infiltrate into public expectations. If the central bank is capable of anchoring expectations, this act will be actually a part of the return to the target. When making pricing decisions, the public considers the central bank forecast, since it can be expected to be met, so, after all, inflation is going to take the predicted path.

We have already given an explanation in Chap. 2 why inflation targeting may prove to be an actually operating mechanism and why market agents may adjust

monetary policy transmission: sometimes there are no empirical evidences that monetary policy influences share prices, and, moreover, shares do not constitute a common form of store of value in some countries (cf. Vonnák 2007).

[76] For instance, Svensson (2003a) mentions it with regard to the interest rate channel, since, due to sticky inflation expectations, shifts in the key policy rate affect short-term real rates of interest as well. Moreover, expectations also exert a strong influence on longer terms. It is a commonplace of economics that it is exactly this assumption that makes it possible to map inflation expectations on the basis of yield curves (cf. Hull 1997).

[77] If monetary policy succeeds in anchoring inflation expectations to the target, i.e. market agents *believe* that actual inflation is going to be around or at the target, then inflation is going to be actually at the target, and this outcome does not require any shifts in the key policy rate–or such shifts may be more moderated, if any. At this point, we should refer back to the things discussed in the previous subsection, since this mechanism also exemplifies constrained discretion well.

their expectations to the forecasts of the monetary authority in central position. As we have seen there, the *plausible* assumption that market agents are capable of forming unbiased estimations can be objected both in information and methodological terms. We can rely on such an assumption in an abstract model at most. Sometimes, market participants produce quite-well-performing estimations about complex processes on the basis of extraordinarily simple techniques. Such a heuristic is the *briefcase index* often cited in the United States: in the mornings of the Federal Open Market Committee meetings, all eyes were on Mr. Greenspan's briefcase, since market tried to evaluate the forthcoming interest rate decision on the basis of how thick the briefcase was. The faith in this indicator was grounded on the simple experience that interest rates tended to be raised if the briefcase was fat, while a flat case predicted interest rates to get cut (cf. Greenspan 2007). Central banks under inflation targeting are remarkably capable of giving such simple devices into the hands of market agents. Operations of credible central banks encourage the public to form expectations on the basis of the published-announced indicators–accomplishing what was envisioned by new classicals a few decades ago.

References

Aarle B, Garretsen H, Huart F (2003) Monetary and fiscal policy rules in the EMU. Ger Econ Rev 5(4):407–434

Akerlof G (1970) The market for "lemons": quality uncertainty and the market mechanism. Q J Econ 84(3):488–500

Allen M, Baumgartner U, Rajan R (2006) Inflation targeting and the IMF. International Monetary Fund, Washington, DC

Andor L (1998) Pénz beszél – a nemzetközi monetáris és finanszírozási rendszer politikai alapjai. Aula, Budapest

Backus D, Driffill J (1985) Rational expectations and policy credibility following a change in regime. Rev Econ Stud 52(2):211–221

Báger G, Pulay G (2008) Analysis of the macroeconomic risks of budgeting. Public Financ Q 53(3):391–408

Ball L (1997) Efficient rules for monetary policy. NBER working paper No. 5952. National Bureau of Economic Research, Cambridge, MA

Barr N (2004) Economics of the welfare state, 4th edn. Oxford University Press, Oxford

Barro RJ (1976) Rational expectations and the role of monetary policy. J Monetary Econ 2(1):1–32

Batini N, Laxton D (2006) Under what conditions can inflation targeting be adopted? The experience of emerging markets. Central Bank of Chile, Santiago

Bayer J (1998) A politikai gondolkodás története. Osiris, Budapest

Benzes I, Ürögdi A (2008) The new theory and empirics of political business cycles. Public Financ Q 53(2):269–282

Berg C, Jonung L (1998) Pioneering price level targeting–the Swedish experience 1931–1937. J Monetary Econ 43(3):525–551

Bernanke BS (2003) "Constrained discretion" and monetary policy. Papers presented at New York University, New York, 3 Feb 2003

Bernanke BS, Mishkin FS (1997) Inflation targeting–a new framework for monetary policy. NBER working paper No. 5893. National Bureau of Economic Research, Cambridge, MA

Bernanke BS, Laubach T, Mishkin FS, Posen AS (1999) Inflation targeting–lessons from the international experience. Princeton University Press, Princeton
Bihari M, Pokol B (1998) Politológia. Nemzeti Tankönyvkiadó, Budapest
Black J (2002) A dictionary of economics. Oxford University Press, Oxford
Blackburn K, Christensen M (1989) Monetary policy and policy credibility–theories and evidence. J Econ Lit 27(3):1–45
Blinder AS (1999) Central bank credibility–why do we care? How do we build it? NBER working paper No. 7161. National Bureau of Economic Research, Cambridge, MA
Blinder AS, Fischer S (1979) Inventories, rational expectations, and the business cycle. NBER working paper No. 381. National Bureau of Economic Research, Cambridge, MA
Bod PÁ (2003) Gazdaságpolitika – intézmények, döntések, következmények. Aula, Budapest
Bod PÁ (2013) Heterodox gazdaságpolitikák korszaka? Köz-gazdaság 8(2):89–101
BoE (2002) Inflation report (February 2002). The Bank of England, London
BoE (2003) Inflation report (February 2003). The Bank of England, London
Boros J (2009) A megismerés talánya. Áron, Budapest
Boskin MJ (1988a) Tax policy and economic growth–lessons from the 1980s. J Econ Perspect 2(4):71–97
Boskin MJ (1988b) Consumption, saving and fiscal policy. Am Econ Rev 78(2):401–407
Botos B (2009) Válságkezelés és protekcionizmus. Külgazdaság 53(5–6):91–105
Brown-Collier EK, Collier BE (1995) What Keynes really said about deficit spending. J Post Keynesian Ec 17(3):341–355
Campbell A, Converse PE, Miller WE, Stokes DE (1960) The American voter. Wiley, New York
Cassirer E (1921) Kants Leben und Lehre. Bruno Cassirer, Berlin. Hungarian edition: Cassirer E (2001) Kant élete és műve (trans: Mesterházi M). Osiris, Budapest
Choi CY (2009) Relationship between inflation and relative price variability–revisited. University of Texas, Arlington
Choi CY (2010) Reconsidering the relationship between inflation and relative price variability. J Money Credit Bank 42(5):769–798
Csaba L (2008) The new kind of macroeconomic populism. Public Financ Q 53(4):601–616
Csaba L (2011) A magyar átalakulás és fejlődés néhány általánosítható elméleti tanulsága. Közgazdasági Szemle 58(10):813–831
Cukierman A (2000) Are contemporary central banks transparent about economic models and objectives and what difference does it make? Paper presented at Deutsche Bundesbank/CFS conference "Transparency in Monetary Policy", Frankfurt am Main, 16–17 Oct 2000
Deane P (1978) The evolution of economic ideas. Cambridge University Press, London. Hungarian edition: Deane P (1997) A közgazdasági gondolatok fejlődése (trans: Szepesi G). Aula, Budapest
Debelle G (1999) Inflation targeting and output stabilisation. Reserve Bank of Australia, Sydney
Debelle G, Lamont O (1997) Relative price variability and inflation–evidence from U.S. cities. J Polit Econ 105(1):132–152
Dobias P (1980) Wirtschaftspolitik–Einführung in ihre Grundlagen und Hauptprobleme. Paderborn: Ferdinand Schöningh. Hungarian edition: Dobias, Peter (1988) Gazdaságpolitika–alapvetések (trans: Penyigey K). KJK, Budapest
Downs A (1957) An economic theory of democracy. Harper, New York
Eijffinger SCW, Geraats PM (2006) How transparent are central banks? Eur J Polit Econ 22 (1):1–21
Erdős T (2003) Fenntartható gazdasági növekedés, különös tekintettel a rendszerváltást követő magyar gazdaságra. Akadémiai, Budapest
Erhard L (1957) Wohlstand für Alle. Econ-Verlag, Düsseldorf. Hungarian edition: Erhard L (1993) Jólétet mindenkinek (trans: Zalán P). Konrad Adenauer Alapítvány, Budapest
ERoP (1982) Economic report of the president 1982. United States Government Printing Office, Washington, DC
Feldstein M (1986) Supply side economics–old truths and new claims. Am Econ Rev 76(2):26–30

Feldstein M (1994) American economic policy in the 1980s–a personal view. In: Feldstein M (ed) American economic policy in the 1980s. National Bureau of Economic Research, Cambridge (MA), pp 1–79

Fischer S (1977) Long-term contracts, rational expectations, and the optimal money supply rule. J Polit Econ 85(1):191–205

FRBoSF (2003) How did the FED change its approach to monetary policy in the late 1970s and early 1980s? Federal Reserve Bank of San Francisco, San Francisco

Friedman M (1948) A monetary and fiscal framework for economic stability. Am Econ Rev 38 (3):245–264

Friedman M (1953) The methodology of positive economics. In: Friedman M (ed) Essays in positive economics. University of Chicago Press, Chicago, pp 3–43

Gál Z (2010) Pénzügyi piacok a globális térben – a válság szabdalta pénzügyi tér. Akadémiai, Budapest

Galbács P (2011) A globális pénzügyi és reálgazdasági válság kezelésének hatása a lisszaboni folyamatra. Külgazdaság 55(1–2):46–64

Galbács P (2015) The age of symbolic economic policies? Soc Econ 37

Giddens A (1993) Sociology, 2nd edn. Polity, Cambridge. Hungarian edition: Giddens A (1997) Szociológia (trans: Acsádi J, Babarczy E, Nagy M, Nagy Z, Tóth L). Osiris, Budapest

Goldfeld SM (1982) Rules, discretion, and reality. Am Econ Rev 72(2):361–366

Goodfriend M (1993) Interest rate policy and the inflation scare problem: 1979–1992. Federal Reserve Bank of Richmond. Econ Q 1(4):1–23

Greenspan A (2001) Transparency in monetary policy. Paper presented at the Federal Reserve Bank of St. Louis, Economic Policy Conference, St. Louis, 11 Oct 2001

Greenspan A (2007) The age of turbulence–adventures in a new world. Penguin Press, New York

Hahn F (1984) Equilibrium and macroeconomics. MIT Press, Cambridge, MA

Harrop M, Miller WL (1987) Elections and voters–a comparative introduction. Macmillan, London

Hawking SW (1993) Black holes and baby universes and other essays. Bantam, New York. Hungarian edition: Hawking SW (1999) Einstein álma és egyéb írások (trans: Ungvárai Nagy Z, Ungvárai J). Vincze, Budapest

Hayek FA (1988) The fatal conceit–the errors of socialism. University of Chicago Press, Chicago. Hungarian edition: Hayek FA (1992) A végzetes önhittség–a szocializmus tévedései (trans: Pásztor E). Tankönyvkiadó, Budapest

Hicks JR (1974) The crisis in Keynesian economics. Basil Blackwell, Oxford. Hungarian edition: Hicks JR (1978) Érték és tőke – A keynesi gazdaságtan válsága (trans: Hegedüs J, Lengyel G). KJK, Budapest

Horváth Á (ed) (2006) Monetary policy in Hungary. Central Bank of Hungary, Budapest

Hull JC (1997) Options, futures, and other derivatives, 3rd edn. Pearson Prentice-Hall, New Jersey. Hungarian edition: Hull JC (1999) Opciók, határidős ügyletek és egyéb származtatott termékek (trans: Antal J). Panem, Budapest

Issing O (2006) Communication and transparency–How do central banks meet these demands? Paper presented at European Central Bank, Franfurt am Main, 18 Jan 2006

Judd JP, Rudebusch GD (1998) Taylor's rule and the FED: 1970–1997. FRBSF Econ Rev 28(3):3–16

Kaldor N (1970) The new monetarism. In: Kaldor N (1978) Further essays in applied economics. Duckworth, London, pp 1–21

Kalecki M (1935) The mechanism of business upswing. In: Kalecki M (1969) Studies in the theory of business cycles 1933–1939. Basil Blackwell, Oxford, pp 26–33. Hungarian edition: Kalecki M (1980) A tőkés gazdaság működéséről – válogatott tanulmányok 1933–1970 (trans: Balogh A). KJK, Budapest, pp 64–73

Kamlet MS, Mowery DC, Su TT (1988) Upsetting national priorities? The Reagan administration's budgetary strategy. Am Polit Sci Rev 82(4):1293–1307

Kant I (1787) Kritik der reinen Vernunft, 2nd edn. Johann Friedrich Hartknoch, Riga. Hungarian edition: Kant I (2009) A tiszta ész kritikája (trans: Kis J).Atlantisz, Budapest

Keech WR (1991) Politics, economics, and politics again. J Polit 53(3):597–611

Keynes JM (1936) The general theory of employment, interest and money. Palgrave Macmillan, London. Hungarian edition: Keynes JM (1965) A foglalkoztatás, a kamat és a pénz általános elmélete (trans: Erdős P). KJK, Budapest
King M (2002) No money–no inflation; the role of money in the economy. Bank Engl Q Bull (Summer):162–177
King M (2005) Monetary policy–practice ahead of theory. Mais Lecture at Cass Business School, London, 17 May 2005
Király J (2009) A tornádó és a hurrikán – a 2007. év válságos hatásai. In: László M (ed) A jelen a jövő múltja. Járatlan utak – járt úttalanságok. Akadémiai, Budapest, pp 295–332
P. Kiss G (1998) A költségvetés tervezése és a fiskális átláthatóság problémái. Central Bank of Hungary, Budapest
Komlós A (1965) A szimbolizmus. Gondolat, Budapest
Kopányi M (2003) Mikroökonómia. KJK–KERSZÖV, Budapest
Körösényi A, Tóth C, Török G (2005) A magyar politikai rendszer. Osiris, Budapest
Kutasi G (2007) Budget deficit and the achievement of fiscal equilibrium–institutional approach. Public Financ Q 52(1):37–52
KVA (2004) Finn Kydland and Edward Prescott's contribution to dynamic macroeconomics: the time consistency of economic policy and the driving forces behind business cycles (Laudation). The Royal Swedish Academy of Sciences, Stockholm
Kydland FE, Prescott EC (1977) Rules rather than discretion–the inconsistency of optimal plans. J Polit Econ 85(3):473–491
Lámfalussy S (2000) Financial crises in emerging markets: an essay on financial globalisation and fragility. Yale University Press, New Haven
László F (ed) (2002) Monetary policy in Hungary. Central Bank of Hungary, Budapest
Lucas RE (1972) Expectations and the neutrality of money. J Econ Theory 4(2):103–124
Lucas RE (1973a) Some international evidence on output-inflation tradeoffs. Am Econ Rev 63 (3):326–334
Lucas RE (1973b) Econometric policy evaluation–a critique. Carn-Roch Conf Ser Public Policy 1 (1):19–46
Lucas RE (1990) Supply-side economics–an analytical review. Oxf Econ Pap 42(2):293–316
Lucas RE (1995) Nobel lecture: monetary neutrality. J Polit Econ 104(4):661–682
Magas I (2009) Cyclical fluctuations and crises in the US economy, 1929–2008. Public Financ Q 54(2–3):348–368
Maróth M (2002) A görög filozófia története. PPKE, Budapest
Marx K (1867) Das Kapital, Band I – Kritik der politischen Ökonomie. Verlag von Otto Meissner, Hamburg
McCallum J (1988) Is increased credibility stabilizing? J Money Credit Bank 20(2):155–166
McCallum BT, Nelson E (2004) Targeting versus instrument rules for monetary policy. NBER working paper No. 10612. National Bureau of Economic Research, Cambridge, MA
Mishkin FS (1996) The channels of monetary transmission–lessons for monetary policy. NBER working paper No. 5464. National Bureau of Economic Research, Cambridge, MA
Mizen P, Albert C, Goodheart E (2003) Central banking, monetary theory and practice. Edward Elgar, Cheltenham/Northampton
Modigliani F (1988) Reagan's economic policies–a critique. Oxf Econ Pap 40(3):397–426
Muth JF (1961) Rational expectations and the theory of price movements. Econometrica 29(3):315–335
Neicheva M (2007) Non-Keynesian effects of government spending–some implications for the stability and growth pact. MPRA, Munich
Neményi J (2009) A monetáris politika szerepe Magyarországon a pénzügyi válság kezelésében. Közgazdasági Szemle 56(5):393–421
Papp J (2008) Freeing the hamstrung economy. Public Financ Q 53(4):675–707
Plato (380 BCE) De republica. English edition: Plato (2006) The republic (trans: Allen RE). Yale University Press, New Haven

Plotinus (250 AD) The six enneads (trans: Mackenna S, Page BS). MIT. http://classics.mit.edu/Plotinus/enneads.html. Accessed 27 Dec 2014
Polanyi M (1964) Science, faith and society. The University of Chicago Press, Chicago
Porter DP, Smith VL (2003) Stock market bubbles in the laboratory. J Behav Finance 4(1):7–20
Rasche RH, Williams MM (2005) The effectiveness of monetary policy. FRB of St. Louis working paper No. 2005-048B. Federal Reserve Bank of St. Louis, St. Louis
Russel B (1980) The problems of philosophy. Oxford University Press, Oxford. Hungarian edition: Russel B (1996) A filozófia alapproblémái (trans: Bánki D). Kossuth, Budapest
Rzońca A, Ciżkowicz P (2005) Non-Keynesian effects of fiscal contraction in new member states. European Central Bank working paper No. 0519. European Central Bank, Frankfurt am Main
Sargent TJ (1986) Rational expectations and inflation, 1st edn. Harper & Row, New York. Hungarian edition: Sargent TJ (2005) Infláció és racionális várakozások (trans: Szabó-Bakos E). Nemzeti Tankönyvkiadó, Budapest
Sargent TJ, Wallace N (1974) Rational expectations and the theory of economic policy. Federal Reserve Bank of Minneapolis working paper No. 29. Federal Reserve Bank of Minneapolis, Minneapolis
Sargent TJ, Wallace N (1981) Some unpleasant monetarist arithmetic. Fed Reserv Bank Minneapolis Q Rev 5(2):1–17
Schelling TC (1982) Establishing credibility–strategic considerations. Am Econ Rev 72(2):77–80
Schultz C (1996) Announcements and credibility of monetary policy. Oxf Econ Pap 48 (4):673–680
Schwartz A, Friedman M (1963) A monetary history of the U.S. (1867–1960). Princeton University Press, Princeton
Shaw GK (1984) Rational expectations: an elementary exposition. Harvester, Brighton
Sivák J, Vigvári A (2012) Rendhagyó bevezetés [a] közpénzügyek tanulmányozásába. CompLex, Budapest
Smith VL (2003) Nobel lecture: constructivist and ecological rationality in economics. Am Econ Rev 93(3):465–508
Smith VL, Suchanek GL, Williams AW (1988) Bubbles, crashes, and endogenous expectations in experimental spot asset markets. Econometrica 56(5):1119–1151
Sobel J (1985) A theory of credibility. Rev Econ Stud 52(4):557–573
Sőtér I (ed) (1965) A magyar irodalom története IV. Akadémiai, Budapest
Svensson LEO (1996) Price level targeting vs. inflation targeting: a free lunch? NBER working paper No. 5719. National Bureau of Economic Research, Cambridge, MA
Svensson LEO (2000) How should monetary policy be conducted in an era of price stability? NBER working paper No. 7516. National Bureau of Economic Research, Cambridge, MA
Svensson LEO (2001) The zero bound in an open economy–a foolproof way of escaping from a liquidity trap. NBER working paper No. 7957. National Bureau of Economic Research, Cambridge (MA)
Svensson LEO (2003a) Monetary policy and real stabilization. NBER working paper No. 9486. National Bureau of Economic Research, Cambridge, MA
Svensson LEO (2003b) What is wrong with Taylor rules? Using judgment in monetary policy through targeting rules. J Econ Lit 41(2):426–477
Szalkai I (1995) A monetáris irányítás. Bankárképző, Budapest
Szentes T (1999) Világgazdaságtan I. – elméleti és módszertani alapok. Aula, Budapest
Tarafás I (2001) A monetáris politika a nagy válságtól az ingatag pénzpiacokig. Aula, Budapest
Taylor JB (1979) Estimation and control of a macroeconomic model with rational expectations. Econometrica 47(5):1267–1286
Taylor JB (1995) Changes in American economic policy in the 1980s: watershed or pendulum swing? J Econ Lit 33(2):777–784
Tengelyi L (1995) Kant. Áron, Budapest
Tobin J (1972) Friedman's theoretical framework. J Polit Econ 80(5):852–863
Tobin J (1981) The monetarist counter-revolution today–an appraisal. Econ J 91(361):29–42
Tommasi M (1992) Inflation and relative prices–evidence from Argentina. Department of Economics, University of California, Los Angeles

Tondl G (2001) Convergence after divergence? Regional growth in Europe. Springer, Wien/New York

Turay A (1984) Ismeretelmélet. Katolikus Hittudományi Főiskola, Budapest

Varian HR (1999) Intermediate microeconomics. W.W. Norton, New York/London. Hungarian edition: Varian HR (2001) Mikroökonómia középfokon – egy modern megközelítés (trans: Bara Z, Temesi J). KJK–KERSZÖV, Budapest

Veress J (2006) Gazdaságpolitika versus a politika gazdasága. Fejlesztés és Finanszírozás 4(2):53–60

Veress J (2007) Búcsú a "klasszikus" gazdaságpolitikától. Paper presented at the 6th international conference of the Faculty of Economics, University of Miskolc, Miskolc-Lillafüred, 10 Oct 2007

Veress J (2013) A gazdasági és pénzügyi világválság gazdaságpolitikai következményei. Statisztikai Szemle 91(4):345–365

Vigvári A (2008a) Positions and standpoints–or what does the "invisible hand" arrange? Public Finance Q 53(3):545–551

Vigvári A (2008b) Pénzügy(rendszer)tan – alapvető ismeretek rendhagyó megközelítésben. Akadémiai, Budapest

Vonnák B (2007) The Hungarian monetary transmission mechanism: an assessment. Central Bank of Hungary, Budapest

Weber M (1964) Wirtschaft und Gesellschaft. Kiepenheuer & Witsch, Köln/Berlin. Hungarian edition: Weber M (1967) Gazdaság és társadalom (trans: Józsa P). KJK, Budapest

Weeks J (1989) A critique of neoclassical macroeconomics. Macmillan, London. Hungarian edition: Weeks J (1998) A neoklasszikus közgazdaságtan kritikája (trans: Szalai Z). Aula, Budapest

Chapter 7
Looking Back

And as a man, with difficult short breath,
Forespent with toiling, 'scaped from sea to shore,
Turns to the perilous wide waste, and stands
At gaze; e'en so my spirit, that yet failed
Struggling with terror, turned to view the straits,
That none hath past and lived.

(Dante: The Divine Comedy)

7.1 Introduction

With some luck, we have succeeded in drawing a distinct picture of the doctrines labelled as "new classical". The achievement of this school cannot be found in its radicalism but rather in the aspects it enriched economic thought the validity of which can hardly be questioned today. Rationality of expectations, supply-side incentives and the role of expectations in price dynamics: all these were highlighted by new classicals, *inter alia*. Their assumptions and findings cannot be held under life-world conditions in their chemically pure form—however, they could draw attention to the limitations of economic policy. Whatever the genuine purpose of new classicals was, they really succeeded in putting the limited potential of economic policy into the lime-light. Moreover, it follows that, as it was argued above, a too rigorous and negative critique against new classical macroeconomics and, through it, mainstream economic theory would certainly miss the point. Sometimes we can meet with judgements that labelled mainstream economics as completely irrelevant, claiming that it has drifted too far from life-world conditions and processes. This critique is partially right: analyses of new classical macroeconomics were carried out with regard to an ideal world that differs from the environment we live in in many aspects. *However, this theory is not irrelevant but chemically pure*. The evolution of new classical macroeconomics has not come to an end: the problem the new generations face is to loosen this ideal framework. If it is really true (but is it true at all?) that models should be close to actual (life-world) circumstances, then our task is to deploy new classical framework under newer and newer conditions by disintegrating and *concretizing* the laboratory settings in order to find a proper scope for the theory. Perhaps, a new general theory may be needed to complete this task and perhaps, what is more, once our knowledge will be

P. Galbács, *The Theory of New Classical Macroeconomics*, Contributions to Economics, DOI 10.1007/978-3-319-17578-2_7

sufficient to realize the missing links that may connect all the present and specific (and often contradicting) theories and models. I think, *this* would mean the emergence of a new general theory. This book was able to accomplish only a small part of this work. After confessing all the above, it may not be completely unhelpful to line up our most important findings again.

7.2 What Have We Learnt?

7.2.1 Monetary Policy

There were some standard and common assumptions applied traditionally in new classical models. Such assumptions are, e.g. complete market flexibility, rational expectations or regarding shocks moving output from the potential (equilibrium) level as white noises. Under these conditions, any countercyclical intervention of monetary policy is completely ineffective and, naturally, impossible as well.

It is a mindset of mine that the way in which these assumptions are set and are applied in models refers beyond itself. Although the aim of Lucas and his followers was to demonstrate countercyclical inefficiency of monetary policy, this purpose has ended up in consequences reaching far beyond the direct results. Such a consequence was, for instance, the precise consideration of the circumstances of inefficiency (or, in other words, its necessary and sufficient conditions). By knowing accurately the conditions under which monetary policy cannot be applied in countercyclical interventions, we can learn *those* conditions at the same time under which, after all, it can be brought into play. A transition to rational expectations, a special maturity structure of certain agreements, long-lasting real economic cycles or market frictions are all such factors that enhance the countercyclical potential of monetary policy. Moreover, the exact content of countercyclical monetary policy is also clear due to the new classicals. A permanent growth enhancement or monetary-policy-based changes in the growth potential are all ruled out, even if the fundamental problem for Friedman, eventually, was the way monetary policy can affect growth—see Chap. 4. This goal would imply serious inflationary pressures and costs, so monetary policy should abandon these efforts. Even if monetary policy is facilitated by the circumstances to exert countercyclical influence on real economy, then this effort should be the closing of the output gap to zero. Monetary policy is right if it supports the real economy in working at the potential (equilibrium) level. By these considerations, after all, the concept of *countercyclical economic policy* could be clarified further—and, maybe thanks to it, *inter alia*, we could not think anymore that monetary (and fiscal) policy interested in the stimulation and control of real economy would be right if it tried to fix output permanently above the potential level. In inflationary terms, this practice is to be avoided—so, monetary policy should seek after the elimination of the output gap, disregarding its positive or negative sign. *Boosting a growth (or development) process permanently (i.e. in the long-run) is not a goal of countercyclical monetary policy*. The best (i.e. the most efficient) option to enhance economic growth is to influence growth potential by applying supply-side instruments (Chap. 6). And, on these grounds, both the proper scope and the role of monetary policy can be outlined well. Over and above closing the output gap, monetary policy should only strive to control the rate of inflation, that is, to establish a clear (stable) inflation environment that does not bias economic decisions. It is exactly the reason why monetary policy should not aim at growth enhancement, since even moderate achievements would imply such inflationary costs that would evidently jeopardize the central bank when it tries to follow its goal of price stability.

Life-world economic shocks are not white noises. For monetarists, this characteristic stripped economic policy of the *possibility* (i.e. the potential) of countercyclical interventions. Even if economic policy contributes to these multi-period business cycles, it has enough time to correct them. If socio-economic reality is such as Keynes (and somehow Marx) portrayed it, if complete flexibility and efficiency of markets are disturbed by frictions, sticky prices or information deficiencies—then even new classical macroeconomics could not deny the potential of countercyclical economic polices.

Multi-period depressions and boosts make room (and also need) for activist countercyclical economic control. Eventually, it was *not* argued above that monetary policy should be applied in countercyclical fine-tuning. It is impossible since monetary policy actions can take real economic effects only with considerable time lags. Moreover, monetary policy is also incapable of long-term growth enhancement, since the pursuit of this goal is accompanied by undesired (and, actually, destructing) inflationary damage.

The relevancy of alternative expectations theories have to be taken into consideration as well. Rationality still seems to be a good assumption, however, some stickiness in forming expectations (plus in pricing behaviour and in setting other nominal quantities) cannot be precluded. The assumption of rational expectations requires of us to be cautious. If we scrutinize actual life-world conditions, a given macroeconomic system cannot be assumed *a priori* to be built on rational, i.e. forward-looking expectations. Rational expectations hypothesis is a marvellous and outstanding theory and an assumption (axiom) that can be modelled in an easy and clear way, but its pure manifestation under life-world conditions can hardly be expected to emerge. It is an empirical problem, of course. But the question why rational expectations could evolve in certain macro-systems while fail in others requires theoretical considerations (even if they have not been phrased by pure theory) yet. We have the intuitive answer that relates the evolution of (the formation of) expectations to the structural stance of macroeconomy and the development of the market system right at hand. If all this is true, rationally formed expectations need to be established—and even if these efforts are successful (which is definitely necessary in order to stabilize the path of real economy), it should not be forgotten that the rationality of expectations does not necessarily imply the complete flexibility of these expectations or that prices and other nominal magnitudes can be anticipated to react with infinite flexibility. It should have become clear by now that the theory of rational expectations could be a dangerous weapon in unguarded hands, since it may lead to absurd consequences; especially when REH lines up together with other assumptions. Rational expectations hypothesis has an unquestionable scope in modern macro-theories, but this scope needs to be clearly and precisely defined.

New classical macroeconomics could not persuasively demonstrate the impossibility of activist economic policies. It was a long way to go until we were able to phrase this sentence. To demonstrate this theorem, new classical doctrines required rational expectations hypothesis built on very extreme statements. REH was reviewed in Chap. 2. The strong definition Muth elaborated may be the most relevant approach made to expectation formation, though it cannot be regarded as the only one. Actually, the weak definition is no more than an empty phrase that can hardly promise anything for a strict analysis. Friedman's adaptive scheme says nothing about the ultimate cause why agents rely only on past data when forming their expectations and why they do not utilize all the other (abundantly available) sets of information regarding future developments either. So, this mechanism is inconsistent with the fundamental assumption (axiom) of utility maximization—however, it should not be set aside after all, since it can be an important limitation of rational expectations. According to some experiences, the formation of forward-looking expectations does not necessarily imply the infinite flexibility of expectations.

Rationality of expectations does not mean omniscience. Market participants posses neither information in the necessary quantity nor methodological knowledge in the necessary quality, and finding the relevant model is predestined to fail because of the joint effects of stochastic events and the trial-and-error technique. However, asymptotical unbiasedness of expectations with respect to the exploited information explains why individual estimations may show qualitative differences. It was only one step ahead to illuminate the mechanism of inflation targeting, since forecasts of the central bank being in a central position can be conceded as outputs from the quasi-relevant model; and one of the consequences of this mechanism is that market agents take these forecasts over. So, criteria of the hard definition are met eventually. Thus, the hard definition of REH cannot be regarded as a realistic theory, but rather as an idealization-based, abstract construction, though the elaborators attributed the ability of describing life-world conditions to it. Discussion given in Chap. 2 was intended to highlight the discrepancies (among others) that lie between pure theory and life-world conditions.

A further prerequisite of the inefficiency of economic policy was the axiom of continuous market clearing. If nominal magnitudes (prices and wages) are capable of flexible adjustment and if expectations are rational, then monetary policy can generate inflation only. However, it is an extraordinarily strong assumption—to realize it, it is enough to put it in other words: under rational expectations, inefficiency of monetary policy can be demonstrated only if prices and wages are capable of continuous and instant adjustment. Models of Chap. 3 and, in part, 4 suggested that the neutrality of money, i.e. classical dichotomy did not prevail (at least in the short-run) in the Phillips-models elaborated by Friedman and the new classicals. Although, for Friedman, adaptivity of expectations made room for the central bank to apply delusion in order to exert real effects, a central bank cannot have the same potential under rational expectations. So, the potential of monetary policy for a deliberate (systematic) output control did not follow from the non-neutrality of money. However, real business cycles were highly problematic, since, in these models, the impotency of monetary policy was based on the fact that the central bank cannot plan an appropriate and efficient systematic strategy to eliminate these unforeseen white noise shocks. Even if the central bank can affect anything real, it can be achieved only by taking random actions. *No* systematic monetary policy can be built upon surprises, as we have seen. However, as real business cycles were introduced into the models, the lone efficiency of surprising actions cannot be held anymore: known, completely anticipated developments are equally capable of exerting real effects.

It has to be stressed again: it was *not* argued here that monetary policy is the most appropriate and most efficient tool of countercyclical control—while some limited efficiency should be attributed to it. Actually, one of the most fundamental purposes of this work was to highlight: traditional countercyclical instruments of fiscal policy and their effectiveness were not questioned by new classicals to such an extent that their actual application should be precluded as well. Moreover, we would rather talk about a new, long-term-focused aspect and an accordant set of supply-side instruments with which Lucas and his followers have armed fiscal policy. Our other purpose was to find a proper scope for monetary policy. If we showed some inclination to blindly believe in the most simple (actually oversimplified) and vulgarized new classical doctrines, then we could say that the central bank can succeed in affecting real output by deceiving at most, stipulating that such efforts are highly destructive and, hence, to be avoided. However, inflation targeting showed that monetary authority can carry out some limited countercyclical control—and we could draw similar conclusions after scrutinizing the limitations of the inefficiency of monetary policy. Of course, too slow reactions of real economy strip monetary authority of the possibility of fine-tuning. However, it is not justified to talk about a central bank impotency as regards the (expected future) path of inflation. Monetary authority has such instruments that make it possible to ease the threatening inflationary pressures. However, in the context of the credibility of economic policy, we could see as well that the lack of harmony between fiscal and monetary policy may erode this potential. Inflationary costs of long-term enhancement are serious, of course. Thus, closing the output gap should be the goal to pursue and monetary policy is really capable of it, in terms of both the experiences of operative economic policy and new classical theory. If there are long-living cycles, monetary policy can contribute to smoothing them.

7.2.2 *Fiscal Policy*

It was surprising even for me that the most interesting and overthrowing theorems of the new classicals were born with regard to the sphere of fiscal policy, bearing direct relation to operative economic policy. The public belief claiming the primacy of monetary policy within monetarism and new classical macroeconomics are too deep-rooted. The shallowness of this view becomes clear if the mainstream notion (using "mainstream" in a broad sense) of monetary policy is reconsidered. Monetary policy is not capable of exerting *efficient* and *powerful* influence on real economy. Yet for Friedman, a temporary real effect was given that occurs at the expense of some additional inflation emerging even in the short-run. New classical macroeconomics, in cases when its standard assumptions can be held, denied even the possibility of systematic countercyclical monetary policies. *Monetary policy can influence only prices and the price level at best.* Money is only a veil thrown atop the real system—if the quantity of money or, as a result, the general price level changes, it does not affect the real system. But what effect is it that cannot influence anything real? What does the faith in the primacy of monetary policy mean—that monetary policy which is capable of influencing only this veil? This monetary policy cannot affect the real system under the standard assumptions, similarly to fiscal policy—while the items that can be influenced are completely unimportant concerning the real economy. At least *almost* unimportant—monetary policy is successful when it is indiscernible. Monetary policy can do the most for the harmonic functioning of real economy through maintaining the stability of prices and enhancing the soundness of macroeconomic environment. I do not think, it has to be stressed that this recommendation, a commonplace today, was planted in public knowledge by monetarism. However, this role is passive-negative, since the pledge of success is a monetary policy that keeps away from activist efforts as far as possible.

Notwithstanding, the scope and role of fiscal policy and its potential is a highly complicated problem for new classicals. Long-term real-economic impotency and limited short-term potential of monetary policy is such a consensus (in terms of both the history of economic thought and economic policy) that has evidently constrained the ways in which new classical macroeconomics could evolve. Before new classicals emerged, it had not been questioned either that monetary policy had only limited potential to control real variables. So, the stake was small in this context: even if it succeeded in realizing that monetary policy can be efficient under certain conditions, this efficiency was only of marginal importance compared to the potential of fiscal policy that had been assumed previously. In case of monetary policy, (countercyclical) efficiency and inefficiency are not considerably far from each other, since efficiency is quite moderated even when inefficiency cannot be held because of the system of assumptions or life-world conditions. Within the realm of countercyclical control, monetary policy cannot do a lot even if it is efficient—it plays the main role in inflation policy. The primary role in real economic stabilization, if this role actually exists at all, is played by fiscal policy. In other words, it means that demonstrating the inefficiency of fiscal policy was (or would have been) the real challenge for new classicals.

"Would have been", since it became doubtful for me whether new classical macroeconomics had really assumed fiscal policy to be (originally or simply) ineffective or wanted to portray it as (originally or simply) ineffective. It is true that new classicals stipulated only the unchanging structure of government expenditures as an assumption to underlie their theses, but, at the same time, they opened the door for considering the circumstances leading to the failure of the theory. The critique of the equivalence-theorem could actually reveal some further limiting assumptions any of which would disturb the equivalence. A case in which all the necessary conditions for the Ricardian equivalence simultaneously hold is very unlikely to emerge. So, it follows that the equivalence-theorem rather tends to fail. Permanent income theory specified under REH definitely pointed towards the efficiency of fiscal policy yet in a sense bearing a resemblance with the theory of Keynes. Again, the purpose of new classicals was not simply to argue for the inefficiency of fiscal policy—rather, we should say that the theory specified and catalogued the conditions for this inefficiency. We have already seen in the case of monetary policy that considering the conditions for inefficiency and relieving some of them made monetary policy capable of exerting a limited countercyclical influence over and above its disinflationary goals (or exactly in their interest). Of course, the other extremity is similarly untrue: the theory of Keynes, that with its oversimplifying system underwent innumerable phases of vulgarization, was not an appealing alternative anymore. *Lucas and his followers hardly wanted to occupy a position at any of the endpoints of complete inefficiency or complete efficiency.* The complete inefficiency of fiscal policy was rather an assumption or, in other words, was built on an assumption: *if* the structure of government expenditures does not change and *if* market agents are forming rational expectations, *then* equivalence holds. Instead of these two extremities, new classicals looked for a tinged and well-elaborated intermediate standpoint that was underpinned by arguments properly.

This intermediate standpoint was characterized by the catalogue of the factors limiting the efficiency of fiscal policy. One of the most fundamental conditions for the equivalence-theorem is that the structure of government expenditures remains intact after the initial and expansion-related tax-cut: lost tax revenues are going to be offset by new bonds to be issued. In Chap. 5, some scenarios were reviewed in the cases of which government can only postpone to fill the tax-cut-induced resource gap at most. Under such circumstances, market agents refute to regard the tax-cut as an increase in either wealth or permanent income, so demand effects the government intended to trigger are expected to fail. In this case, aggregate demand dynamics cannot be influenced by fiscal policy instruments.

Our discussion was grounded on the distinction made between simple quantitative *corrections* (contractions, eventually) and *reforms* covering also the qualitative aspects of public sector operations. Simple fiscal corrections do not prevent expansionary tax-cuts from leading to simultaneous resource gaps that could remain hidden from market agents analysing and interpreting budgetary operations. This is the reason why demand-enhancing effects of expansionary tax-cuts scrutinized by the equivalence-theorem fail to emerge. If the play-field of fiscal policy is narrow, the efficiency of short-run countercyclical fiscal control can be regained through qualitative changes (i.e. through reforms, instead of simple contractions). As we have seen, both the economics of reforms and the theory of political business cycles point to the fact that these reforms tend to fail to occur—in other words, execution may exert even the surprise often cited by new classicals. However, it was also stressed that qualitative changes are not the instruments of countercyclical budgetary policy—however, the recovery of the demand-enhancing effects of fiscal policy can be one of the beneficial results of these reforms.

The doubtful inefficiency of fiscal policy and the existence of supply-side incentives draw attention to the fact that the possibilities of short- and long-run control do not preclude each other. The former is enabled exactly by multi-period business cycles and by the potential of fiscal policy to influence expectations. Perhaps the distinction of *short-term control* vs. *long-term control* may exaggeratedly curtail the validity of theories. On the basis of our previous considerations, we may rather draw the conclusion that positive outcomes realized in one time-horizon are far from ruling out efficiency in the other. Keynes did not set the possibility and potential of long-term output-enhancement aside, just did not pay attention to it, since he phrased his purposes in the context of short-run developments. Similarly, new classicals did not persuasively question the possibility and the grounds of this short-run control. Rather, we should say, as we have seen, that new classical macroeconomics clarified and specified the conditions of the *efficiency* of fiscal policy. The success of new classical macroeconomics, accordingly, did not manifest itself in undermining the predecessors but, first, in making rational expectations hypothesis widely conceded and, second, in stressing some long-run effects that had been played down up to then by high theory, and, third, in clarifying and specifying the (necessary) conditions for short-run efficiency of fiscal policy. As a result, these conditions were clearer and, at the same time, tighter than previously.

7.3 Meditations Upon a Pure Economic Theory

In Chap. 1, I have referred to neoclassical-rooted mainstream economics as a *pure theory*. The basis of the argumentation was that mainstream economics applies ideal-typical concepts (axioms), gained through abstraction and idealization, from which it derives the fundamental economic laws through a mathematical apparatus, in a deductive way. Friedman's positivist methodological principles were highly criticized, according to which the realism of the underlying assumptions is not a relevant aspect when judging the performance of a theory. On the contrary, I unequivocally advocated the view that the right way of creating new concepts is abstraction combined with idealization—since it may help us avoid arbitrary conceptualization methods through which we can gain (approximately) correct predictions, but cannot get closer to economic reality. Models based on arbitrary assumptions do not bring us closer to understanding any observed phenomena, though this is the fundamental purpose and aspect of causal explanations (cf. Hausman 2008).

I could hardly deny that my sympathy for new classical macroeconomics intensifies up to prejudice sometimes. The purpose and intention of the critique on which this work was grounded was to *defend* and *justify* the theory. New classical macroeconomics is so imposing and so much characterized by an uncompromising determination and elaboration that it makes it a unique episode in the history of economic thought. I gave voice in abundance to this view (or belief, perhaps) of mine in Chaps. 1 and 6 and maybe somewhere else, too. I think, the relevancy of neoclassical economics has not been questioned fundamentally up to now—though one may find it hard to accept this statement uncritically. The theories motivated by this intention (not only the system of Keynes but also the achievements of the Nobel-prized Vernon Smith in the field of experimental economics) could usually register only moderated success, though their importance and greatness could hardly be questioned.

For me, economics which is rooted in and oriented by the neoclassical theory is a pure theory that gives us only a broad framework by which some (theoretical) outcomes can be generated. Actual life-world processes should be expected to differ from these outcomes—however, a pure theory teaches what the things we scrutinize are *in themselves*. Not only neoclassical economics is a pure theory. Newton's mechanics or Euclid's geometry were also pure theories. For example, one of the most fundamental reviews of Euclidean geometry written in Hungarian phrases the following with regard to the axiomatic system of absolute geometry:

> "The names of the elements of sets $\mathcal{E}$, $\mathcal{P}$, $\mathcal{L}$ and, additionally, d and m are undefined (unmeaningful) words attached to our language; and though we refer to them through words that have meanings in everyday language (such as points, planes, lines, distance, measure of the angle), we must not attribute to them any qualities taken from physical space." (Kovács 2004)

Axioms of geometries are described in the following again:

> We get to the concept of space from experience through abstraction. [...] If the physical attributes of objects are abstracted [...] and only the form of the parts of the space they occupy is considered, then we get to the concept of geometric objects. If we imagine objects that do not exist in reality at all, such as a pillar reaching to infinity, then their forms, viz. the geometric objects they provide can still be considered. [...] Objects are confined by surfaces. [...] Surfaces do not have thickness. If a surface is cut into pieces, these pieces are constituted by lines. [...] The parts of these lines are still labelled as lines, so, for instance, an edge of a table is also a line. Lines can be depicted through a thin wire or yarn as well. Lines have no thickness or width. If a line is cut into pieces, the resulting pieces are delimited by points. A point can be depicted by the prick of a needle or any object with a very small size. A point has no expansure at all." (Hajós 1999)

No further commentaries are needed, I think. Pure theories are not concerned with directly observed reality—we will never see an Euclidean point or a Newtonian pendulum, and it is similarly sure that the laws of planetary motion theorized by Newton only approximate the motion of actual planets. Similarly, neoclassical economics separates the most fundamental attributes of the phenomenon under investigation and sets them as axiom(s); and, the same as other pure sciences, operates elements non-existing in reality. If we want to throw neoclassical economics away *on these grounds*, then we have to do the same with ancient mathematics or Newtonian physics as well. The relevancy of such theories are unquestionable, as their ideal-typical concepts grasp such fundamental features of phenomena that are unavailable for empirical sciences. Men are *human beings* in the empirical world and not a machine specialized for making economic decisions, as economics tries to depict us through its own man concept. It is out of question that the theories of highly prestigious physics also apply ideal-typical concepts—the idea of friction-free surfaces or completely rigid objects are such concepts.

As it was mentioned above, the relevancy of neoclassical doctrines has not been questioned successfully even up to now. During the last few decades, a number of theories had emerged from which this denial was expected both by professionals and the historians of economic thought. Actually, Keynes (often mentioned as the first paradigm-shifting economist) "only" *complemented* neoclassical economics. We could hardly talk about a denial, since Keynes (1936), for instance, built the quantity theory of money in its rough form into his system, and mentioned some neoclassical doctrines as sub-cases of his own theory. As opposed to neoclassical theory, Keynes mainly described how reality works. For me, it is undoubted that Keynes did *not* try to refute the equilibrium-tendencies built in macroeconomic systems—since questioning theoretically the tendencies towards macro-equilibrium would end up in a paradox conclusion that involuntary unemployment is voluntary after all. Keynes only stressed that the equilibrium postulated by neoclassical economics is disturbed by various factors under life-world conditions. The prevailing paradigm has not changed—it is so much true that even Keynes himself claimed that some neoclassical axioms should not be questioned. For example, he wrote the following in his General theory regarding the marginal product of labour:

> [...] *The wage is equal to the marginal product of labour.* [...] subject, however, to the qualification that the equality may be disturbed [...] if competition and markets are imperfect.

Additionally, he phrased the following view of the other *fundamental* (in his word) assumption of neoclassical economics:

> *The utility of the wage when a given volume of labour is employed is equal to the marginal disutility of that amount of employment.* [...] subject to the qualification that the equality for each individual unit of labour may be disturbed by combination between employable units analogous to the imperfections of competition which qualify the first postulate.

Emergence of neoclassical axioms under pure market conditions is disturbed by factors of actual socio-economic environment. For me, the same holds for the rationality-theory of Vernon Smith. While the system of neoclassical axioms and homo oeconomicus built and characterized on those grounds show what a man *would be* like if his behaviour was governed and framed only by the principles of formal rationality and if his only job was to make economic decisions, experimental economics describes what a man is like in *reality* (as far as a laboratory can be considered as reality). No other relation is possible, since while neoclassical economics gets to its traditional man concept on the grounds of its axioms in a deductive way, then experimental and behavioural economics create their own concepts on an empirical basis, following an inductive logic (cf. Tversky and Kahneman 1974, 1986; Mérő 2004). I do not want to dispute the merits of these schools through the considerations above, since these theories taught more about human economic behaviour than any other theoretical constructs. I just wanted to highlight that neoclassical axioms were not challenged, since they deliberately were *not* intended to grasp actually existing human beings. So, they are unfalsifiable on empirical grounds. Sometimes, we can come across with some banal but still powerful arguments claiming that the neoclassical (mainstream) paradigm still clearly prevails. For instance, there are some popular textbooks that scrutinize the fundamental problems of economics in a way that the problems and the accordant chapters can be permuted almost at will—that is to say, it is completely indifferent which textbook the students rely on (Rosenberg 2009).

In Chap. 1, some time was also devoted to the statement according to which results of an idealizing process are not only unfalsifiable but unverifiable as well. Although modern physics succeeded in complementing and reconsidering Newtonian mechanics, and more and more precise numerical results are available on these grounds, all this does *not invalidate* but *complements* the achievements of Newton. Invalidation will be an option only when we can find a more appropriate ground than the postulation of gravity. It reveals on second thoughts that this is exactly the relation between the gravitational theories of Newton and Einstein: Newton's model was not *complete*; the generalizing amendments should be considered as Einstein's achievements. It is typical, for instance, that both theories lead to the same results under earthly conditions, though there are such extreme circumstances under which Newton's model already fail. Under these circumstances, it is Einstein's amendments that are relevant (cf. Singh 2004; on the perihelion shift of Mercury and the divergent results from these theories see Hédervári 1986; Einstein 1934). Additionally, the generalizing-complementing character should be stressed even in spite of the fact that Einstein completely gave up the Euclidean geometry that had underlain the theory of Newton. This is nothing more than changing and *replacing* the basic assumptions (of fundamental importance and leading to distant consequences) that results in a whole new theory (cf. Feynman 1964). The same is true of neoclassical economics and, moreover, the theories of Newton and Einstein may also recall the relation between Keynes (1936) and the quantity theory of money to us. Quantity theory (and, simultaneously, the case of a general equilibrium with full employment) is a special sub-case of the theory of Keynes and, as such, it is a useful and valid theory for describing certain macroeconomic settings after all, where the replacement of presumptions is represented, e.g. by the differences of the views on the velocity of money. The fundamental relation is also a complementation in the case of neoclassical economics. It could hardly be denied that the concept of *homo oeconomicus* extremizes existing (though maybe not prevailing) features of human beings and that an *exclusive* occurrence of these features *would* lead to the predictions deduced by the theory. However, reality in its completeness is not of such kind and it is similarly evident and undeniable. So, neoclassical economics is not the only science that organizes itself following similar logical guidelines. As we have seen, *there were and still there are sciences besides neoclassical economics that do not provide realistic descriptions of the directly experienced reality*. It was exactly the circumstance that was suggested by Husserl as the original sin of European sciences (cf. Husserl 1976).

My view on the essence and the role of abstraction may be criticized—to which I have to react here, in advance, before these critiques would be phrased at all, since this issue is in very close relation to the interpretation of neoclassical economics as a pure theory. In this context, I have to cite Max Weber's methodological principles. Weber (1904) says the following—far from being unimportant as to our central problem, viz. neoclassical economics:

> We have in abstract economic theory an illustration of those synthetic constructs which have been designated as *"ideas"* of historical phenomena. It offers us an ideal picture of events on the commodity-market under conditions of a society organized on the principles of an exchange economy, free competition and rigorously rational conduct. [...] Substantively, this construct in itself is like a *utopia* [...]. [...] Its relationship to the empirical data consists solely in the fact that where market-conditioned relationships of the type referred to by the abstract construct are discovered or suspected to exist in reality to some extent, we can make the *characteristic* features of this relationship pragmatically *clear* and *understandable* by reference to an *ideal-type*.

Here, Weber's example was the ideal-type of "city economy". The concept of city economy does not embrace all the features of all city economies that have ever existed. Following the way of creating an ideal-type, we extremize (*accentuate*, in Weber's terminology) *one* feature or *some* features (being relevant for the current investigation) of an actually existing and observed phenomenon and arrange it/them into a unified (and well-ordered) thought construct. The concept of the city economy created this way cannot be found anywhere and any time in reality, so it is unverifiable in empirical terms. In other words: we will never find a city economy described with a short ideal-typical definition. Actual city economies only approximate this ideal form more or less. An ideal-typical concept emerging as a result of abstraction *cannot be found in its conceptual purity anywhere in reality and cannot be verified empirically*; its occurrence is hindered by other features not fading away under actual conditions. We are all *homo oeconomici* somehow, though not only that, and our elementary acts corresponding to the economic man are always blenched by the circumstance that, e.g. we are social beings as well with behaviour subject to moral norms that prevent the hard and brutal characteristics of economizing evidently present in our actual behaviour from prevailing in their purity. A homo oeconomicus can hardly be expected to walk in the local park (unless as a pathological case), but the concept is still a result of abstraction, coming from the accentuation of Weber-Fechner's law. I consistently emphasized the principle that *the proper way of creating ideal-typical concepts is abstraction*. This aspect developed to be the guideline of investigation, since it enabled us to check the first step of theorizing: the system of axioms and concepts. On the grounds of this principle, I could phrase the sentence: *homo oeconomicus cannot be criticized exactly for the lack of a clear empirical verifiability*. It was also this principle that led me to the critique of rational expectations hypothesis and, at the same time, of new classical macroeconomics, since formal rationality was assumed by new classicals to actually operate even in reality. So, it has to be stressed: checking the correctness of abstractions, viz. criticizing the arbitrary way of conceptualizing was a relevant aspect of this work.

In this context, it is worth stressing the *purpose* of abstraction (idealization). For me (hoping that I could avoid contradicting Weber) the guiding aspect of idealization is subjective, the direction of conceptualization is always established by the interest of the researcher and science. *Homo politicus* was invented by political science, *homo psychologicus* by psychology, and *homo sociologicus* by sociology (as equally abstract creatures) in a way that is similar to the procedure of neoclassical economics (cf. Andorka 1997). The ultimate goal and the leading aspect of these creating acts were to accentuate some phenomena standing in the lime-light of a given science by setting other features aside. So, *homo oeconomicus* is an abstract creature that is destined to make fundamental economic laws and their consequences analyzable through its "behaviour" (or *operation*, rather).

The consequent distinction I made between *exaggerated* and *unreal* assumptions also relates to this context of pure theories. I would like to introduce my view through an example. In his seminal methodological paper, Milton Friedman argues for the principle that if we get to a theory that is consistent with reality as far as its predictions are concerned, this theory has to be preserved even if the underlying assumptions cannot be found in reality. Friedman's example was the behaviour of leaves: if their density is to be modelled, and we have a model performing well in terms of predictions while leaves are assumed to be rational agents capable of mathematical optimizing, then our model cannot be objected (Friedman 1953). For me, the assumption Friedman mentioned is wrong because it follows from an act of arbitrary alchemy and is not based on abstraction. On the contrary, the assumption of agents of REH forming expectations according to the objective probability distribution *is not wrong* but *too tight*. The axiom is right: if market agents really possessed all the methodological skills and sets of information (in other words: if they were homo oeconomici), that *would be* needed for forming such expectations, then nothing would prevent new classical macroeconomics from providing actually a realistic description of reality (as it was intended). However, human beings are not like REH assumes. By the way, this picture would be acceptable if it was provided by a pure theory—and, moreover, should not be criticized *on these grounds*: if we talked about an agent of simulations, running on a computer screen, who was created as an analogy to homo oeconomicus, the idea of the agent forming such expectations would be acceptable. However, if the theory is aimed at giving a realistic portray of reality, this *exaggerated* assumption cannot be held, since it leads immanently to the failure of the theory, as far as the intention of a realistic description of reality is concerned. So, in the course of a thought experiment (that led to a moderated version of the strong definition of REH), I started from a notion of agents who do not form expectations on their own but take these expectations over from a specialized institution—and the estimation itself cannot be unbiased, but only approximates the objective probability distribution as much as possible. The distinction I made between *wrong* and *exaggerated* assumptions was based on *this* principle. It is a twofold investigation of assumptions: the first question is whether a given assumption is based on abstraction; and the second is whether this assumption is in accordance, e.g. with the purposes and ambitions of the theory that applies it. For me, loosening the assumptions is not aimed at making wrong presuppositions right. A wrong assumption cannot be fixed only replaced (and is to be replaced—the famous F-twist cannot be accepted as a standpoint with grounded scientific aspirations). However, the tightening and the loosening of right abstractions result in such model families the investigation of which is highly capable of providing new lessons.

The question whether my critique of new classical doctrines is carried out to decide whether the assumptions are *irrelevant* or *exaggerated* is very difficult to answer. *Equilibrium is grounded on the assumption of equilibrium*, i.e. on a presumed equilibrium or an equilibrium that is presumed. I think this is the most significant thought of the new classical system. It was argued here that the presumption of equilibrium-tendencies *as an abstraction* cannot be criticized: if price flexibility is not hindered, equilibrium emerges. *This* is the fundamental mechanism of markets—perhaps the law of markets. Nineteenth century neoclassical models by separating actual circumstances from artificial model environment could apply the pure theory in the right way. The only danger was raised by the *imperative* connotation. For new classicals, the assumption of a general macroeconomic equilibrium is *exaggerated* since they promised to give a realistic description of reality. In other words, the purpose of the theory and the assumption it applies are not in accordance.

It is interesting that while the demonstration of the emergence of equilibrium states is based on a well-ordered set of presumptions, i.e. the emergence of an equilibrium necessarily follows from the assumptions, then the arguments for the benefits of the emergence and subsistence of equilibrium lift to almost metaphysical heights sometimes—in a sense as Prof. Mátyás talked about the metaphysical equilibrium of demand and supply with regard to Say's law (Mátyás 2002). For example, Hayek (1988) thinks he revealed the manifestation of the evolutionary logic in the mechanisms of the free market. The bottom-line of his reasoning is that, due to market freedom, a spontaneously occurring and evolving order (capitalism itself) emerges that cannot be established through deliberate actions. In other words, capitalism cannot be designed. For Prof. Mérő (2004), an economy is a thing that *emerges* but is not developed and the progression of which is governed by forces of general evolution. Free market steps into the reasoning where we have to regard it as a medium articulating and manifesting the evolutionary logic. Market mechanisms follow special rules with which any external interference should be avoided. Of course, referring to the evolution and evolutionary logic is dangerous but not necessarily wrong, since we can list evident examples from natural sciences. Economic policy intervention as an interference with a naturally evolving order that can be easily represented through the example of an ecosystem living and evolving in a Petri dish. The evolving process can be disturbed if the researcher, for example, changes the acidity/basicity of the aqueous solution in the dish (i.e. the environment), the temperature or the nutrient supply. In this case, such conditions may emerge through which the scientist can *externally* modify the processes, dooming some species and favouring others—or creating new species. He may find the new setting he established to serve some (scientific, technical, moral, etc.) interests better than others. I am not sure that government interventions are effects that

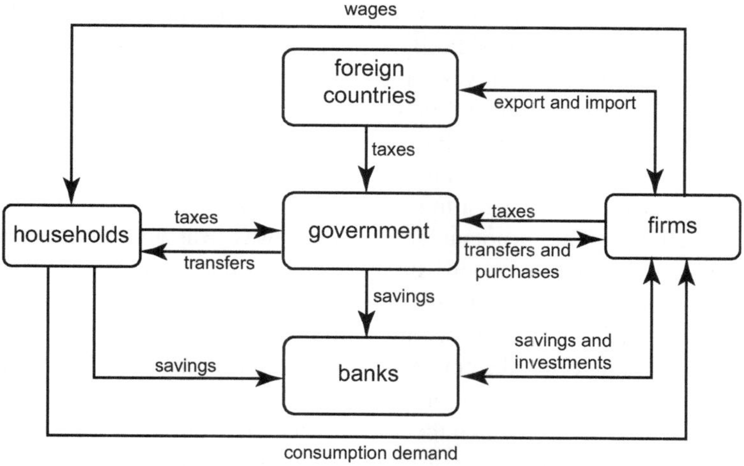

Fig. 7.1 Income and money flows in a government-augmented open economy (Nagy 2002)

really come from somewhere outside of the economy. Government as a *market agent* is a part of the system that is a common-place today on the grounds of lessons from mixed economies (for further considerations see Stiglitz 1988 or Barr 2004). Even the most elementary conceptions regard government (or public sector, as a whole) as a component of the system and not as an agent intervening from outside. Moreover, it does not follow from the bureaucracy model of Kornai (1984) either that public sector stands somewhere outside of the economy—even if it stands *above* (all other) market agents through its powerful position (Fig. 7.1).

Government, central budget, public sector (or whatever) is a component of the economy, so the parable of the biologist *externally* interfering with the evolutionary logic hobbles. Rather, government (if we insist on biological examples) is a species that, if emerges, shows an aggressive behaviour striving after to control the actions of other agents, but economy as a system still goes its own (evolutionary) way—it is just another spontaneous way and not the path of evolution of a government-free economy. The architectural example of Vernon Smith (2003) is also exact. The appearance of a town, the setting of the buildings can evolve without any deliberate control in a way that it shows a well-ordered structure. An order may evolve even without a planner—and this spontaneous order can be really disturbed from the outside world. But government is not of this kind. Its controlling efforts come from somewhere inside of the system. For me, any effort towards controlling economy seems to be one of the manifestations of the natural logic of economy—similarly, interveners can be interpreted also as components of the economy (e.g. as entities of a special species). Of course, it does not follow that there is no market without government—but when the regulator-controller government appears, the evolutionary logic of economy remains in operation, since only the set of agents has broadened. A government modifies processes and mechanisms of an economy and the governing evolutionary logic only from an insider position—and does not switch them off from the outside. By the way, government itself is the child of the socio-economic system: it was set up by the agents of this system. To put it in other words, it was born inside of the structure and, moreover, its evolution can be characterized by a continuous adjustment and dynamic changes in its roles and duties. The existence of the government is subject to the logic of evolution. Maybe, outlining an evolutionary constitutional theory and history or a public service theory as (sub)disciplines is not an idea to sweep away. But anyway, if government is not a child of the socio-economic structure, if it did not evolve within the system, then who established it from outside? The Great Creator? I do not think that those who argue for the logic of universal evolution want to refer to an external creator exactly when talking about the emergence of government *as such*. The essence and the actual content of evolution is an uninterrupted adjustment process to environment from the necessity of which the government could not escape either. Nothing can illustrate it better than the debate on the potential, the instruments and the duties of economic policy—after all, this work also has a reading towards this direction. But not only government as a market agent obeys this logic—so does the legislator.

The relation between economic theories and reality is still one of the fundamental questions of the philosophy of economics—and idealization can really be regarded as a strategy of being connected to reality. Economists build theoretical models from the components gained through idealization and abstraction (or through other techniques), but the roles and scope of these models are not clarified. It is evident that, through the analyses of artificial worlds (models) filled with completely rational agents operating in completely flexible markets, economists could considerably enhance our knowledge on the mechanisms of *these* models and the behaviour of *these* agents. However, it is a different question how these analyses could support a deeper understanding of directly experienced relations, circumstances and structures. This is the reason why the correctness of abstraction gets importance, since models grounded on *unreal* assumptions (as it was discussed in detail in Chap. 1) do not help us with *learning* and *understanding* the actual life-world conditions. The success of predictions creates another problem. A model, if a proper set of assumptions is unimportant for us, may even provide good estimations of the future dynamics of a relevant (macroeconomic) variable—while *the ultimate driving forces* behind this dynamics cannot be revealed on the basis of this model. It is highly important to consider this problem, since the investigations in economics we studied above were aimed directly at regarding the features of the models as *surrogate systems* and not as actual life-world structures. We can even say that analyzing these models *stands proxy for* the investigation and description of life-world circumstances and developments. This is the very reason why in the previous chapters special attention was devoted to abstraction, creating and applying pure theories, and stressing the importance of this application. If we are not careful when creating and combining the elementary components of our models, the results will be almost completely useless with no significance *beyond the direct numerical content*. In such a case, we know nothing more about the actual life-world conditions and processes than the fact that they are somehow different but the broad framework, mentioned above, of actual mechanisms will *not* be available. According to Mäki (2009), theoretical models in economics are unavoidable because we do not have reliable direct access to some deep facts of economic reality. In order that life-world conditions could be learned on the grounds of findings from models, these theoretical models have to be *representative* of those life-world conditions somehow—this representative character is guaranteed by abstraction and idealization. Our theoretical models as surrogate systems are such intermediating devices that enable us to have indirect access to reality. To put it in other words, these surrogate systems convey information about real socio-economic systems. It is still a question, of course, what the purpose of the examination of theoretical models is, if *this* is not. This is the case when our theoretical models are regarded as *substitute systems*—these are examined for their own sake, i.e. analysis is *not* for gaining information about real systems. The resemblance to directly observed life-world socio-economic systems is an aspect of marginal importance, since, after all, the study of these substitute systems *substitutes for* the interest in reality and questioning this reality. Evidently, in such a case, creating the building blocks through idealization is not required either.

Mäki (2009) regards theoretical models as describing imaginary situations—imaginary in the sense they are not built on components discovered or observed in reality and, moreover, in the sense that the settings these models manifest and describe contain only a few (more or less, but hardly all) components from real life-world situations. Accordingly, a part of the building blocks really comes from idealization (these blocks may be even *too* idealized sometimes), while the other part is built in without having real-actual counterparts. In this, economic models are unreal indeed in a sense that appeared in Chap. 1. The only question is where the border-line between arbitrary, non-idealized, non-reality-based elements and abstraction/idealization-grounded components is. Where is the demarcation line that isolates arbitrary conceptualizing alchemy from the technique that is built on abstraction and that starts from directly observable phenomena? According to Mäki, for example, perfect information, zero transaction costs, closed economy, or the generally applied *ceteris paribus* clause are all unreal (in other words: false idealizing) assumptions, since they are characteristics of models (model worlds) used as surrogate models and not of the real world. However, it is dubious for me that these examples are really arbitrarily created *false* (as Mäki says) assumptions and not the results of idealization (abstraction). For instance, in case of the assumption of zero transaction costs, something is disregarded that really exists under life-world conditions—nothing is added to the model that is non-existent in reality. Perfect rationality is the same: scarcity of time, limited brain capacities, costs of the information process are all factors that prevent us from having perfect and complete knowledge in reality. But if these disturbing circumstances can be disregarded, then, in an abstractive context, homo oeconomicus can be rational. By the way, it bears some resemblance to what Karl Polanyi (1957) writes about money at a point: anything can be money but, actually, only a few things used to be money. In theoretical terms, as an ideal-utopian creature, a human being can be omniscient and fully rational, but, in reality, this state can be approximated only. The big picture becomes even more complicated by the fact that natural sciences also apply such assumptions in abundance—but is assuming a frictionless plane or a perfectly rigid physical object an unreal component or the result of abstraction and idealization? Of course, I do not want to suggest that new classical macroeconomics as one of the consummations of mainstream economics is flawless. The notion of the labour union leader as the incarnation of the economy-wide auctioneer raises suspicions even in the most fanatic mainstream economists—but, actually, abstraction is extant only in that non-organized workers are assumed away. The direct tendencies towards equilibrium can be doubted of course, but this theory is still a *pure theory* to which actual life-world conditions can be compared. This reasoning is dangerous, I know, since, eventually, any assumption seems to be defendable on these grounds—we can stipulate anything, since this is "only" a pure theory after all and searching for the determining factors underlying observed discrepancies can be started immediately. On second thoughts, it can be realized that our line of reasoning radically differs from the methodology of positive economics as Friedman elaborated it. For him, the goal justifies the means: the crucial point is if predictions are (approximately) true—while the assumptions on the grounds of which these results are gained are completely irrelevant. However, we are talking about something else: abstraction, idealization is a technique easy to use, almost completely in accordance with researchers' (individual) taste—the ultimate purpose is to accentuate some features (elements) while disregarding others. The assumption of tendencies towards equilibrium is *not* undefendable in a utopian context, in an idealized model environment, since, in the case of a comprehensive, all-embracing organization, this is eventually the exact way the labour union leader may succeed in making everybody employed. By reason of the principle of the short side, equilibrium employment makes the maximum level of employment. To mention a counterexample, the notion of agents forming adaptive expectations is weird, since the most important underlying motive of expectations formation is a forward-looking attitude, and it is exactly the element that is missing from Friedman's model. Is it really an abstraction if even the most important component of the observed mechanism or phenomenon is disregarded? After all, it could be, if these characterises are unimportant for us in the context of the current investigation. However, Friedman assumed his agents forming expectations concerning future outcomes to show inclination and to have ability to utilize all pieces of information, apart from those referring to relevant future developments.

I have no doubts, my reasoning fails to give the ultimate and satisfying answers as regards the scope of idealization and abstraction in the methodology of economics—others, who criticize, regard economics as an embryonic and confused science being indisposed and seeking ways and means. Hausman (2008) declares the assumption of complete rationality and perfect information clearly false (*illegitimate*, in his word). Maybe, our science cannot be characterized by the rigor and consistency of physics or mathematics, even if these disciplines are as rigorous and consistent as they try to make themselves appeared. If compared to these sciences, economics may not really be full-fledged and full-blown yet. Some might think that it could not be at all. But if we remember the achievements through which economics contributed to our knowledge about socio-economic reality, we might think that the measures and criteria by which disciplines are judged are wrong.

My faith is strong: *we are making science*. If forms of life known on Earth are taken as a basis, will we ever be capable of perceiving a form of life the evolution of which followed a completely different path? Is the search for extraterrestrial life not predestined to fail as long as we hold definite ideas about what life means, what its conditions are and what its typical forms look like (which may not be fully known even on Earth)? Is our science concept not wrong if it leads us to question the scientific quality of economics? Maybe, the measures of our scientific models are too rigorous if economics cannot pass them.

References

Andorka R (1997) Bevezetés a szociológiába. Osiris, Budapest
Barr N (2004) Economics of the welfare state, 4th edn. Oxford University Press, Oxford
Einstein A (1934) Mein weltbild. Querido, Amsterdam, Hungarian edition: Einstein A (2000) Hogyan látom a világot? (trans: Szécsi F). Gladiátor, Budapest
Feynman RP (1964) The character of physical law. MIT, Cambridge, MA, Hungarian edition: Feynman RP (2005) A fizikai törvények jellege (trans: Gajzágó É). Akkord, Budapest
Friedman M (1953) The methodology of positive economics. In: Friedman M (ed) Essays in positive economics. University of Chicago Press, Chicago, pp 3–43
Hajós G (1999) Bevezetés a geometriába. Nemzeti Tankönyvkiadó, Budapest
Hausman DM (2008) Philosophy of economics. In: Zalta EN (ed) The Stanford encyclopedia of philosophy, Fall 2008 edition. http://plato.stanford.edu/entries/economics. Accessed 13 Apr 2012
Hayek FA (1988) The fatal conceit–the errors of socialism. University of Chicago Press, Chicago, Hungarian edition: Hayek FA (1992) A végzetes önhittség–a szocializmus tévedései (trans: Pásztor E). Tankönyvkiadó, Budapest
Hédervári P (1986) Ismeretlen (?) naprendszerünk. Kossuth, Budapest
Husserl E (1976) Die Krisis der europäischen Wissenschaften und die transzendentale Phänomenologie. Martinus Nijhoff, The Hague, Hungarian edition: Husserl E (1998a–b) Az európai tudományok válsága I–II (trans: Berény G, Mezei B, Egyedi A, Ullmann T). Atlantisz, Budapest
Keynes JM (1936) The general theory of employment, interest and money. Palgrave Macmillan, London, Hungarian edition: Keynes JM (1965) A foglalkoztatás, a kamat és a pénz általános elmélete (trans: Erdős P). KJK, Budapest
Kornai J (1984) Bürokratikus és piaci koordináció. Akadémiai, Budapest
Kovács Z (2004) Geometria – az euklideszi geometria metrikus megalapozása. Kossuth Egyetemi Kiadó, Debrecen
Mäki U (2009) Realistic realism about unrealistic models. In: Kincaid H, Ross D (eds) The Oxford handbook of philosophy of economics. Oxford University Press, Oxford, pp 68–98
Mátyás A (2002) A korai közgazdaságtan története. Aula, Budapest
Mérő L (2004) Az élő pénz – a gazdasági vállalkozások eredete és az evolúció logikája. Tericum, Budapest
Nagy A (ed) (2002) Bevezetés a közgazdaságtanba I–II. BGF PSZFK, Budapest
Polanyi K (1957) The semantics of money-uses. In: Polanyi K (ed) (1971) Primitive, archaic, and modern economies. Beacon Press, Boston, pp 175–203

Rosenberg A (2009) If economics is a science, what kind of a science is it? In: Kincaid H, Ross D (eds) The Oxford handbook of philosophy of economics. Oxford University Press, Oxford, pp 55–67

Singh S (2004) Big bang, the most important scientific discovery of all time and why you need to know about it. Fourth Estate, London, Hungarian edition: Singh S (2006) A nagy bumm: minden idők legfontosabb tudományos felfedezésének története (Szécsényi-Nagy G). Park, Budapest

Smith VL (2003) Nobel lecture: Constructivist and ecological rationality in economics. Am Econ Rev 93(3):465–508

Stiglitz JE (1988) Economics of the public sector, 2nd edn. W.W. Norton, New York, Hungarian edition: Stiglitz JE (2000) A kormányzati szektor gazdaságtana (Mezei G). KJK–KERSZÖV, Budapest

Tversky A, Kahneman D (1974) Judgement under uncertainty–heuristics and biases. Science 185(4157):1124–1131

Tversky A, Kahneman D (1986) Rational choice and the framing of decisions. J Bus 59(4): S251–S278

Weber M (1904) Die Objektivität sozialwissenschaftlicher und sozialpolitischer Erkenntnis. Archiv für Sozialwissenschaft und Sozialpolitik 19(1):22–87. English edition: Weber M (1949) Objectivity of social science and social policy (trans: Shils EA, Finch HA). In: Shils EA, Finch HA (eds) The methodology of the social sciences. Free Press, New York, pp 49–112

MIX
Papier aus verantwortungsvollen Quellen
Paper from responsible sources
FSC® C105338

If you have any concerns about our products,
you can contact us on
ProductSafety@springernature.com

In case Publisher is established outside the EU,
the EU authorized representative is:
Springer Nature Customer Service Center GmbH
Europaplatz 3, 69115 Heidelberg, Germany

Printed by Libri Plureos GmbH
in Hamburg, Germany